WELSH FERNS

CLUBMOSSES, QUILLWORTS AND HORSETAILS

A DESCRIPTIVE HANDBOOK

SEVENTH EDITION
BY
G. HUTCHINSON
AND
B. A. THOMAS

AMGUEDDFEYDD AC ORIELAU CENEDLAETHOL CYMRU
NATIONAL MUSEUMS & GALLERIES OF WALES

Cardiff 1996

© Adran Botaneg, Amgueddfeydd ac Orielau Cenedlaethol Cymru
© Department of Botany, National Museums & Galleries of Wales

Cyhoeddir gan
Adran Botaneg
Amgueddfa ac Oriel Genedlaethol, Caerdydd
Parc Cathays
CAERDYDD CF1 3NP

Published by
Department of Botany
National Museum & Gallery, Cardiff
Cathays Park
CARDIFF CF1 3NP

FIRST PUBLISHED	1940
SECOND EDITION	1948
THIRD EDITION	1954
FOURTH EDITION	1962
FIFTH EDITION	1969
SIXTH EDITION	1978
SEVENTH EDITION	1996

ISBN 07200 04 35 7
Printed in Great Britain by J & P Davison, Pontypridd, Mid Glamorgan.

PREFACE TO THE SEVENTH EDITION

Since the publication in 1940 of *Welsh Ferns* by Hyde and Wade, increased knowledge and nomenclatural changes have necessitated a number of new editions, the last appearing in 1978. The present seventh edition represents the most radical revision, involving not only updates in nomenclature but also incorporating new material and methods of presentation.

Welsh species, as previously, are described in detail and briefer accounts of the remaining British taxa are given, thus increasing the overall usefulness of the work. Quick reference has been made easier by rearranging the genera and species into alphabetical order and placing the hybrids at the end of the accounts of each genus instead of interspersing them among the specific descriptions. The only exception to this order is the grouping of the clubmosses, quillworts and horsetails together at the beginning. A synopsis of the classification of the British pteridophytes puts the genera into an evolutionary perspective.

Greatly inproved clarity has resulted from several changes, especially by replacing lists of herbarium specimens by maps showing the distribution of ferns both in Wales and in the wider context of Europe. A number of new diagrams and charts illustrate technical terms and features of critical importance. The glossary has been expanded with over 100 additional entries including geographical terms that cover the relevant areas. Many of the terms found in earlier editions have been expanded and the plural for those of Latin origin included. The new and modified material has greatly improved the ease of use of the book whilst still retaining the tradition of the original work.

TREVOR G. WALKER
President
British Pteridological Society

The publication of this edition of Welsh Ferns was supported by a grant from the Bentham-Moxon Trust.

CONTENTS

	Page
INTRODUCTION	1
ACKNOWLEDGEMENTS	3
PTERIDOPHYTES	3
LIFE-HISTORIES	5

The Male-fern (p. 5), Chromosomes (p. 11), Polyploidy (p. 12), Hybridization (p. 12), Apogamy (p. 13), Homospory and Heterospory (p. 14)

IDENTIFICATION	15
THE HERBARIUM OF THE NATIONAL MUSEUM & GALLERY, CARDIFF	15
HINTS TO COLLECTORS	16
EXPLANATION OF VICE-COUNTIES AND 10-KM SQUARE (HECTAD) MAPS	17
WORLD DISTRIBUTIONS	23
SYNOPSIS OF CLASSIFICATION OF PTERIDOPHYTA IN THE BRITISH ISLES	24
ARRANGEMENT OF GENERA AND SPECIES IN THE TEXT.	30
KEY TO GENERA OF PTERIDOPHYTA	30
CLUBMOSSES	33
LESSER CLUBMOSSES	42
QUILLWORTS	46
HORSETAILS	49
FERNS	69

The Stem (p. 69), Hairs and Scales (p. 69), The Leaf (p. 71), Fertile Leaves (p. 75)

OTHER ALIEN SPECIES	200
BIBLIOGRAPHY	220
SOME FERN SOCIETIES	226
GLOSSARY	227
INDEX	239

INTRODUCTION

The first edition of *Welsh Ferns*, published in 1940, was written by the then Keeper of Botany, H.A.Hyde, and the Assistant Keeper, A.E.Wade in response to the perceived need for an inexpensive text. Notes were given on the distribution of species in Wales together with lists of specimens in the National Museum of Wales herbarium. There were few changes in the next two editions, but the fourth edition, published in 1962 reflected an increased knowledge of both fern taxonomy and distribution in Wales. The fifth edition (1969) was largely rewritten by the new Keeper, S.G.Harrison, to include the clubmosses, quillworts and horsetails and the sixth (1978) gave many more details of hybrids.

This new edition incorporates several major changes. The systematic arrangement of the main text in the first six editions followed, with a few exceptions, the scheme of classification proposed by Professor R. E. G. Pichi-Sermolli, in *Uppsala Univ. Årsskrift*, 6, 70-90 (1958). However, in this edition we have followed the more practical approach of arranging the genera, and the species within them, in alphabetical order. The only exception to this order is the grouping of the clubmosses, quillworts and horsetails together at the beginning (p. 33). Nevertheless, a full synopsis of classification is given earlier in the book (p. 24). There are a few additional aliens, associated with gardens and estates, mentioned briefly at the end of the main alphabetical sequence (p. 200). The names of authors of the Latin names of plants are all abbreviated according to Brummitt, R. K. & Powell, C. E. (1992): *Authors of Plant Names*.

Every Welsh species is described and, in order to increase the usefulness of the work, brief descriptions of the remaining British species have also been added. A few alien species, which may be encountered as garden escapes, are also included. Each specific description is followed by a note on the kind or kinds of habitat in which the plant is found.

Listing of herbarium specimens, as given in the previous editions, is no longer practical, nor is it a suitable method of portraying the distribution of taxa. Instead, 10-km square (hectad) maps are given for all but the rarest species. They are updated versions, including a few minor corrections, of the maps given in Hutchinson and Thomas (1992), incorporating new records received up to December 31st 1995, being compiled from published records, information from county recorders and new records from our own field work and our herbarium. Outline distribution maps are also given for the distribution of taxa in Europe. These are based on the *Atlas Florae Europaeae* and the *Atlas of North European Vascular Plants*. For further details of these see under World Distributions, p. 23.

It is hoped that this book will be used frequently for the identification of specimens in the field. For this purpose a key to the genera immediately precedes the start of the descriptions, and keys to the species are placed under the headings of the respective genera. These keys, however, are only of use as pointers and the reader should always check the identifications by reference to the complete descriptions in the text and preferably to authenticated specimens.

All measurements are given in metric units (occasionally in microns (µm): 1,000µm=1mm). Scales in centimetres and inches are printed **inside the front cover**. The dimensions of rhizomes are given as length diameter; only the height of fronds is stated; the size of the blade is quoted as length breadth. The normal range of a particular dimension is usually indicated and the size attained by unusual specimens is sometimes also added in brackets, but even these do not always represent the extremes. For quick reference, definitions of terms used in the text can be found in the glossary, with extra details for many given in the introductory sections in the book.

ACKNOWLEDGEMENTS

We put on record our greatful thanks to the previous authors H. A. Hyde, A. E. Wade and S. G. Harrison for their efforts in producing the previous editions of *Welsh Ferns* on which this edition is based. Many of the illustrations which have also appeared in the previous editions were drawn by Evelyn A. Jenkins. We thank D. Davies and K. L. Davies for translating into Welsh those names which are extra to *Enwau Cymraeg ar Blanhigion - Welsh Names of Plants* (D. Davies & Jones, A., 1995). We thank all Welsh vice-county recorders, and other interested members of the Botanical Society of the British Isles and the British Pteridological Society for adding to the updated base-maps, for pointing out erroneous records and answering numerous queries. C. D. Preston allowed access to the record-files at the Biological Record Centre, Monks Wood Experimental Station, provided a set of pteridophyte maps resulting from the B.S.B.I Monitoring Scheme (1987-1988) and answered numerous queries about the records. A. C. Jermy allowed us to examine specimens in the cryptogamic herbarium, Natural History Museum, London (**BM**). Thanks also to the Botanical Society of the British Isles for giving permission to reprint the list of vice-counties and accompanying map, and to use the pteridophyta maps (Hutchinson & Thomas, 1992) for updating and amending.

Thanks are due to B.S.B.I. referees for determinations and help with the taxonomy, especially R. H. Roberts (*Polypodium*) and C. N. Page (*Equisetum* and the *Pteridium aquilinum* complex). C. R. Fraser-Jenkins helped in a similar way with *Dryopteris affinis*. M. Gibby kindly helped with the chromosome counts. A base-map for the gametophyte stage of *Trichomanes speciosum* was supplied by F. J. Rumsey. We also thank the numerous recorders who sent in specimens for determination, J. G. Gavan for preparing the original Welsh distribution maps, K. L. Davies for help in producing the European maps, A. M. Townsend for preparing the line-drawings additional to the previous editions and D. M. Spillards for preparing the World map. The cover design is based on an original painting by D. E. Evans.

PTERIDOPHYTES

The ferns, clubmosses, quillworts and horsetails constitute one of the major divisions of the plant kingdom, the PTERIDOPHYTA (Greek: pteris - fern; phyton - plant.). They are Vascular Cryptogams, sharing with the seed-plants or PHANEROGAMS the possession of well-developed conducting and supporting tissues but differing from them in lacking seeds. The plant, as ordinarily understood in all classes of the Pteridophyta, forms small reproductive bodies called sporangia. These produce minute spores which germinate to form a relatively simple and insignificant body called a prothallus (the gametophyte generation). Each prothallus bears the sex organs; male or female or both. The female egg, after fertilization by the male antherozoid (sperm), develops to form the recognisable plant (the sporophyte generation), which, when mature, will produce spores. There is, therefore, an alternation of sporophyte and gametophyte generations of plants completing what we call a life-cycle. There are variations to this alternation of generations as some pteridophytes reproduce themselves freely by vegetative means, such as rhizomes and bulbils, while others complete their life-cycles without sexual fusion.

The relationship between the two alternating phases in the life-history of ferns in general may be exemplified by reference to the familiar Male-fern (*Dryopteris filix-mas* (L.) Schott) on page 5.

Living Pteridophyta can be divided into three classes, Filicopsida or Polypodiophyta, Lycopsida or Lycopodiopsida, and Sphenopsida or Equisetopsida. Some botanists divide the Filicopsida (Ferns) further, recognising as classes the Ophioglossopsida (*Botrychium*, *Ophioglossum*), Osmundopsida (*Osmunda*), and Polypodiopsida for the rest of our 'true ferns'.

Ferns are distinguished by having leaves that are always more strongly developed than the stem. The leaves are often very large and much divided, and usually rolled up crozier-like when young and bud-like. In most ferns the fertile leaves, i.e. those which bear sporangia, are not otherwise distinguishable from the sterile leaves and they are not confined to any particular part of the shoot.

Clubmosses living in the British Isles have a rather moss-like appearance with small, simple leaves, which are usually spirally arranged on the stem. Their sporangia, whether all alike (in Lycopodiaceae) or of two kinds (in Selaginellaceae) are borne on or near the base of the fertile leaves (sporophylls). The sporophylls may be like or unlike the foliage leaves and either in distinct terminal cones (strobili) or occurring in fertile zones at intervals along the stem. In *Selaginella*, but not in the other British clubmosses, there is a minute membranous flap of tissue (ligule) at the base of the upper surface of the leaf or above the sporangia on the fertile leaves. The sporangia in the Lycopodiaceae all contain the same kinds of spores, whereas those in the Selaginellaceae contain either megaspores (in megasporangia) or smaller and more numerous microspores (in microsporangia).

Quillworts have subulate leaves which are long in proportion to the size of the rather small, corm-like axis on which they are borne. A sporangium is embedded in the broad basal part of the fertile leaves. As in *Selaginella* there is a minute ligule on the upper surface of the leaf immediately above the sporangium. Megaspores and microspores are formed in different sporangia as in *Selaginella*.

Horsetail leaves are small, but they are quite distinctive in the way they are joined below to form a sheath round the stem. The British species all have distinct terminal cones. The sporangia are borne on peltate (mushroom-shaped) sporangiophores which are quite different from the fertile leaves of the other groups.

LIFE-HISTORIES

The Male-fern (*Dryopteris filix-mas*)

The plant consists essentially of a short stocky stem surmounted by a basket-like tuft of large leaves, or fronds. The stem, or stock (Fig. 1), grows in length by the activity of an apical growing point; when very young the stem is small, but as it grows older it increases in size so that its general shape becomes that of an inverted cone. The stem only very occasionally branches. The growing point is concealed by the young, as yet still rolled-up, leaves, and the rest of the stem by leaf bases, spirally arranged on its surface, and thin wiry roots borne at the bases of the stipes (leaf stalks).

The fronds (leaves) of the Male-fern (Fig. 107, B) are divided like a feather into segments, called pinnae, and each pinna is similarly divided almost down to its main vein, or midrib, into a number of lobes or segments. If a frond from a vigorous plant is examined in the late summer it will be seen that, in the upper half of the frond at least, each segment carries on its lower surface a number of raised brown patches called sori (sing. sorus). These are arranged in two parallel rows one on either side of the midrib and nearer to it than to the edge or margin of the frond (Fig. 2). Each sorus stands above a veinlet which supplies it with nourishment.

The essential structures in a sorus are the stalked capsules called sporangia (Fig. 3) which grow from a swelling called the receptacle. The sporangia, in the young stage at least, are covered and protected by a kidney-shaped scale called the indusium. The head of each sporangium, the capsule, is shaped like a bi-convex lens, within which develop the dark-coloured spores. The margin of the capsule, i.e. round the edge of the lens, is almost completely surrounded by a series of cells having thickened inner and radial walls, which together form a structure called the annulus. There may be thinner walled, basal cells, between the base of the annulus and the stalk while remaining thin-walled marginal cells stand on either side of a point (the stomium) at which the sporangium, when ripe, will open to liberate the spores. This process, called dehiscence, takes place in dry air. The annulus cells dry up and the whole annulus shrinks and gradually bends back on itself (Fig. 4). Finally the tension on the water in the annulus cells becomes too great and a gas bubble appears in each of them. The whole annulus then recovers its shape with a sudden jerk slinging the spores out into the air.

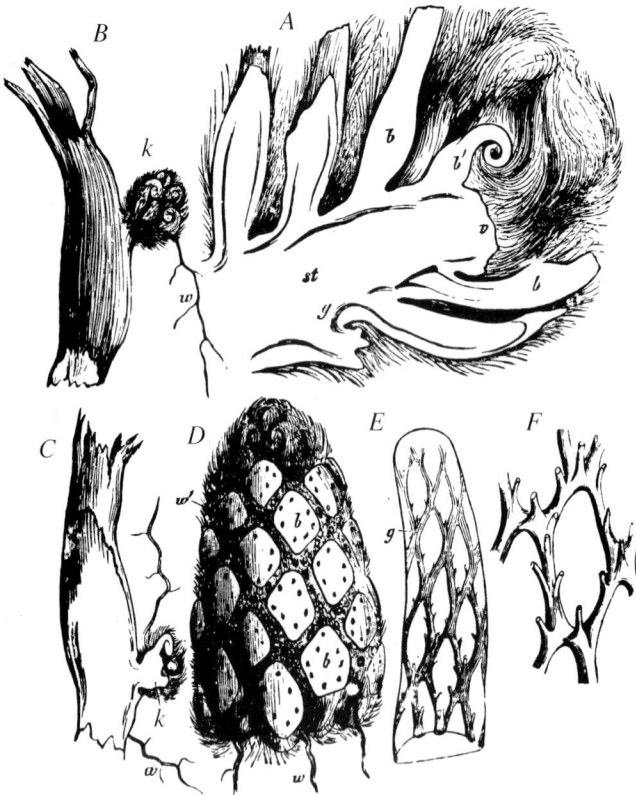

Fig. 1. Male-fern (*Dryopteris filix-mas*). *A*. Stem (or stock) in longitudinal section: *v*, apex with growing point; *st*, stem (or stock); *b*, leaf stalks (or stipes); *b'*, one of the still enfolded leaves (or fronds); *g*, vascular strands. *B*. Leaf stalk (or stipe) bearing at *k* a bud with a root (*w*) and several fronds. *C*. A similar leaf stalk cut longitudinally. *D*. Stock from which the mature leaves have been cut away to their bases leaving only those of the terminal bud: *w,w'*, some of the numerous roots which fill the spaces between the leaves. *E*. Stock from which the rind has been removed to show the network (*g*) of vascular bundles (the plant's water- and food-supply system). *F*. A single mesh of the network enlarged, showing the smaller bundles which pass up into the leaf-bases. (After Sachs).

The most usual number of spores in each fern sporangium is 64, but in the Male-fern complex, chromosome imbalance may produce 16, 32 or 48; the number is usually a multiple of the 16 original spore mother cells.

Each spore consists of a single cell enclosed in a thick wall. Given moisture and a suitable temperature the wall bursts and the contents

LIFE-HISTORIES (MALE-FERN)

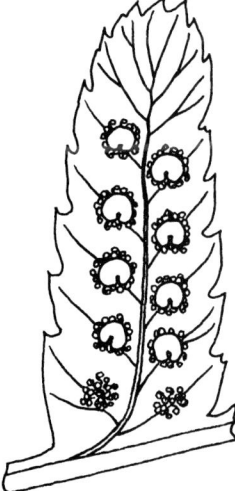

Fig. 2. Male-fern (*Dryopteris filix-mas*). A single fertile segment bearing kidney-shaped (reniform) sori (*c.*×4).

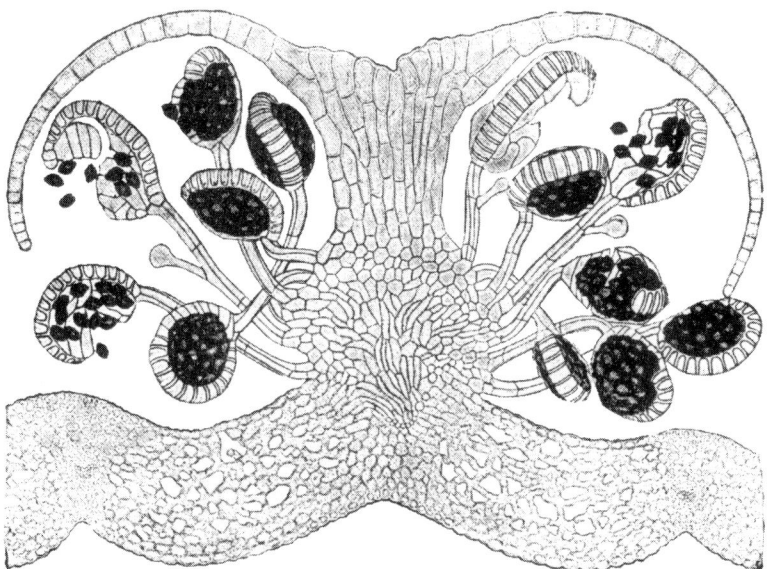

Fig. 3. Single sorus of Male-fern (*Dryopteris filix-mas*) in vertical section with the underside of the blade uppermost. The lower part of the sections passes through a small portion of the leaf blade and shows the lower and upper epidermis and in between them the assimilating tissue with its many air spaces. The sporangia are seen attached by their stalks to the receptacle and (at this stage) covered by the indusium (which seen here in vertical section looks like the head of a horned sheep).

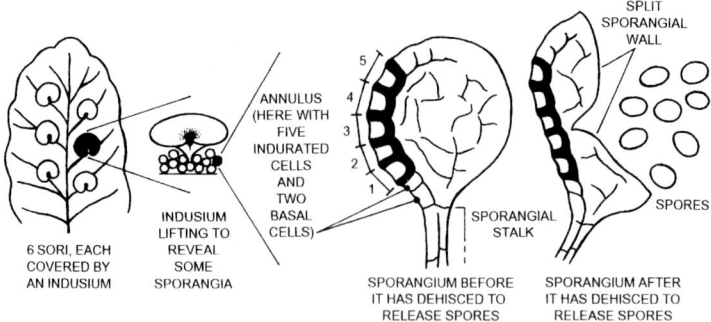

Fig. 4. Magnification of a sorus of a homosporous fern to show eventual dehiscence of sporangium and release of spores.

protrude. The one original cell divides to form the prothallus which is at first thread-like, then later widens out until finally it becomes heart-shaped (Fig. 5). The fully formed prothallus consists of a single layer of cells, except

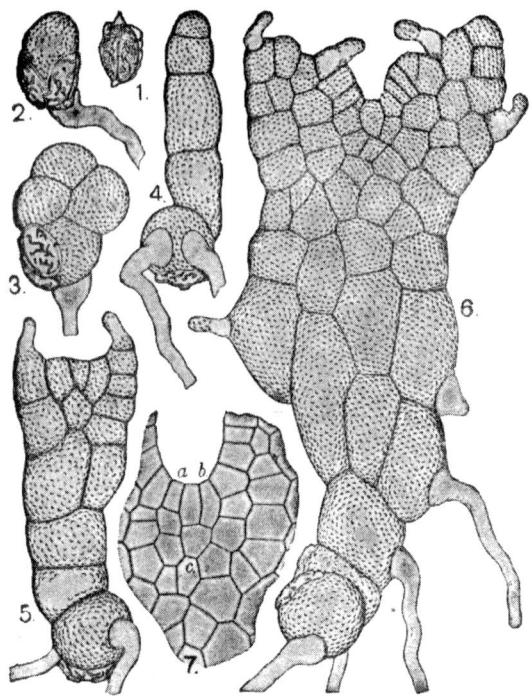

Fig. 5. Successive stages in the germination of a spore of Male-fern (*Dryopteris filix-mas*) to form a prothallus (*c.*×200). (After Kny).

below the notch of the heart where it swells out into a cushion several cells thick. The entire independent plantlet is attached to the soil and supplies itself with moisture and nutrient salts by means of hair-like root-cells called rhizoids. The remaining vegetative cells all contain chlorophyll and the prothallus is, therefore, able to assimilate carbon from the air and, with sunlight, build up all the foods it requires. However, unlike the fern plant, the prothalli of most species can live only under conditions of continuous moisture. Moreover, the sexual reproductive processes demand the presence of external fluid water.

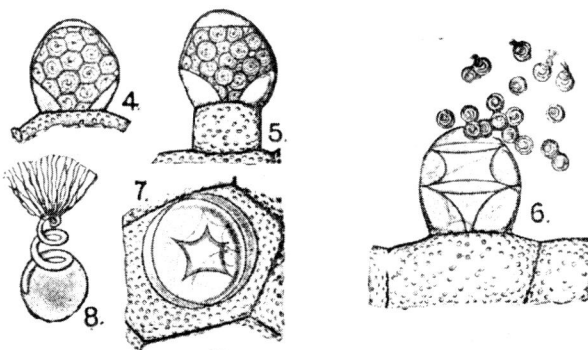

Fig. 6. Antheridia and sperms of Male-fern (*Dryopteris filix-mas*). 4&5. Developing antheridia containing sperm cells (*c.*×350). 6&7. Ruptured antheridia and liberated sperms (*c.*×350). 8. Single sperm (*c.*×1,050). (After Kny).

The sexual organs are antheridia (male organs) and archegonia (female organs). Young prothalli often produce antheridia, although both antheridia and archegonia are normally borne on the lower surface of mature prothalli; the antheridia in the basal region or at the sides, the archegonia on the central cushion. Each antheridium (Fig. 6) consists of a little knob projecting above the surface. When mature it contains a group of sperm cells which are liberated as free swimming sperm in the presence of external water. The archegonium (Fig. 7) when mature comprises a large egg cell situated below the surface of the prothallus, together with a short tube of cells called the neck which projects beyond the surface. In the presence of fluid water the two central neck canal cells turn into mucilage and the neck bursts open, thereby chemically attracting sperm which swim through the water towards the archegonium. Sperm enter the canal and one finally fuses with the egg. The fertilized egg (zygote) starts growing immediately into a young fern plant (Fig. 8) which, though at first dependent for its nourishment on the prothallus, soon becomes independent. The prothallus then dies and rots

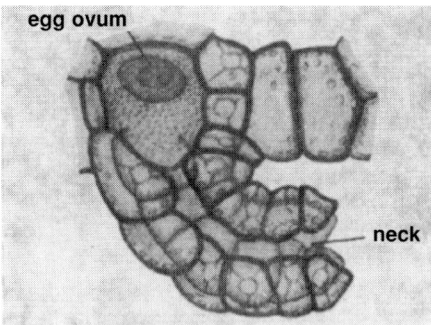

Fig. 7. Mature archegonium, its neck open to receive the sperm cells (c.×350).

away. Prothalli of some species are very long-lived. This is especially true of the prothalli of the Killarney Fern (*Trichomanes speciosum*) which can form very large colonies, even in areas where the parent sporophyte does not live (p.193).

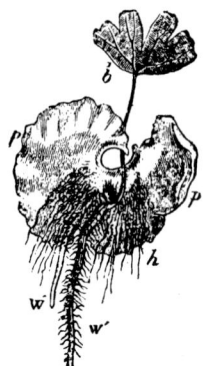

Fig. 8. Early stage in the development of a young fern plant (sporophyte), still attached to the prothallus (gametophyte). *pp'*. Prothallus. *b*. First leaf. *w,w'*. First and second roots. *h*. Rhizoids of the prothallus. (c.×20). (After Sachs).

The fern life-cycle is thus seen to be made up of two alternating phases or generations. One, called the sporophyte, begins with the zygote and culminates in the formation of spores. The other, called the gametophyte begins with the spore and culminates in the fusion of the two sexual cells, or gametes (Fig. 9). This handbook uses sporophyte characters to identify the species.

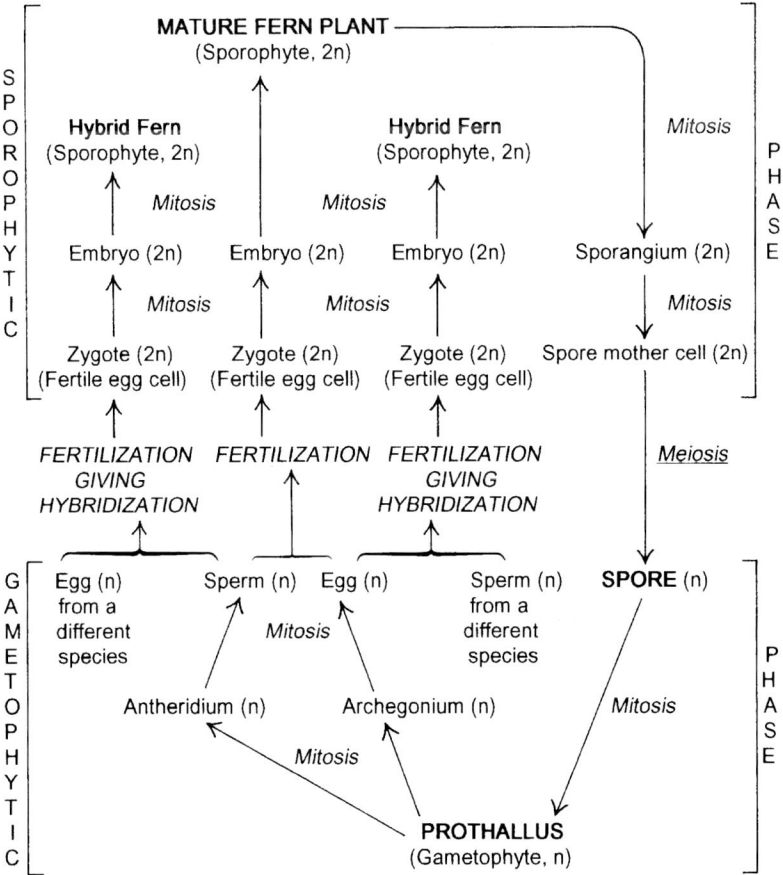

Fig. 9. Diagram illustrating the life cycle of a typical homosporous fern, with routes to hybridization.

Chromosomes

The sporophyte and the gametophyte of ferns differ not only in their form and behaviour but also in the structure of their cells. The nucleus of each cell of the prothallus may be seen, at least when the cell is about to divide, to contain a fixed number of thread-like structures called chromosomes (in Dwarf Male-fern, *Dryopteris oreades*, cited here as a simple example, the number is 41). The chromosomes in any particular species are all similar although each has its own individuality. In the course of cell-division each chromosome divides along its length, one longitudinal half going to one daughter cell, and the other half to the other daughter cell. This process is called mitosis. Since the hereditary potentialities of each cell (the genes) are carried by the chromosomes, all those potentialities are passed to each and every cell of the prothallus, including the sperms and egg cells. At fertilization, a sperm with (in *D. oreades*) 41 chromosomes fuses with an egg also possessing 41 chromosomes. The fertilized egg or zygote, thus contains two sets of 41 chromosomes or 82 in all.

The zygote immediately divides by mitosis giving rise to two cells and the process is repeated again and again until the fern plant (the sporophyte) becomes mature. Each and every cell of the sporophyte therefore contains 82 chromosomes.

When the spores are about to be formed, each spore mother cell in the developing sporangium gives rise to four spores. The nuclear divisions which take place during spore-formation are different from mitosis and together are termed meiosis and result in four nuclei (one to each spore) each containing 41 chromosomes. When a spore germinates, its single cell divides mitotically and such divisions continue throughout the growth of the gametophyte plant thus formed so that this plant also contains, as already described, the reduced number of chromosomes (41) in each of its cells.

Other species of ferns possess various numbers of chromosomes, but in general the fern 'plant' (the sporophyte or diploid generation) of a particular species contains the unreduced number of chromosomes (referred to as 2n) appropriate to that species, while the gametophyte (or haploid generation) contains exactly half as many (n).

Polyploidy

The number 41, or some multiple of it is found in all the normal species of *Dryopteris* native to Wales, e.g. in Male-fern 2n=164. It is said to be the base-number or basic chromosome number. Other fern genera contain species in which 2n may be up to six times the base-number of the group concerned. Such species are referred to variously as triploids (in which 2n=3 times the base-number), tetraploids (2n=4 times the base-number), pentaploids (5) or hexaploids (6), and in general as polyploids.

Hybridization

If two or more prothalli grow side by side there is a possibility that a sperm from one may fertilize an egg cell of another. It is even possible for cross fertilization to occur between prothalli of different species. This is hybridization and the hybrid sporophyte which arises from such a union is likely to have some resemblances to the sporophytes of both species (Fig. 9). Many supposed natural hybrids between fern species have been described, e.g. in the genera *Dryopteris* and *Asplenium*. Some of these have also been reproduced artificially in the laboratory and their hybrid nature has been confirmed by microscopic observation of their chromosomes. Several hybrids have also been described in the genus *Equisetum*. Hybrids are probably more numerous than has generally been supposed and should be looked for, especially in mixed populations of the potential parents.

Interspecific hybrids have the special sign (×) placed between the generic name and the specific epithet.

Apogamy

In a few species of fern (e.g. *Dryopteris affinis*), the prothallus has the same number of chromosomes as the parent plant. The young sporophyte grows from the gametophyte not as the result of fertilization but (by a process called apogamy) as a sort of bud and itself produces spores by a series of nuclear and cell divisions from which the number of chromosomes emerges unchanged (Fig. 10).

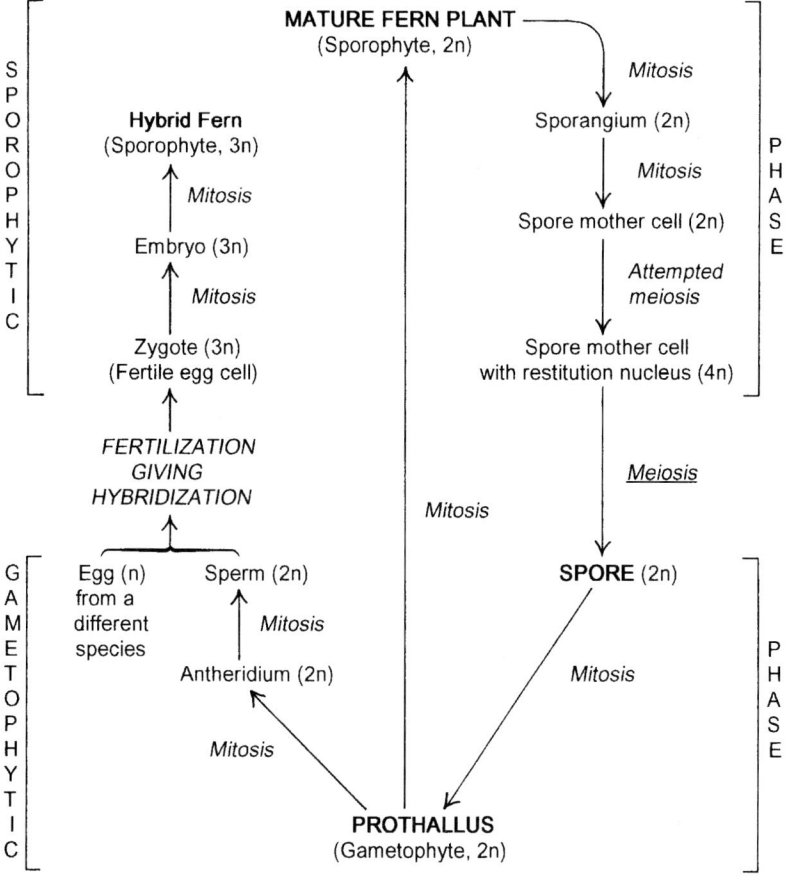

Fig. 10. Diagram illustrating the life cycle of a typical apogamous fern, with route to hybridization.

There are several different types of nuclear division in the sporangia of such ferns which can result in apogamy but, as a result of all of them, the gametophytes do not produce functional archegonia, only antheridia. There is, therefore, no possibility of self fertilization. However, there can be intergametophytic hybridization involving a diploid sperm, from an apogamously produced gametophyte, and a haploid egg cell, from a normal gametophyte. Such a fusion would result in a triploid sporophyte.

Homospory and Heterospory

Homospory: The description of the life-cycle of ferns given in the preceding section is true for the great majority of families of Pteridophyta. The spores are all alike and capable of developing into independent prothalli, which usually form both kinds of sexual organs. The spores are said to be homospores and the plants exhibit homospory. The horsetails and those clubmosses belonging to the Lycopodiaceae family are all homosporous and have similar life-cycles to the Male-fern. Their prothalli are, however, rather different. Those of the horsetails may grow into small thin antheridial gametophytes and others into larger thicker archegonial gametophytes. The latter subsequently produce antheridial lobes to become bisexual. Species of *Lycopodium* have either subterranean tuberous prothalli that are usually associated with a fungus or green prothalli that grow on the surface.

Heterospory: There are, however, four families represented in Wales which have rather different life-cycles: the Pillworts (p. 164), the Water Ferns (p. 107), the Lesser Clubmosses (p. 42) and the Quillworts (p. 46). These families differ widely in many respects but they all form two different kinds of spores (microspores and megaspores) in different sporangia. These spores are said to be heterospores and the plants exhibit heterospory. A microspore is relatively small and on germination gives rise to a male prothallus which is not capable of existing independently for any length of time, but which forms an antheridium almost immediately. A megaspore, by comparison, is relatively large containing abundant nourishment. It germinates to give a female prothallus which is able to live, for some time at least, an independent existence and forms one or more archegonia. The embryonic plant which results from fertilization of one of the eggs is nourished, at first, at the expense of the food reserve contained in the megaspore.

The differences between the spores are reflected also in the numbers in which they are produced. The microspores are formed in large numbers in each microsporangium but there are far fewer megaspores formed in each megasporangium. There are up to 300 in *Isoetes,* 4 in *Selaginella selaginoides* but only one in each megasporangium of *Azolla* and *Pilularia*.

IDENTIFICATION

The main aim of this book is to help its readers to identify all the species of ferns, clubmosses, quillworts and horsetails native to Wales. Those few species and hybrids not found in Wales but native to other parts of the British Isles have also been included, but with shorter descriptions. Attention must be focused on features which, at a summary glance, the beginner might overlook completely. These features often provide somewhat subtle points of distinction, and it is essential that they should be described in language which is both concise and exact. For this purpose it is necessary to use a number of special botanical words (such as glabrous, meaning devoid of hairs) and to limit the use of certain other words to special meanings, e.g. entire which means 'devoid of any sort of toothing' when used to describe the margin of a leaflet. For quick reference, the definitions of these technical terms can be found in the glossary, with extra details of many given in the introductory sections of the book and at the start of each class of pteridophyte.

THE HERBARIUM OF THE NATIONAL MUSEUM & GALLERY, CARDIFF
(NATIONAL MUSEUM OF WALES, NMW)

A Brief History

The Department of Botany at the National Museum & Gallery, Cardiff (National Museum of Wales, **NMW**) first took shape in 1919 but the history of its herbarium may be traced from very small beginnings some ninety years earlier. Charles Conway of Pontrydyryn formed a small collection of dried plants during the eighteen-thirties which, after his death, was acquired by the then Cardiff Museum. The Herbarium thus begun was added to by John Storrie, who in 1888 gave his own specimens to the Museum. Around the turn of the century important additions were received from two well-known botanists of the period, Arthur Bennett and the Rev. H. J. Riddelsdell. In 1912, when the entire contents of the Cardiff Museum were taken over by the newly constituted National Museum of Wales, the Herbarium was estimated to contain some 3500 sheets plus an unknown number of unmounted specimens.

So far the collection had not advanced beyond the requirement of a local museum, but with the establishment of the Department of which it forms an essential part, the Herbarium grew very rapidly and now contains 518,000 specimens. This expansion has been guided by the stated aim of the Department as part of the Museum, namely the complete illustration of the flora of Wales and its relationship to the rest of the world by means of collections that are international in scope, importance and quality.

In order to facilitate the comparison of Welsh specimens with corresponding material from the rest of the British Isles and Europe, the Welsh collections have been treated as an integral part of the European collections. Specimens of each species are arranged geographically, those from Wales being placed first followed by those from the rest of the British Isles and Europe. For plants outside Europe and those that have been cultivated, there is an extensive Foreign and Cultivated collection. Despite the name of the Museum changing slightly in 1995 the international herbarium abbeviation is retained as **NMW**.

The Pteridophyte Collection at NMW

The collection of ferns, clubmosses, quillworts and horsetails, comprises over 30,000 specimens from nearly 200 collectors; these have been acquired from a variety of sources. The most notable early collections were from W. A. Shoolbred and H. J. Wheldon from South Wales, and T. G. Rylands and J. E. Griffith from North Wales. European specimens were received from H. S. Thompson and D. A. Jones. M. S. Percival's material was the main collection from the rest of the world.

Since the publication of the last edition of Welsh Ferns in 1978 the pteridophyte herbarium has been expanded considerably with the arrival of university herbaria from Bristol, Cardiff and Queen Mary College, London. The main private collections have included those from C. R. Fraser-Jenkins, with his early plants from South Wales and the rest of the British Isles, as well as many from remote corners of the world. Other noteworthy collections have been from F. Rose, W. S. Lacey and W. E. Hughes. There are also many smaller donations of specimens as well as specimens arising from departmental research, exchange and purchase.

HINTS TO COLLECTORS

Collecting ferns presents no special difficulties, but care must be taken to gather fronds which are well matured. They need not necessarily be the largest, but should show ripe sori. Immature fronds are usually misleading and should be avoided, or if collected should be accompanied by mature ones. In the other pteridophytes, fertile stems are often not available but it is usually possible to identify vegetative material, except in *Isoetes*, where megaspores are needed for identification. The scales which clothe the base of the stipe often provide characters of importance in identification, so the fronds should be collected complete with the stipe, i.e. they should be detached from the root-stock at their base. This can be easily accomplished with most ferns by gripping the base of the stipe and applying a firm downward pressure, so breaking it away from the root-stock.

It is not necessary to collect complete fern plants, and they should never be uprooted. It is against the law for an unauthorized person to uproot any wild plant without the permission of the landowner.

There is further legislation to protect the very rare and endangered or vulnerable species in the British Isles. Essentially the legislation makes it illegal to interfere in any way with the species (listed below) or their habitats or to be involved in trading them. What species to include on the lists is regularly reviewed.

In **Great Britain** the following pteridophytes are Protected under the *Countryside and Wildlife Act, 1981*: *Equisetum ramosissimum* (Branched Horsetail); *Cystopteris dickieana* (Dickie's Bladder-fern); *Ophioglossum lusitanicum* (Least Adder's-tongue); *Trichomanes speciosum* (Killarney Fern); *Woodsia alpina* (Alpine Woodsia) and *Woodsia ilvensis* (Oblong Woodsia).

In the **Republic of Ireland** the following are protected under the *Flora Protection Order, 1987*: *Asplenium obovatum* subsp. *lanceolatum* (Lanceolate Spleenwort); *Asplenium septentrionale* (Forked Spleenwort); *Cryptogramma crispa* (Parsley Fern); *Gymnocarpium robertianum* (Limestone Fern); *Pilularia globulifera* (Pillwort) and *Trichomanes speciosum* (Killarney Fern).

In **Northern Ireland** the following are protected under the *Wildlife (NI) Order, 1985*: *Lycopodiella inundata* (Marsh Clubmoss); *Gymnocarpium dryopteris* (Oak Fern); *Pilularia globulifera* (Pillwort); *Polystichum lonchitis* (Holly-fern) and *Trichomanes speciosum* (Killarney Fern).

EXPLANATION OF VICE-COUNTIES AND 10-KM SQUARE (HECTAD) MAPS

Vice-counties

Records of plant distribution throughout the British Isles have been kept for many years using the vice-county system devised initially for Great Britain by H. C. Watson. He divided the whole area for botanical purposes into 112 divisions, which often coincided exactly or approximately with the old shire-counties, though some large counties had to be subdivided to give areas of more or less uniform size. Watson called his divisions 'vice-counties' and numbered them serially from West Cornwall, including Scilly, (v.c.1) to Shetland (v.c.112). A more detailed explanation of these has been published in more recent times (Dandy, 1969). The Channel Islands, given the abbreviation v.c.S ('S' for Sarniense) and the 40 Irish vice-counties, given the abbreviations v.c.H1-H40 ('H' for Hibernia), were added later (Figs. 11 and 12). From time to time the Watsonian vice-counties have been

18 EXPLANATION of VICE-COUNTY and 10-KM SQUARE (HECTAD) MAPS

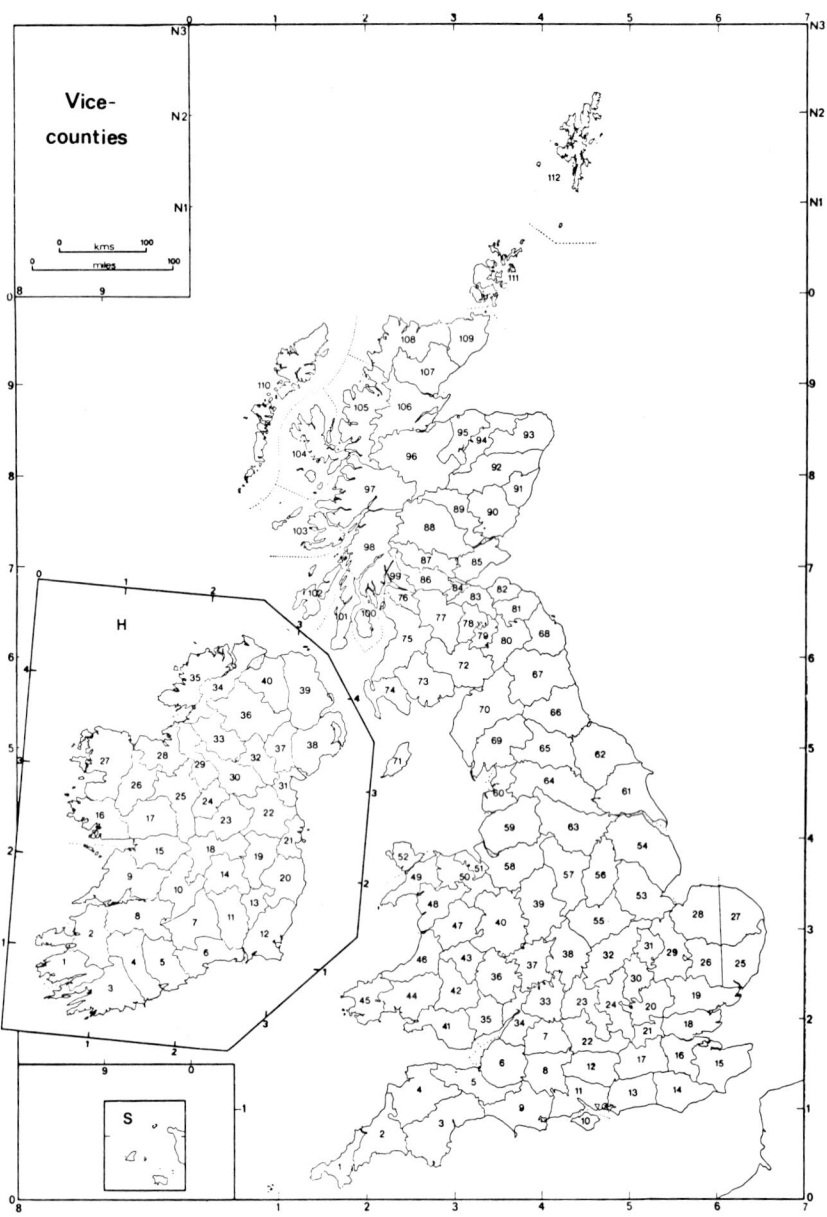

Fig. 11. Map of British Isles showing vice-counties.

Names of Vice-counties

ENGLAND I
1   W. Cornwall
1b   Scilly
2   E. Cornwall
3   S. Devon
4   N. Devon
5   S. Somerset
6   N. Somerset
7   N. Wilts.
8   S. Wilts.
9   Dorset
10   Wight
11   S. Hants.
12   N. Hants.
13   W. Sussex
14   E. Sussex
15   E. Kent
16   W. Kent
17   Surrey
18   S. Essex
19   N. Essex
20   Herts.
21   Middlesex
22   Berks.
23   Oxon
24   Bucks.
25   E. Suffolk
26   W. Suffolk
27   E. Norfolk
28   W. Norfolk
29   Cambs.
30   Beds.
31   Hunts.
32   Northants.
33   E. Gloucs.
34   W. Gloucs.
35   (see Wales)
36   Herefs.
37   Worcs.
38   Warks.
39   Staffs.
40   Salop

WALES
35   Mons.
41   Glam.
42   Brecs.
43   Rads.
44   Carms.
45   Pembs.
46   Cards.
47   Monts.
48   Merioneth
49   Caerns.
50   Denbs.
51   Flints.
52   Anglesey

ENGLAND II
53   S. Lincs.
54   N. Lincs.
55   Leics.
55b Rutland
56   Notts.
57   Derbys.
58   Cheshire
59   S. Lancs.
60   W. Lancs.
61   S.E. Yorks.
62   N.E. Yorks.
63   S.W. Yorks.
64   Mid-W. Yorks.
65   N.W. Yorks.
66   Co. Durham
67   S. Northumb.
68   Cheviot
69   Westmorland
70   Cumberland

ISLE OF MAN
71   Man

SCOTLAND
72   Dumfriess.
73   Kirkcudbrights.
74   Wigtowns.
75   Ayrs.
76   Renfrews.
77   Lanarks.
78   Peebless.
79   Selkirks.
80   Roxburghs.
81   Berwicks.
82   E. Lothian
83   Midlothian
84   W. Lothian
85   Fife
86   Stirlings.
87   W. Perth
88   Mid Perth
89   E. Perth
90   Angus
91   Kincardines.
92   S. Aberdeen
93   N. Aberdeen
94   Banffs.
95   Moray
96   Easterness
97   Westerness
98   Main Argyll
99   Dunbarton
100 Clyde Is.
101 Kintyre
102 S. Ebudes
103 Mid Ebudes
104 N. Ebudes

105 W. Ross
106 E. Ross
107 E. Sutherland
108 W. Sutherland
109 Caithness
110 Outer Hebrides
111 Orkney
112 Shetland

IRELAND
H01 S. Kerry
H02 N. Kerry
H03 W. Cork
H04 Mid Cork
H05 E. Cork
H06 Co. Waterford
H07 S. Tipperary
H08 Co. Limerick
H09 Co. Clare
H10 N. Tipperary
H11 Co. Kilkenny
H12 Co. Wexford
H13 Co. Carlow
H14 Laois
H15 S.E. Galway
H16 W. Galway
H17 N.E. Galway
H18 Offaly
H19 Co. Kildare
H20 Co. Wicklow
H21 Co. Dublin
H22 Meath
H23 Westmeath
H24 Co. Longford
H25 Co. Roscommon
H26 E. Mayo
H27 W. Mayo
H28 Co. Sligo
H29 Co. Leitrim
H30 Co. Cavan
H31 Co. Louth
H32 Co. Monaghan
H33 Fermanagh
H34 E. Donegal
H35 W. Donegal
H36 Tyrone
H37 Co. Armagh
H38 Co. Down
H39 Co. Antrim
H40 Co. Londonderry

CHANNEL ISLANDS
S

Fig. 12. Vice-county names in abbreviated form, with their vice-county number.

EXPLANATION of VICE-COUNTY and 10-KM SQUARE (HECTAD) MAPS

divided further, the distinct part being allocated a 'b' after the number of the vice-county. Currently there are two sub-divisions: v.c.1b, Scilly, from West Cornwall, and v.c.55b, Rutland, from Leicester.

In Wales, the vice-county boundaries follow closely the political boundaries of the 13 old shire-counties as they were before the Local Government reorganisation of 1974. They start with Monmouthshire (v.c.35) and then run in numerical order from Glamorgan (v.c.41) through to Anglesey (v.c.52) (Fig. 13). Each vice-county has a recorder appointed by the Botanical Society of the British Isles (B.S.B.I.) who is responsible for collating the vascular plant records (ferns, fern allies and flowering plants) and to whom all records should be sent. Details of current vice-county recorders for Wales can be obtained from Department of Botany, National Museum & Gallery, Cardiff CF1 3NP.

10-kilometre square (or hectad) distribution maps

These follow the Ordnance Survey's National Grids for Great Britain and in recent times for Ireland. The grid system for the whole of Ireland is tilted at a slightly clockwise angle from the Great Britain grid system. Previously another system operated for Ireland (e.g. *Atlas of the British Flora*, 1962), which was a continuation westwards of the grid system for Great Britain. The Isle of Man follows the Great Britain National Grid but the Channel Islands' records are plotted in the 10-km (hectad) squares of the Universal Transverse Mercator grid.

If a pteridophyte has been recorded from a 10-km square (hectad) it is represented by a symbol in the square corresponding to the most recent record. Symbols used for the maps are as follows:-

- ● 1970 - 1995
- ◐ 1950 - 1969
- ◓ 1930 - 1969 untraced *Atlas of the British Flora* record
- ○ pre-1950
- ◆ Introduced 1970 - 1995
- ◇ Introduced pre-1970

The *Atlas of the British Flora* was first published in 1962 (Perring & Walters eds) with a *Critical Supplement...* in 1968 (Perring ed.) both using 10-km square (hectad) maps of the British Isles to display plant distribution. Two date zones were used: 1930 onwards and pre-1930. The B.S.B.I. Monitoring Scheme (1987-1988) updated the 'Atlas' for a limited number of squares in the British Isles - every third 10-km square (hectad) being chosen in a grid across the British Isles. Publication of the results was in 1995 (Palmer, M. A. & Bratton, J. H. eds). Currently, work is in progress with a view to publishing an 'Atlas 2000' for the whole of the British Isles.

EXPLANATION of VICE-COUNTY and 10-KM SQUARE (HECTAD) MAPS   21

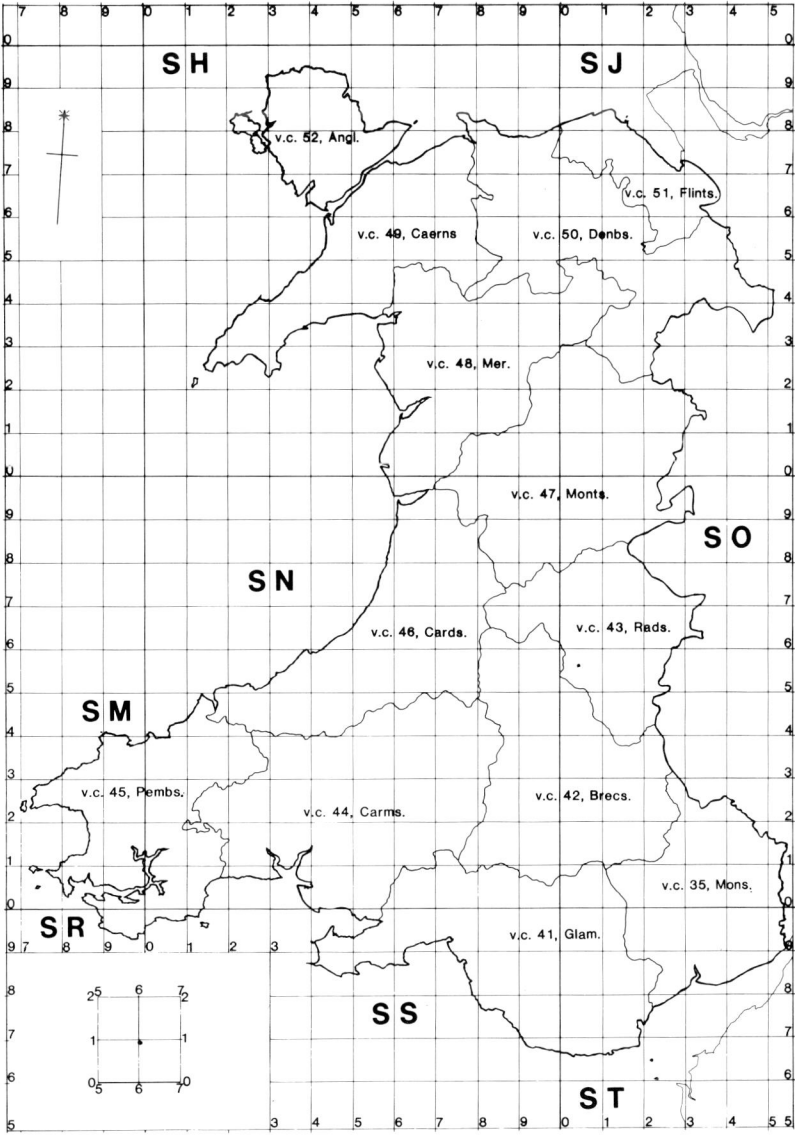

Fig. 13. Map of Wales showing vice-county boundaries and 10-km squares (hectads).

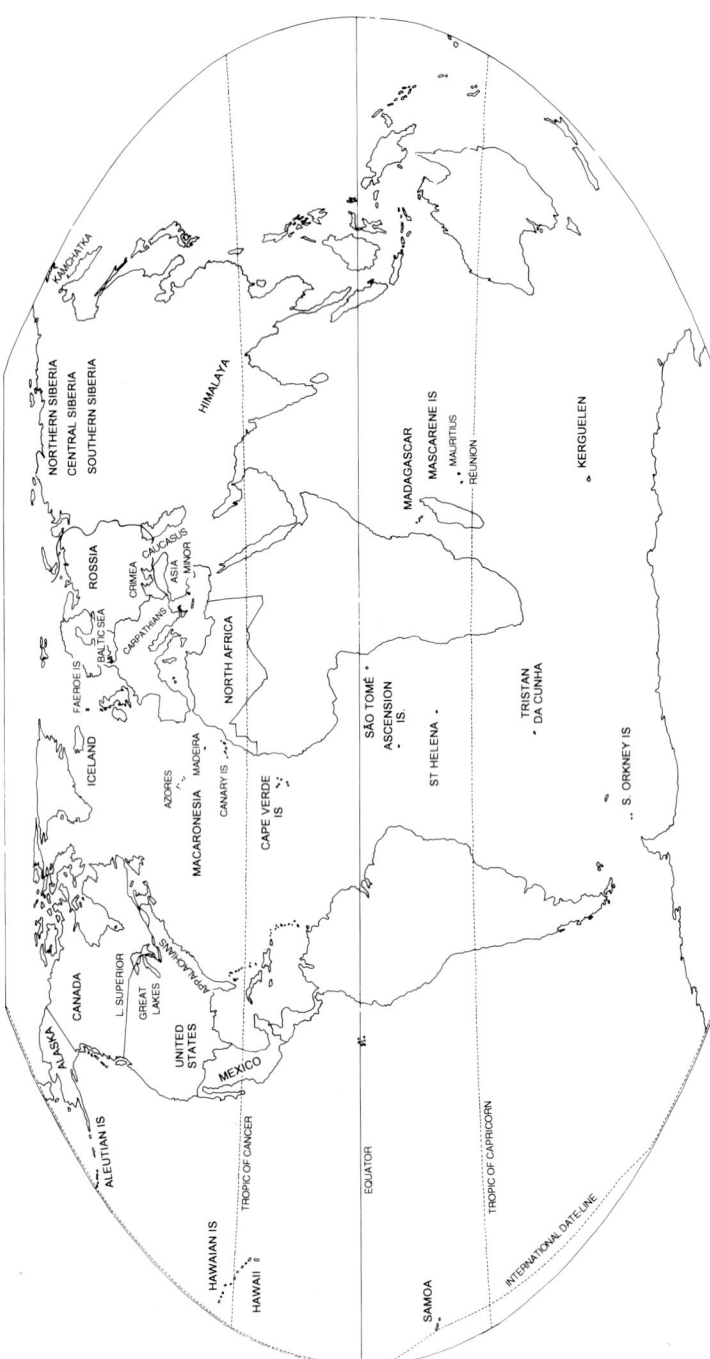

Fig. 14. Map of the World showing some of the place-names mentioned in the text.

WORLD DISTRIBUTIONS

Many species of ferns are polymorphic so that in different parts of the world they may exist as different subspecies or varieties from those which occur in the British Isles. The world distributions given, therefore, refer to the species as a whole. In a short paragraph, only a general idea of the world distribution of a species can be given. Some species outside Europe originally thought by earlier pteridologists to be the same as the European species are now treated as different species, e.g. the *Huperzia selago* from Greenland is *H. appalachiana* Beitel & Mickel. A World Map showing some of the less well known localities referred to in the text is opposite. Definitions of some of the collective terms used in the text for various parts of the world are given in the glossary.

The European maps follow Hultén, E. & Fries, M. (1986): *Atlas of North European Vascular Plants*. Vol. 1. and Jalas, J. & Suominen, J. eds (1972): *Atlas Florae Europaeae*. Vol. 1.

The European area of the former USSR is referred to in the text as Rossia and its divisions are listed in brackets afterwards (Fig. 15): Northern (N), Baltic (B), Central (C), South-western (W), Crimea (K), South-eastern (E). These divisions essentially follow those of Komarov, V. L. *et al.* eds (1934-1964). [*Flora of the USSR*].

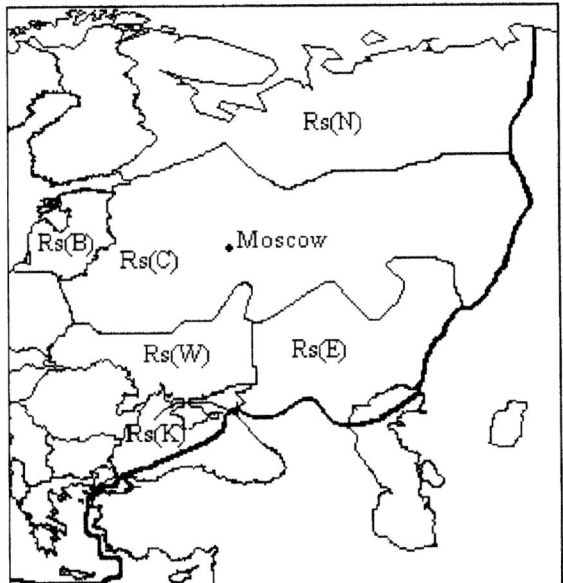

Fig. 15. Map of Rossia essentially following the floristic divisions of V. L. Komarov *et al.* eds (1934-1964). [*Flora of the USSR*].

SYNOPSIS OF CLASSIFICATION OF PTERIDOPHYTA IN THE BRITISH ISLES

This systematic list consists of species considered to have been correctly identified for the British Isles and essentially follows the classification in *Flora Europaea* (second edition) (Tutin, *et al.*, 1993) Vol. 1. Those species considered to have been correctly identified for Wales are in bold. Those recorded in error for the British Isles which are described in the alphabetical arrangement are in italics. All alien species and totally alien genera and families in the synopsis are asterisked (*). Some additional aliens associated with gardens and estates can be found, with brief comments, at the end of the main alphabetical sequence.

Class **Lycopsida**

Family **Lycopodiaceae** ....Clubmoss family
    Genus **Huperzia** ....Fir Clubmosses .... p. 36
        Huperzia selago ....Fir Clubmoss............................. p. 36
    Genus **Lycopodiella** ....Marsh Clubmosses .... p. 38
        Lycopodiella inundata ....Marsh Clubmoss .... p. 38
    Genus **Lycopodium** ....Clubmosses .... p. 39
        Lycopodium annotinum ....Interrupted Clubmoss ...... p. 40
        L. clavatum ....Stag's-horn Clubmoss ...... p. 40
    Genus **Diphasiastrum** ....Alpine Clubmosses .... p. 33
        Diphasiastrum complanatum subsp. issleri.......................
        ....Issler's Clubmoss .... p. 35
        D. alpinum ....Alpine Clubmoss .... p. 33
Family **Selaginellaceae**.........Lesser Clubmoss family
    Genus **Selaginella**...............Lesser Clubmosses .... p. 42
        Selaginella selaginoides ....Lesser Clubmoss .... p. 44
        S. kraussiana* ....Kraus's Clubmoss .... p. 43
Family **Isoetaceae**...............Quillwort family
    Genus **Isoetes** ....Quillworts .... p. 46
        Isoetes lacustris ....Quillwort .... p. 48
        I. echinospora ....Spring Quillwort .... p. 46
        I. histrix ....Land Quillwort .... p. 47

Class **Sphenopsida**

Family **Equisetaceae** ....Horsetail family
    Genus **Equisetum** ....Horsetails .... p. 49
        Equisetum hyemale...............Rough Horsetail .... p. 55
        E. ramosissimum*...................Branched Horsetail .... p. 59
        E. variegatum. ....Variegated Horsetail ...... p. 63
        E. fluviatile ....Water Horsetail .... p. 54
        E. palustre ....Marsh Horsetail .... p. 57

E. sylvaticum ....Wood Horsetail .... p. 60
E. pratense...........................Shady Horsetail .... p. 59
E. arvense...........................Field Horsetail .... p. 52
E. telmateia ....Great Horsetail .... p. 61

              Hybrids in Equisetum

E. hyemale × E. ramosissimum
  = E. × moorei ....Moore's Horsetail .... p. 67
E. hyemale × E. variegatum
  = E. × trachyodon ....Mackay's Horsetail .... p. 68
E. fluviatile × **E. palustre**
  = **E. × dycei** ....Dyce's Horsetail .... p. 67
E. fluviatile × E. telmateia
  = E. × willmotii ....Willmot's Horsetail .... p. 67
E. palustre × **E. telmateia**
  = **E. × font-queri** ....Font-Quer's Horsetail ...... p. 68
E. sylvaticum × E. telmateia
  = E. × bowmanii ....Bowman's Horsetail .... p. 68
E. pratense × E. sylvaticum
  = E. × mildeanum ....Milde's Horsetail .... p. 68
E. arvense × **E. fluviatile**
  = **E. × litorale** ....Shore Horsetail................ p. 65
E. arvense × E. palustre
  = E. × rothmaleri ....Ditch Horsetail .... p. 66

              Class **Filicopsida**

Family **Ophioglossaceae** ......Adder's-tongue family
    Genus **Ophioglossum** ....Adder's-tongues .... p. 154
        Ophioglossum lusitanicum ......Least Adder's-tongue ...... p. 155
        O. azoricum ....Small Adder's Tongue...... p. 155
        O. vulgatum ....Adder's-tongue .... p. 156
    Genus **Botrychium** ....Moonworts .... p. 112
        Botrychium lunaria..............Moonwort .... p. 112
Family **Osmundaceae** ....Royal Fern family
    Genus **Osmunda** ....Royal Ferns..................... p. 160
        Osmunda regalis ....Royal Fern .... p. 161
Family **Adiantaceae** ....Maidenhair Fern family
    Genus Anogramma ....Jersey Ferns .... p. 80
        Anogramma leptophylla ....Jersey Fern .... p. 80
    Genus **Adiantum**.................Maidenhair Ferns .... p. 78
        Adiantum capillus-veneris.......Maidenhair Fern .... p. 78
        A. pedatum* ....American Maidenhair Fern p. 79
    Genus **Cryptogramma.** ......Parsley Ferns .... p. 114
        Cryptogramma crispa. ....Parsley Fern .... p. 114

Family **Pteridaceae*** ....Brake Fern family
    Genus **Pteris*** ....Brake Ferns.............................. p. 189
        Pteris incompleta*...................Spider Brake .... p. 189
        P. tremula*...........................Tender Brake .... p. 190
        P. cretica* ....Brake Fern .... p. 189
        P. vittata* ....Ladder Brake .... p. 190
Family **Hymenophyllaceae** ...Filmy-fern family
    Genus **Hymenophyllum** ......Filmy-ferns............................. p. 148
        Hymenophyllum tunbrigense...Tunbridge Filmy-fern ...... p. 149
        H. wilsonii ....Wilson's Filmy-fern ...... p. 151
    Genus **Trichomanes**............Killarney Ferns .... p. 193
        Trichomanes speciosum ....Killarney Fern................ p. 193
        T. venosum* ....Veiled Bristle-fern .... p. 195
Family **Polypodiaceae** ....Polypody family
    Genus **Polypodium** ....Polypodies .... p. 167
        Polypodium cambricum ....Southern Polypody .... p. 170
        P. cambricum 'Cambricum' ...Welsh Polypody .... p. 171
        P. vulgare ....Polypody..................... p. 173
        P. interjectum ....Intermediate Polypody ... p. 172
        Hybrids in Polypodium
        P. cambricum × P. vulgare
        = **P. × font-queri** ....Font-Quer's Polypody...... p. 176
        P. cambricum × P. interjectum
        = **P. × shivasiae** ....Shivas' Polypody .... p. 175
        P. interjectum × P. vulgare
        = **P. × mantoniae.** ....Manton's Polypody......... p. 176
    Genus Phymatodes* ....Kangaroo Ferns .... p. 164
        Phymatodes diversifolia* ....Kangaroo Fern .... p. 164
Family Dicksoniaceae* ....Tree-fern family
    Genus Dicksonia* ....Australian Tree-ferns .... p. 120
        Dicksonia antarctica* ....Australian Tree-fern ...... p. 121
    Genus Cyathea* ....Silver Tree-ferns .... p. 115
        Cyathea dealbata* ....Silver Tree-fern .... p. 116
Family **Hypolepidaceae** ......Bracken family
    Genus **Pteridium**. ....Brackens .... p. 184
        Pteridium aquilinum. ....Bracken .... p. 184
        P. aquilinum subsp. aquilinum ......Common Bracken ... p. 187
        P. aquilinum subsp. atlanticum ......Atlantic Bracken...... p. 187
        P. aquilinum subsp. fulvum ....Perthshire Bracken ... p. 188
        P. pinetorum ....Pinewood Bracken .... p. 188
        P. pinetorum subsp. pinetorum ....p. 188
        P. pinetorum subsp. osmundaceum .... p. 188

Family **Thelypteridaceae** ......Marsh Fern family
    Genus **Thelypteris** ....Marsh Ferns .... p. 190
        Thelypteris palustris ....Marsh Fern .... p. 191
    Genus **Oreopteris** ....Lemon-scented Ferns .... p. 158
        Oreopteris limbosperma.........Lemon-scented Fern ...... p. 158
    Genus **Phegopteris** ....Beech Ferns............................. p. 162
        Phegopteris connectilis ....Beech Fern .... p. 162
Family **Aspleniaceae** ....Spleenwort family
    Genus **Asplenium** ....Spleenworts .... p. 81
        Asplenium marinum ....Sea Spleenwort............... p. 86
        A. trichomanes.....................Maidenhair Spleenwort ... p. 93
        A. trichomanes subsp. trichomanes
            Delicate Maidenhair Spleenwort .... p. 97
        A. trichomanes subsp. quadrivalens
            Common Maidenhair Spleenwort .... p. 96
        A. trichomanes subsp. pachyrachis
            Lobed Maidenhair Spleenwort .... p. 95
        A. trichomanes-ramosum ......Green Spleenwort............ p. 97
        A. fontanum* ....Smooth Rock-spleenwort ... p. 85
        A. obovatum subsp. lanceolatum Lanceolate Spleenwort ... p. 87
        A. adiantum-nigrum. ....Black Spleenwort .... p. 82
        A. onopteris ....Irish Spleenwort .... p. 88
        A. cuneifolium* ....Serpentine Black Spleenwortp. 85
        A. septentrionale ....Forked Spleenwort .... p. 92
        A. ruta-muraria ....Wall-rue .... p. 89
        A. ceterach ....Rustyback ....p. 84
        A. scolopendrium ....Hart's-tongue Fern .... p. 90
            Hybrids in Asplenium
        A. trichomanes subsp. quadrivalens × **subsp. trichomanes**
        = **A.** × **lusaticum** ....Hybrid Maidenhair Spleenwort
        ....p. 104
        A. trichomanes subsp. pachyrachis × **subsp. trichomanes**
        = **A.** × **staufferi** .... p. 103
        A. obovatum subsp. lanceolatum × A. trichomanes
        = A. × refractum .... p. 100
        A. obovatum subsp. lanceolatum × A. scolopendrium
        = A. × microdon ....Guernsey Fern ....p. 100
        A. adiantum-nigrum × A. obovatum
        = A. × sarniense ....Guernsey Spleenwort ...... p. 99
        A. adiantum-nigrum × A. onopteris
        = A. × ticinense............ ......Hybrid Black Spleenwort ... p. 99
        A. adiantum-nigrum × **A. septentrionale**
        = **A.** × **contrei** ....Caernarvonshire Fern ...... p. 100
        A. adiantum-nigrum × A. scolopendrium
        = A. × jacksonii ....Jackson's Fern .... p. 99

A. septentrionale × A. trichomanes subsp. trichomanes
= A. × alternifolium............ Alternate-leaved Spleenwort
....p. 102
A. ruta-muraria × A. trichomanes subsp. quadrivalens
= A. × clermontiae ....Lady Clermont's Spleenwort
....p. 101
A. ruta-muraria × A. septentrionale
= A. × murbeckii..................Murbeck's Spleenwort ... p. 101
A. scolopendrium × A. trichomanes subsp. quadrivalens
= A. × confluens ....Confluent Maidenhair Spleenwort
....p. 101

Family **Woodsiaceae** ....Lady-fern family
   Genus **Athyrium** ....Lady-ferns .... p. 104
      Athyrium filix-femina............Lady-fern .... p. 105
      A. distentifolium var. distentifolium
....Alpine Lady-fern .... p. 105
      A. distentifolium var. flexile......Newman's Lady-fern ...... p. 105
   Genus **Cystopteris** ....Bladder-ferns. .... p. 117
      Cystopteris fragilis ....Brittle Bladder-fern......... p. 118
      C. dickieana ....Dickie's Bladder-fern ...... p. 117
      C. montana ....Mountain Bladder-fern ... p. 119
   Genus **Woodsia**..................Woodsias ....p. 195
      Woodsia ilvensis ....Oblong Woodsia............ p. 198
      W. alpina ....Alpine Woodsia .... p. 196
   Genus **Gymnocarpium** ......Oak Ferns .... p. 143
      Gymnocarpium dryopteris......Oak Fern..................... p. 144
      G. robertianum ....Limestone Fern .... p. 146
   Genus **Matteuccia*** ....Ostrich Ferns .... p. 152
      Matteuccia struthiopteris* ......Ostrich Fern .... p. 153
   Genus **Onoclea*** ....Sensitive Fern .... p. 153
      Onoclea sensibilis* ....Sensitive Fern............... p. 153

Family **Dryopteridaceae** ......Buckler-fern family
   Genus **Polystichum** ....Shield-ferns............................ p. 177
      Polystichum lonchitis ....Holly-fern .... p. 179
      P. munitum* ....Western Sword-fern ...... p. 181
      P. aculeatum.......................Hard Shield-fern............ p. 177
      P. setiferum ....Soft Shield-fern .... p. 181
            Hybrids in Polystichum
      P. lonchitis × P. setiferum
        = P. × lonchitiforme ....Atlantic Hybrid Shield-fern p. 184
      P. aculeatum × P. lonchitis
        = P. × illyricum ....Alpine Hybrid Shield-fern.. p. 183
      P. aculeatum × P. setiferum
        = **P. × bicknellii** ....Lowland Hybrid Shield-fern
....p. 183

SYNOPSIS of CLASSIFICATION of PTERIDOPHYTA in the BRITISH ISLES   29

    Genus **Cyrtomium*** ....House Holly-ferns .... p. 116
        Cyrtomium falcatum* ....House Holly-fern .... p. 116
    Genus **Dryopteris** ....Buckler-ferns .... p. 121
        Dryopteris filix-mas ....Male-fern .... p. 135
        D. affinis ....Scaly Male-fern .... p. 124
        D. affinis subsp. affinis .... Yellow Scaly Male-fern... p. 126
        D. affinis subsp. borreri .... Common Scaly Male-fern p. 128
        D. affinis subsp. cambrensis ... Narrow Scaly Male-fern... p. 129
        D. oreades..........................Dwarf Male-fern. .... p. 137
        D. submontana....................Rigid Buckler-fern .... p. 138
        D. dilatata..........................Broad Buckler-fern......... p. 132
        D. expansa ....Northern Buckler-fern ... p. 134
        D. remota ....Scaly Buckler-fern .... p. 138
        D. aemula ....Hay-scented Buckler-fern   p. 123
        D. carthusiana ....Narrow Buckler-fern. ...... p. 130
        D. cristata ....Crested Buckler-fern. ...... p. 131
              Hybrids of Dryopteris
        D. filix-mas × D. oreades
          = **D. × mantoniae** ....Manton's Male-fern ...... p. 143
        D. affinis × D. filix-mas
          = **D. × complexa** ....Hybrid Male-fern .... p. 140
        D. affinis × D. oreades .... p. 141
        D. dilatata × D. filix-mas = D. × subaustriaca .... p. 143
        D. dilatata × D. expansa
          = **D. × ambrosiae** ....Gibby's Buckler-fern ...... p. 142
        D. aemula × D. oreades
          = D. × pseudoabbreviata..........Mull Fern .... p. 140
        D. aemula × D. dilatata. .... p. 140
        D. carthusiana × D. filix-mas
          = D. × brathaica ....Brathay Fern .... p. 142
        D. carthusiana × D. dilatata
          = **D. × deweveri** ....Hybrid Narrow Buckler-fern
                                                                                    .p. 141
        D. carthusiana × D. expansa
          = D. × sarvelae ....Kintyre Buckler-fern ...... p. 142
        D. carthusiana × D. cristata
          = D. × uliginosa ....Hybrid Fen Buckler-fern   p. 141
Family **Davalliaceae*** ....Hare's-foot Fern family
    Genus **Davallia*** ....Hare's-foot Ferns....................... p. 120
        Davallia canariensis* ....Hare's-foot Fern .... p. 120
        D. bullata group* ....Squirrel's-foot Fern group   p. 120
Family **Blechnaceae** ....Hard-fern family
    Genus **Blechnum** ....Hard-ferns .... p. 109
        Blechnum spicant ....Hard-fern.................... p. 110

    B. cordatum* ....Chilean Hard-fern .... p. 109
    B. penna-marina* ....Antarctic Hard-fern......... p. 110
  Genus **Woodwardia*** ....Chain-ferns .... p. 200
    Woodwardia radicans* ....European Chain-fern ...... p. 200
Family **Marsileaceae** ....Pillwort family
  Genus **Pilularia**...................Pillworts .... p. 164
    Pilularia globulifera ....Pillwort .... p. 165
Family **Azollaceae*** ....Water Fern family
  Genus **Azolla*** ....Water Ferns.............................. p. 107
    Azolla filiculoides* ....Water Fern .... p. 107

ARRANGEMENT OF GENERA AND SPECIES IN THE TEXT

In this edition we have followed the more practical approach of arranging the genera, and the species within them, in alphabetical order. **The only exception to this order is the grouping of the clubmosses, quillworts and horsetails together at the beginning and the inclusion of some additional aliens associated with gardens and estates at the end.** Every Welsh species is described and in order to increase the usefulness of the work brief descriptions of the remaining British species have also been added. A few alien species, which may be encountered as garden escapes, are also included.

KEY TO GENERA OF NATIVE WELSH PTERIDOPHYTA

1 Aquatic   **2**
  Not aquatic (although some in wet habitats)   **4**
2 Plant free-floating, leaves moss-like in appearance
              Azolla (strictly, an alien) p. 107
  Plant not usually free-floating, leaves rush-like, subulate or filiform **3**

3 Rhizomes long, slender, visible, bearing uniformly narrow leaves at intervals   **Pilularia** p. 164
  Rhizomes long, not visible, bearing jointed stems with whorls of scale leaves   **Equisetum fluviatile** p. 54
  Rhizomes short, stout, corm-like, leaves narrow but broader based in a rosette   **Isoetes** p. 46
4 Leaves simple, rush-like, subulate or filiform   **5**
  Leaves not rush-like   **6**
5 Rhizomes long, slender, bearing narrow uniformly filiform leaves at intervals   **Pilularia** p. 164
  Rhizomes short, stout, corm-like, subterranean. Leaves narrow but broader based, in a rosette   **Isoetes** p. 46

KEY to GENERA of NATIVE WELSH PTERIDOPHYTA

6   Plant more or less moss-like, stem clothed with small, simple, subulate, or scale-like leaves   **7**
Plant not moss-like, or if seeming so, then leaves not simple   **11**

7   Leaves either 4-ranked, of 2 kinds, and serrulate; or all alike, spirally arranged and sparsely slender-toothed. Ligule present (inconspicuous). Sporangia of 2 kinds, with either 4 large spores or many small spores   **Selaginella** p. 42
Leaves all alike, entire or serrulate. Sporangia all alike   **8**

8   Stems ascending, with stout dichotomous branches. Roots confined to the short basal region. Sporangia axillary   **Huperzia** p. 36
Stems creeping; rooting at intervals, with short lateral branches, more or less dichotomous or apparently monopodial. Sporangia in terminal spikes   **9**

9   Leaves on vegetative branches opposite and decussate, often dimorphic   **Diphasiastrum** p. 33
Leaves spirally inserted; imbricate, spreading or curved upwards   **10**

10   Leaves lanceolate, flat; sporophylls ovate, acuminate, with broad, scarious, toothed margins   **Lycopodium** p. 39
Leaves linear-subulate, curved upwards; sporophylls similar to the foliage leaves   **Lycopodiella** p. 38

11   Stem simple with or without branches attached in whorls up the stem. Leaves small, fused to form small toothed sheaths round the stem above most nodes   **Equisetum** p. 49
Stems and leaves not as above   **12**

12   Leaves filmy in texture (Hymenophyllaceae, but see also *Adiantum*, p. 78)   **13**
Leaves firm in texture   **14**

13   Receptacle bristle-like, projecting from the indusium   **Trichomanes** p. 193
Receptacle included within the indusium   **Hymenophyllum** p. 148

14   Fertile frond or fertile portion of frond distinct from sterile frond   **15**
Fertile and sterile fronds similar in appearance   **19**

15   Young fronds straight in the bud (Ophioglossaceae)   **16**
Young fronds circinate in the bud   **17**

16   Sterile leaf and fertile spike both simple   **Ophioglossum** p. 154
Sterile leaf pinnate, fertile spike branched   **Botrychium** p. 112

17   Only upper pinnae fertile, lower ones sterile and leafy   **Osmunda** p. 160
Fertile frond entirely distinct in appearance and bearing sori throughout   **18**

18   Frond three or four times pinnate   **Cryptogramma** p. 114
Frond 1-pinnate   **Blechnum** p. 109

19   Sori marginal, covered by the incurved edge of the frond   **20**
Sori superficial   **21**

20 Sori in a continuous marginal line covered by a continuous indusium
 Pteridium p. 184
 Sori separate, covered by distinct lappets of the frond margin
 Adiantum p. 78
21 Indusium present (in the young condition at least) 22
 Indusium absent or apparently absent 29
22 Sori nearly circular 23
 Sori oblong or linear (may be very short) 28
23 Indusium inferior 24
 Indusium superior 25
24 Indusium surrounding the base of the sorus and entirely laciniate
 Woodsia p. 195
 Indusium almost entire, at first hoodlike **Cystopteris** p. 117
25 Indusium circular **Polystichum** p. 177
 Indusium reniform 26
26 Sori large, indusium persistent at least until the spores are being shed
 Dryopteris p. 121
 Sori small, indusium small, shed well before the spores 27
27 Frond glandular beneath, stipe $1/5$ as long as the blade
 Oreopteris p. 158
 Frond without glands, stipe about as long as the blade
 Thelypteris p. 190
28 Sori oblong or reniform **Athyrium** p. 104
 Sori linear **Asplenium** p. 81
29 Sori linear, frond densely covered with scales beneath
 Asplenium ceterach p. 84
 Sori circular or oval, frond glabrous or somewhat hairy or scaly beneath
 30
30 Fronds becoming disarticulated from the rhizome **Polypodium** p. 167
 Fronds becoming disarticulated, if at all, some distance above the rhizome at a joint on the stipe 31
31 Fronds becoming disarticulated at or about half-way* **Woodsia** p. 195
 Fronds not becoming disarticulated 32
 *(*Woodsia* is repeated here in order to allow for the possibility that its indusium may not always be recognized as such.)
32 Frond pinnate, with pinnatifid pinnae **Phegopteris** p. 162
 Frond bi- to tri-pinnate **Gymnocarpium** p. 143

LYCOPSIDA - CLUBMOSSES, LESSER CLUBMOSSES and QUILLWORTS

CLUBMOSSES

All are placed in the order Lycopodiales and in the only Family, the Lycopodiaceae. All are homosporous and have no ligules. There are eight or more genera with four occurring in the British Isles. Some botanists place *Huperzia* and the tropical *Phlegmariurus* in a separate family: the Huperziaceae.

Family LYCOPODIACEAE - Clubmoss family

Genus DIPHASIASTRUM Holub

Alpine Clubmosses - Cnwpfwsoglau Alpaidd

Stems dorsiventral, creeping, branching frequently with the ultimate groups of branchlets usually in tufts. Leaves simple, small, dimorphic, in 4 ranks, the lateral leaves keeled. Cones terminal on cylindrical dichotomous branchlets. Sporophylls different from the leaves. Homosporous. Sporangium usually unilocular, kidney-shaped, dehiscing by a transverse slit. Spores numerous. Prothallus commonly subterranean and saprophytic, with mycorrhiza. Sperm biflagellate. About 22 species. North Temperate to Sub-Arctic regions.

KEY TO SPECIES

Plant glaucous. Ventral leaves trowel-shaped with free apical portion *c*.1.2mm long. Lateral leaves fused to stem for $1/2$ their length. Cones sessile. Sporophylls at least 2 × the length of the sporangia   **D. alpinum**

Plant yellow-green. Ventral leaves elliptical-lanceolate with free apical portion *c*.0.8mm long. Lateral leaves fused to stem for up to $2/3$ their length. Cones pedunculate. Sporophylls $1 1/2$-2 × the length of the sporangia
D. complanatum

DIPHASIASTRUM ALPINUM (L.) Holub

(*D. complanatum* subsp. *alpinum* (L.) Jermy; *Lycopodium alpinum* L.; *Diphasium alpinum* (L.) Rothm.)

Alpine Clubmoss; Cnwpfwsogl Alpaidd

Stems usually of two kinds; the main axis, 15-50cm long, creeping, usually subterranean or concealed beneath vegetation, simple or sparingly

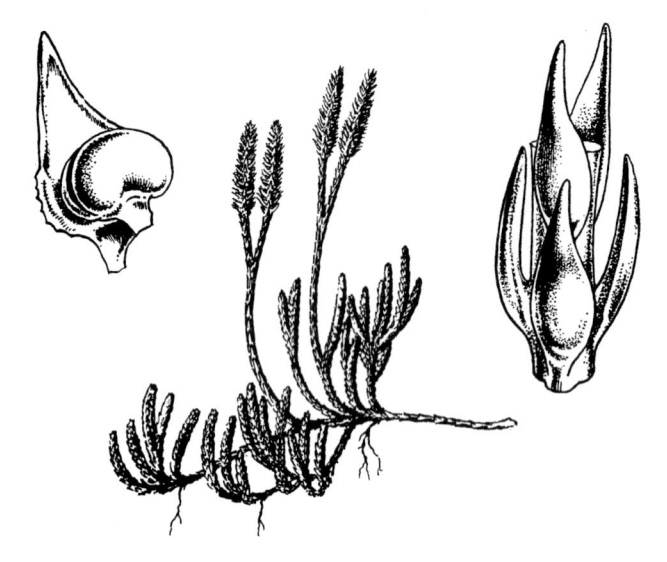

Fig. 16. Alpine Clubmoss (*Diphasiastrum alpinum*). Left: sporophyll (×6). Centre: plant (×1/2). Right: portion of stem (×3).

branched, with distant leaves; the lateral branches, more or less erect, often branched from near the base, usually with 3-5 dichotomies, forming dense tufts of cylindrical or slightly flattened branchlets 3-10cm high, with their leaves close together. **Leaves** glaucous, entire, 2-4mm long, about 1mm wide, decurrent, lanceolate, acute or acuminate, loosely appressed, usually rounded on the back, distant on the main axis, densely arranged on the branchlets, on which the lateral pairs are sometimes larger and more spreading than the ventral row and have a well-developed decurrent keel, giving these branchlets a flattened appearance. Flattened and cylindrical branchlets are often to be found on the same plant. **Cones** distinct, 1-2cm long, sessile and terminal, on cylindrical, dichotomous branchlets. **Sporophylls** at least twice as long as the sporangia, with the basal half ovate or broad-ovate, tapering suddenly at first, then more gradually towards the acute or acuminate apex, with narrow, scarious, denticulate margins. **Sporangia** yellow, reniform. **Spores** ripening June-August, generally several weeks earlier than *L. clavatum* in the same district. 2n=46.

First recorded, 1650; How, *Phyt. Brit.* 77. 'On the top of Snowdon'.

Mountain grassland and high moors. Locally common in the uplands of the north and west of Great Britain; southwards to Derbyshire although formerly to S. Devon. In the mountainous areas of Ireland, from Wicklow northwards, but mostly round the coastal counties. Formerly Kerry.

Throughout Europe south to the Pyrenees, Northern Apennines, Bulgaria, Rossia (N,C,W), Iceland and Faeroes. Japan. North America: NW and NE. Greenland.

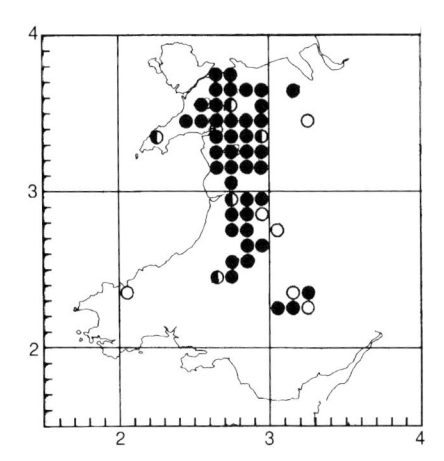

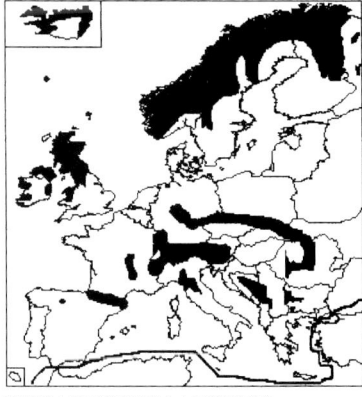

DIPHASIASTRUM ALPINUM

DIPHASIASTRUM COMPLANATUM (L.) Holub subsp. *ISSLERI* (Rouy) Jermy

(*D. issleri* (Rouy) Holub); *Lycopodium issleri* (Rouy) Lawalrée; *L. complanatum* auct. brit.; *L. alpinum* var. *decipiens* Syme; *Diphasium issleri* (Rouy) Holub)

Issler's Clubmoss; Cnwpfwsogl Issler

Like *D. alpinum*, but **leaves** larger and more conspicuously dimorphic, the lateral leaves much broader and flatter than the ventral leaves, giving a flattened appearance to the branchlets, which are usually longer than those of *D. alpinum* and bear a superficial resemblance to the European *D. complanatum*. **Cones** on erect elongated peduncles. **Sporophylls** 1½-2 × length of sporangia, ovate, abruptly acuminate. The chromosome number of British material has not yet been determined.

It has probably arisen through hybridization between *D. alpinum* and *D. complanatum*. Jermy and Camus (1991) used the epithet *decipiens* (*sensu* Syme) for the British plants. They suggest that the backcross of this form with *D. alpinum* may occur in Great Britain (e.g. the Malvern Hills, Worcs.).

D. complanatum is reputed to have been found in Hants., Gloucester, Worcester, N. Devon, N. Wales and the Scottish Highlands, but some of the plants referred to this taxon in British herbaria seem to be etiolated forms of *D. alpinum*. Etiolated and normal branchlets can sometimes be found on the

same specimen. The material from N. Wales previously determined as '*D*. × *issleri*' is now considered to be definitely an etiolated form of *D. alpinum*.
Central Europe westwards to central France and Great Britain.

Diphasiastrum complanatum subsp. *complanatum* is not known from the British Isles. It is found in N. and central Europe extending locally to Spain. World-wide it is a circumboreal subspecies, including Greenland. Whereas in subsp. *issleri* the ventral leaf width is $1/3$ the branch width, and the free apex of lateral leaves is $c.1/2$ as long as the leaf, in subsp. *complanatum* they are $1/5$ and $c.1/6$ respectively.

Genus HUPERZIA Bernh.

Fir Clubmosses - Cnwpfwsoglau Ffeinid

Main stems ascending or erect, dichotomously branched into branches of equal length. Leaves simple, small, spirally arranged. Sporangia kidney-shaped, dehiscing by a transverse slit, pedunculate, in the axils of leaf-like sporophylls, borne in fertile zones towards the tops of the branches. Homosporous. Spores numerous. Prothallus commonly subterranean and saprophytic, with mycorrhiza. Sperm biflagellate. About 393 species, pan-temperate and pan-tropical.

HUPERZIA SELAGO (L.) Bernh. ex Schrank & C.Mart.

(*Lycopodium selago* L.)

Fir Clubmoss; Cnwpfwsogl Mawr

Stems 5-25cm long, with an ascending or erect, unbranched basal portion usually 1-5cm long, on which the dichotomously branched, adventitious, endogenous roots are borne, often intermixed with persistent older leaves. **Branches** more or less erect, usually with 2 dichotomies, the terminal branchlets 1-6(-12 or more)cm long. **Leaves** spirally arranged, 8mm long and about 1.5mm wide, linear to ovate-lanceolate, acute, imbricate or spreading, often slightly incurved, entire or minutely serrulate, dull green, often bearing in their axils flattened trident-like buds (bulbils or gemmae). **Sporangia** reniform, pedunculate, in the axils of leaf-like sporophylls, becoming conspicuously yellow-brown in late summer; not forming terminal cones but usually aggregated in fertile zones alternating with sterile ones, near the ends of the branchlets. **Spores** ripening June-August. **Prothalli** subterranean, saprophytic, large and cylindrical; but rarely found. The readily detached gemmae seem to be the usual means of reproduction. 2n=264.

First recorded, 1670; Ray, *Cat. Plant. Ang.*, 214. 'On Snowdon, Caderidris, and the other high Mountains of Wales'.

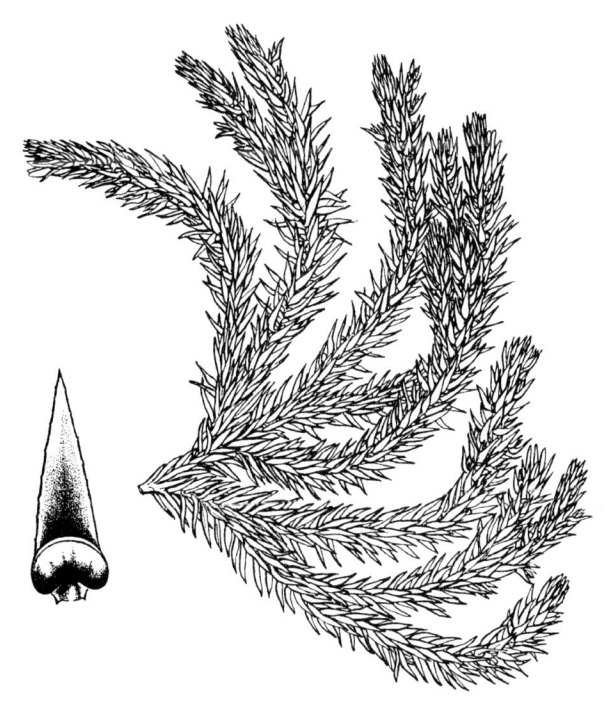

Fig. 17. Fir Clubmoss (*Huperzia selago*). Plant (natural size). On the left: sporophyll (×6).

Heaths, moors, mountain grassland and rock ledges. Locally common in mountainous areas, but now very rare in lowland regions, throughout most of the N. and W. of the British Isles.

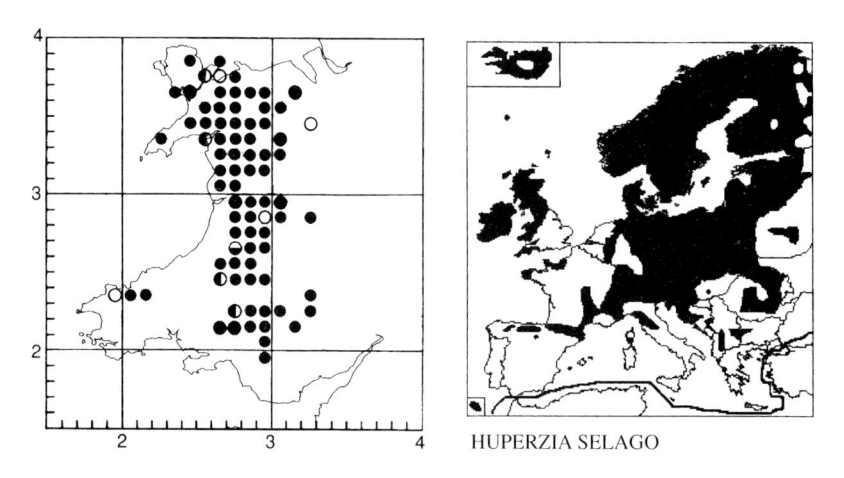

HUPERZIA SELAGO

Europe: Northern regions and central and southern mountains, Rossia (K,E), Iceland, Faeroes, Spitzbergen, Azores. Madeira. Northern Asia, although needs confirming in east-central Siberia. Himalaya. Japan. North America: throughout much of Canada to NE USA. Subspecies or allied taxa in scattered parts of the southern hemisphere. The *H. selago* recorded from Greenland is really *H. appalachiana* Beitel & Mickel.

Genus LYCOPODIELLA Holub

Marsh Clubmosses - Cnwpfwsoglau y Gors

Stems creeping and rooting, branching not conspicuously dichotomous, owing to unequal development. Leaves simple, small, spirally arranged, subulate, often curving upwards. Sporophylls similar but more spreading, broadened and toothed at the base, with axillary sporangia, borne in an erect, terminal cone. Homosporous. Sporangia unilocular, kidney-shaped, dehiscing by a transverse slit. Spores numerous. Prothallus commonly subterranean and saprophytic, with mycorrhiza. Sperm biflagellate. About 15 species. Most of Europe and Siberia, N. America, Africa, Asia and Australasia.

LYCOPODIELLA INUNDATA (L.) Holub

(*Lycopodium inundatum* L.; *Lepidotis inundata* (L.) Opiz)

Marsh Clubmoss; Cnwpfwsogl y Gors

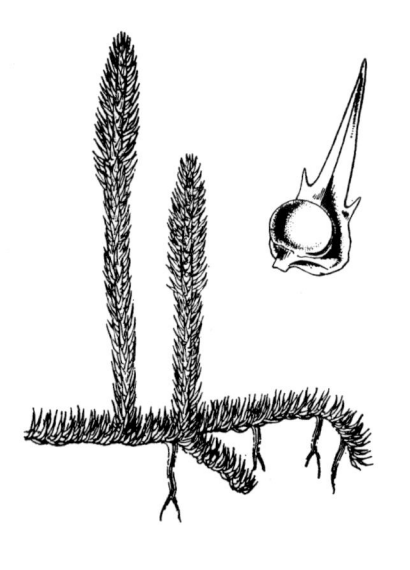

Stems 5-15(-20)cm long, prostrate, sparingly branched; usually with a single erect fertile branch. Previous year's growth usually dying away during the winter.

Fig. 18. Marsh Clubmoss (*Lycopodiella inundata*). Plant (natural size). Above, right: sporophyll (×6).

Leaves 6mm long and about 1mm wide, often curved, linear-subulate, entire, acute, imbricate or spreading, spirally arranged but those on the prostrate stems curved upwards. **Cones** 1-3cm long, terminal on erect, solitary fertile branches, 2-10cm long. **Sporophylls** similar to the foliage leaves but more spreading, narrow triangular, with one or two teeth usually near their broad base. **Spores** ripening June-September. 2n=156.

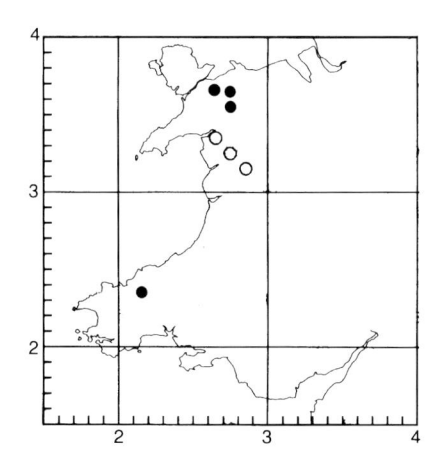

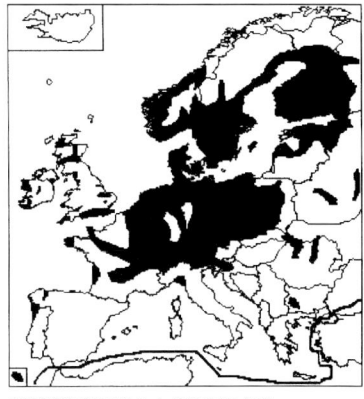

LYCOPODIELLA INUNDATA

Boggy moorland and wet heaths. Rare.

First recorded, 1695; in Camden's *Britannia* (ed. Edmund Gibson), 701. 'Capel Ceirig'.

Protected in Northern Ireland under the *Wildlife (NI) Order, 1985*.

Throughout Europe, including Rossia (N,B,C,W,E), with the exception of the Mediterranean region. Azores. NW and NE North America.

Genus LYCOPODIUM L.

Clubmosses - Cnwpfwsoglau

Stems usually creeping, with short branches. Leaves spirally arranged or in whorls, linear or lanceolate, spreading or appressed. Sporophylls different from the leaves, ovate, with toothed margins arranged in a more or less distinct terminal cone. Homosporous. Sporangium unilocular, kidney-shaped, dehiscing by a transverse slit. Spores numerous. Prothallus commonly subterranean and saprophytic, with mycorrhiza. Sperm biflagellate. About 50 species, pan-temperate and pan-tropical.

KEY TO SPECIES

Leaves acute but not hair-pointed. Cones sessile **L. annotinum**

Leaves and sporophylls with a long, white hair-point. Cones on long peduncles, which bear distant, small, scale-like leaves **L. clavatum**

LYCOPODIUM ANNOTINUM L.

Interrupted Clubmoss; Cnwpfwsogl Meinfannau

Stems 30-60cm long, procumbent, with several or many ascending branches. **Leaves** on the main stem 3-5mm long, those on the branches 5-7(-10)mm, about 1mm wide, linear-lanceolate, acuminate, usually sparsely toothed, dull green, spirally arranged, spreading, denser on the branches than on the main stem. **Cones** distinct, terminal, 1.5-3cm long, on erect branches 6-15(-25)cm long. **Sporophylls** quite different from the leaves, ovate, acuminate, with broad, scarious, toothed margins. **Spores** ripening June-August. 2n=68.

First recorded in 1690; Ray, *Syn.* edn I, 16, 'Above Llin y Cwn'. Griffith wrote of it, in 1895:-'I am afraid this is extinct in Carnarvonshire. It used to grow on the Glyders, but I have failed to find it for years now.' Now extinct in Wales.

Local on moors and mountain grassland, in central and northern Scotland and one site in northern England.

Europe southwards to the Pyrenees, northern Apennines and southern Carpathians. Siberia. Himalaya. North America: across the continent and southwards to Virginia and New Mexico. Greenland.

LYCOPODIUM CLAVATUM L.

Stag's-horn Clubmoss; Common Clubmoss; Cnwpfwsogl Corn Carw

Stems 30-100cm long, procumbent, with numerous branches, mostly procumbent. **Leaves** 3-7mm long, 0.75-1.5mm wide, spirally arranged, bright green, linear, minutely serrulate, acuminate, ending in a long white hair-point. **Fertile branches** erect, 5-15(-25)cm long, with 1-3 cones, 1-5cm long, on long peduncles. **Sporophylls** ovate, with broad scarious denticulate margins and long white hair-points. **Sporangium** on the adaxial surface of the sporophylls. Peduncles bearing remote, yellowish, linear-subulate leaves, quite distinct from both sporophylls and foliage leaves. **Spores** ripening June-September. 2n=68.

First recorded, 1670; Ray, *Cat. Plant. Ang.* 214. 'On Snowdon, and the other mountains of Wales'.

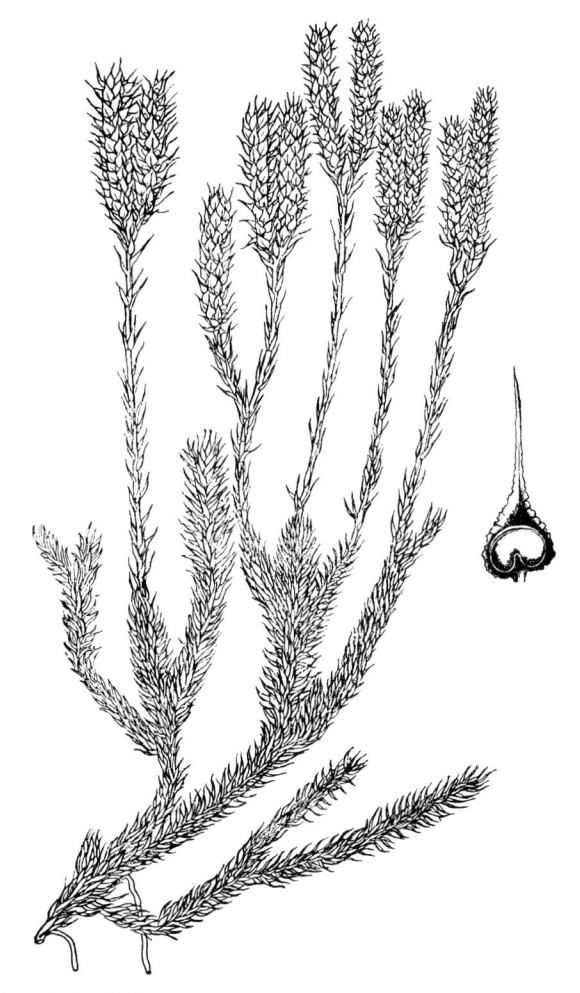

Fig. 19. Stag's-horn Clubmoss (*Lycopodium clavatum*). Plant (natural size). Right: sporophyll (×6).

Still locally common on pastures, heaths and moorland in mountainous regions of the British Isles, but rare and decreasing in lowland areas.

An almost cosmopolitan species growing in humid temperate and boreal regions of the northern hemisphere and on high mountains in the tropics and in the south temperate zone. A highly variable species with an almost continuous range of forms. Tetraploid forms with aggregated cones have been found in Japan.

Europe: Northern and central regions, extending locally into southern and eastern regions, Rossia (N,B,C,W,E), Iceland and Faeroes. Caucasus. Northern east Asia. Japan. SE Asia. Africa: South-western Cape northwards

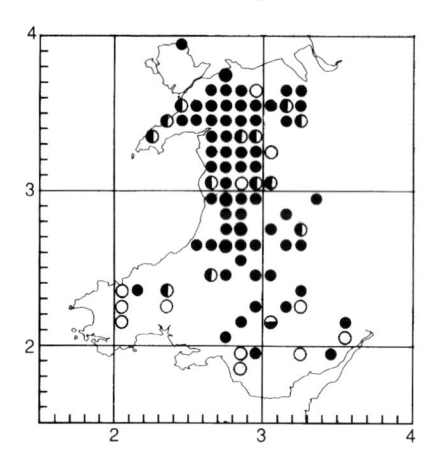

LYCOPODIUM CLAVATUM

through East Africa to the tropical mountains. Mascarene Islands. Madagascar. Kerguelen. North America: across the continent and southwards to Oregon and North Carolina. West Indies. South America: along the Andes southwards to Bolivia, NW Argentina and central Brazil.

LESSER CLUBMOSSES

Family SELAGINELLACEAE - Lesser Clubmoss family

Genus SELAGINELLA P.Beauv.

Lesser Clubmosses - Cnwpfwsoglau Lleiaf

A genus with about 700 terrestrial or epiphytic species, mainly in tropical and warm temperate regions but a few species extend into arctic and alpine zones.

Stems elongated, creeping or erect; the creeping stems producing leafless branches (rhizophores) which bear roots at their tips. Leaves small, ligulate, spirally arranged or in 4 ranks and dimorphic. Sporophylls leaf-like, arranged in terminal cones. Sporangia near the base of the upper surface of the sporophylls, usually unilocular, dehiscing by a transverse slit, the megasporangia usually borne towards the base of the cone, the microsporangia near its apex. Microspores numerous. Megaspores usually 4 or fewer per sporangium, but some exotic species may produce up to 42. The male prothallus is retained in the microspore until almost mature. It consists of one small prothallial cell and an antheridium producing 128 or 256 biflagellate sperm which swim in the surface film of water to reach the female prothallus. The female prothallus is also retained within the megaspore. Fertilization sometimes occurs before the megaspore is released.

The genus *Selaginella* has been divided into sections and subsections, subgenera or genera by various botanists. Two of these are represented in the wild flora of the British Isles, one native and one naturalized species. A cosmopolitan genus, but naturalized in New Zealand.

KEY TO SPECIES

Leaves dimorphic, 4-ranked   **S. kraussiana**
Leaves uniform, spirally arranged   **S. selaginoides**

SELAGINELLA KRAUSSIANA (Kunze) A.Braun
Kraus's Clubmoss; Cnwpfwsogl Krauss

Stems creeping, articulated at the nodes, bistelic in the internodes. **Leaves** ovate-lanceolate, acute, asymmetrical at the base, minutely toothed, the lateral leaves 2-3mm long, the leaves on the upper side of the stem little

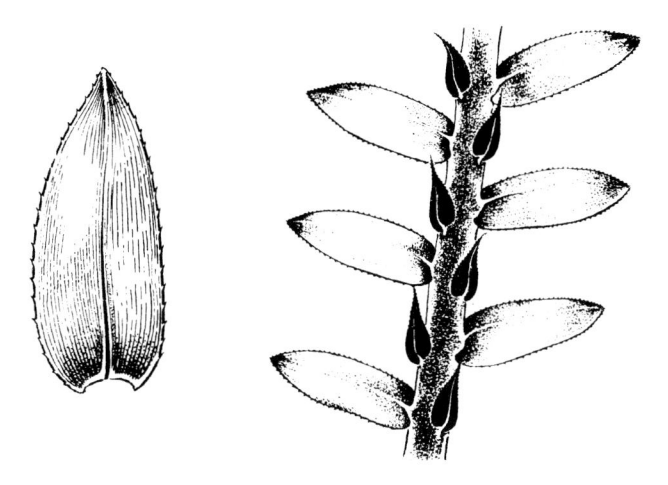

Fig. 20. Kraus's Clubmoss (*Selaginella kraussiana*). Left: leaf (×15). Right: portion of stem (×7.5).

more than half as long. **Sporophylls** ovate, cuspidate, keeled, minutely toothed. **Strobili** distinct, sessile, up to 2cm long, *c*.1mm wide. 2n=20.

First recorded, 1947, J. A. Webb, *Proc. Swansea Sci. & Field Nat. Soc. 2*, 258 (1947), 'nr. Mayals, Gower'.

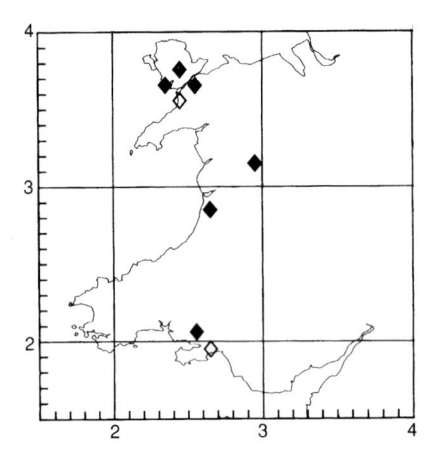

Cultivated in greenhouses, escaped and naturalized in a few sites in the British Isles especially in the west. Naturalized in central and western Europe and now also in SE North America, Jamaica, S. and E. Australia, New Zealand. Native of the Azores, Madeira, Canary Islands, tropical and S. Africa.

SELAGINELLA KRAUSSIANA

SELAGINELLA SELAGINOIDES (L.) Link
Lesser Clubmoss; Cnwpfwsogl Bach

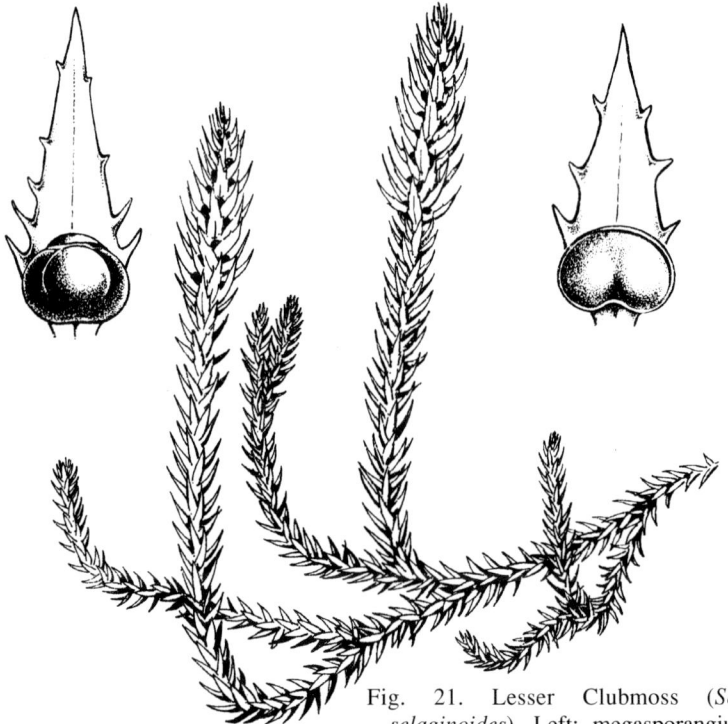

Fig. 21. Lesser Clubmoss (*Selaginella selaginoides*). Left: megasporangium (×10). Centre: plant (×2). Right: microsporangium (×10).

Decumbent stems slender, 3-15cm long, with short, slender, ascending, sterile branches. Fertile branches yellowish, more robust, 3-9cm long. Stele single. **Leaves** 1-3mm long, up to 4mm on fertile branches, spreading or loosely appressed, lanceolate, acute, with 1-5 long slender teeth on each side. **Sporophylls** similar to the leaves, but larger-up to 5mm long. **Cones** sessile, solitary, 1-5cm long, 0.5cm or more in width, not very distinct from the rest of the branch. **Megaspores** whitish, apparently smooth (microscopically papillose), broad obovoid, with a distinct triradiate ridge, borne in tetrads in the megasporangia, which dehisce by a tangential split. **Microspores** yellow, numerous, about $1/20$ the size of the megaspores. **Microsporangia** few, near the apex of the cone, sometimes absent. **Spores** ripening June-August. 2n=18.

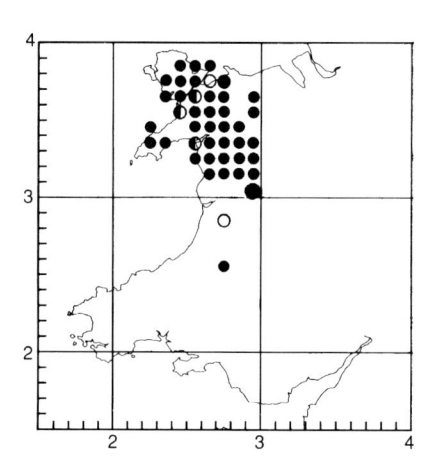

SELAGINELLA SELAGINOIDES

First recorded, 1670; Ray, *Cat. Plant. Ang.* 215, 'Snowdon'.

Locally common in North Wales, northern England, Scotland and many parts of Ireland, mainly in mountainous districts on damp grassland and bogs up to 3,500ft.(1067m), descending down to sea-level in some districts, on sand-dunes.

Most of Europe in the mountainous regions southwards to the Pyrenees, Alps and the Caucasus, Rossia (N,B,C,W), Iceland and Faeroes. Asia southwards to Kamchatka. Japan. North America: across the continent and southwards to British Columbia, Minnesota, Michigan and Maine. Greenland.

QUILLWORTS

Family ISOETACEAE - Quillwort Family

Genus ISOETES L.

Quillworts - Gwair Merllyn

Aquatic or terrestrial plants. **Stems** short, stout, 2- or 3-lobed, largely concealed by numerous roots beneath and by the bases of the leaves above. **Roots** slender, dichotomously branched. **Leaves** in a dense rosette, subulate or filiform usually terete or trigonous, sometimes flattened, broadening at the spathulate, sheathing base. The outer leaves produced early in any season are sporophylls bearing megasporangia, later leaves bear microsporangia, then sterile leaves are formed. Two flushes of fertile leaves have been observed, in spring and in autumn. **Sporangia** sessile, embedded in the base of the sporophyll, below the ligule, usually more or less concealed by a flange-like outgrowth of tissue (the velum). **Megasporangium** traversed by strands of tissue (trabeculae). **Megaspores** numerous, tetrahedral, germinating *in situ* to form a many-celled prothallus, its archegonia exposed through the split top of the megaspore. **Microspores** minute, very numerous (estimated as up to 1 million in each microsporangium). **Male prothallus** consisting of 1 prothallial cell and an antheridum with a 4-celled wall enclosing a central cell which gives rise to 4 multiciliate antherozoids. About 70 species of which 3 are native to the British Isles. A second generic name, *Stylites*, was given to plants found in 1954 in the Peruvian Andes. They are now thought to belong to *Isoetes*.

KEY TO SPECIES

1 Plant terrestrial, in damp, peaty or sandy places; stem dormant and leafless in summer, with persistent leaf bases; leaves *c*.3cm × 1mm (W.Cornwall and Channel Islands) **I. histrix**
Plant submerged aquatic, on lake bottoms; stems never completely leafless without persistent leaf-bases; leaves 4-20(-50)cm × 1.5-5mm **2**

2 Megaspores 440-550µm, with pointed spines; leaves rather flaccid
I. echinospora
Megaspores 530-700µm, with blunt tubercles; leaves stiff **I. lacustris**

ISOETES ECHINOSPORA Durieu

(*I. setacea* Lam.)

Spring Quillwort; Gwair Merllyn Bychan

Stem sub-globose or short cylindrical at first, becoming bi-lobed, *c*.1-3cm long. Roots numerous, *c*.5-15cm long, borne on the base of the stem.

Leaves 4-12(-20)cm × 2-3mm, pale green, rather flaccid, compressed, often curving outwards and downwards, with four longitudinal, septate air canals. **Megaspores** 440-550µm, white or yellowish, with pointed, fragile spines. **Microspores** dust-like, yellowish-brown. **Spores** ripening May-August, the microspores before the megaspores. 2n=22.

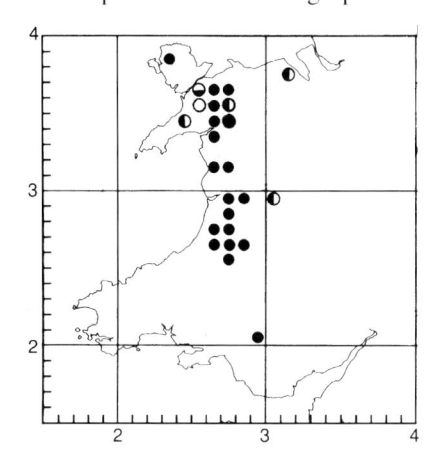

ISOETES ECHINOSPORA

First recorded, 1863: Babington, *Journ. Bot.* I, 4. 'Near Llanberis'.

In more oligotrophic waters than *I. lacustris*, perhaps because of competition, as it also occurs in a few eutrophic lakes in Wales.

Local, mainly in the west (scarcer than *I. lacustris* in the north) of the British Isles. In SW England eastwards into Dorset. W. Ireland.

Europe: north and central regions southwards to Spain and northern Italy. Faeroes. North America: Alaska to Nova Scotia and southwards to Nevada and Virginia. Greenland.

ISOETES HISTRIX Bory

Land Quillwort; Gwair Merllyn y Tir

Terrestrial. Dormant and leafless during the summer; **stem** covered with persistent, short, 2-pronged, glossy, blackish leaf-bases. **Leaves** 3-8cm × 0.5-1mm, more or less flat, rather glossy dark green, with broad membranous margins near the base, and one longitudinal non-septate air canal. **Sporangium** completely covered by the velum. **Megaspores** 400-560µm, with small blunt tubercles often confluent to form a net-like pattern. 2n=20.

W. Cornwall, Channel Islands.

Europe: Mediterranean and Atlantic coasts.

ISOETES LACUSTRIS L.

Quillwort; Gwair Merllyn

Stem sub-globose or short cylindrical at first, becoming bi-lobed, *c*.1-3cm long. **Roots** numerous, *c*.5-15cm long, borne on the base of the stem. **Leaves** 8-20(-50)cm × 1.5-5mm, dark green, rather stiff, erect or curving outwards and downwards, subulate, more or less rounded in section above the whitish base which has flattened membranous margins. Leaves have four

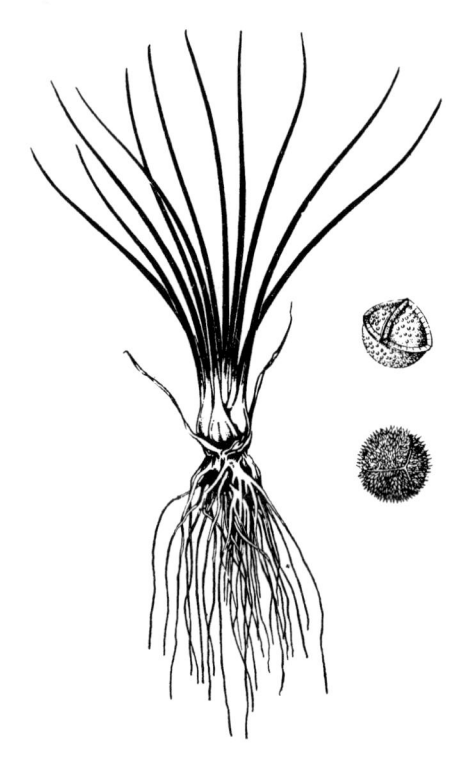

Fig. 22. Quillwort (*Isoetes lacustris*). Plant (×1/3), and megaspore, top right (×20). Bottom right: megaspore of Spring Quillwort (*Isoetes echinospora*) (×20).

longitudinal, septate air canals. **Sporangium** only partly covered by the velum. **Megaspores** usually pale yellowish. rarely white, 530-700µm diameter, covered with blunt tubercles. **Microspores** dust-like, yellowish-brown. **Spores** ripening May-July. 2n=110.

First recorded, 1690: Ray, *Syn*. Ed. 1, 210. Snowdon Phynon vrêch.

In upland lakes and ponds, at altitudes up to 2,700 ft.(823m), usually on stones, clay or sand, at depths from a few cms. to 6m. or more.

Locally frequent, in the north and west of the British Isles to south Devon.

Europe: North and central regions and locally southwards to the Pyrenees, northern Italy and Bulgaria, Rossia (N,B,C), Iceland and Faeroes. Greenland.

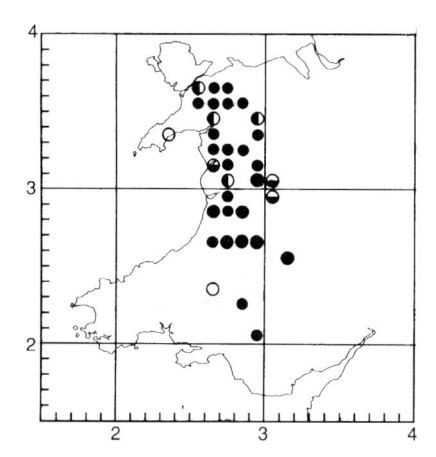

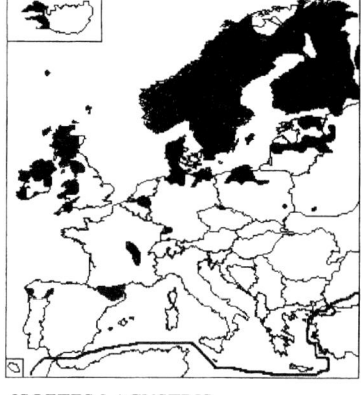

ISOETES LACUSTRIS

Intermediates between *I. lacustris* and *I. echinospora* sometimes occur where they grow together. Further work is needed to resolve the distribution of such hybrids.

SPHENOPSIDA - HORSETAILS

Family EQUISETACEAE - Horsetail Family

Genus EQUISETUM L.

Horsetails - Marchrawn

Perennial herbs with long branching underground rhizomes from which the aerial stems arise at intervals. Stems either all green pigmented, whether sterile or fertile, or of two kinds, fertile stems without chlorophyll, produced early in the season and later replaced by sterile stems, either whitish or green pigmented. Stems grooved, faintly or prominently, hollow except at the nodes, with a ring of smaller vallecular canals in the cortex and, between these and the central cavity, another ring of even smaller 'carinal canals'. Branches, when present, few or numerous, usually much more slender than the stem, arising in whorls from the nodes, simple or with secondary whorls of branchlets, hollow or solid (Fig. 23). Leaves small, corresponding usually with the same number of grooves on the stem. Fertile stems bearing terminal, cylindrical, ovoid or obovoid cones, consisting of a central axis bearing whorls of peltate sporangiophores, close together when young,

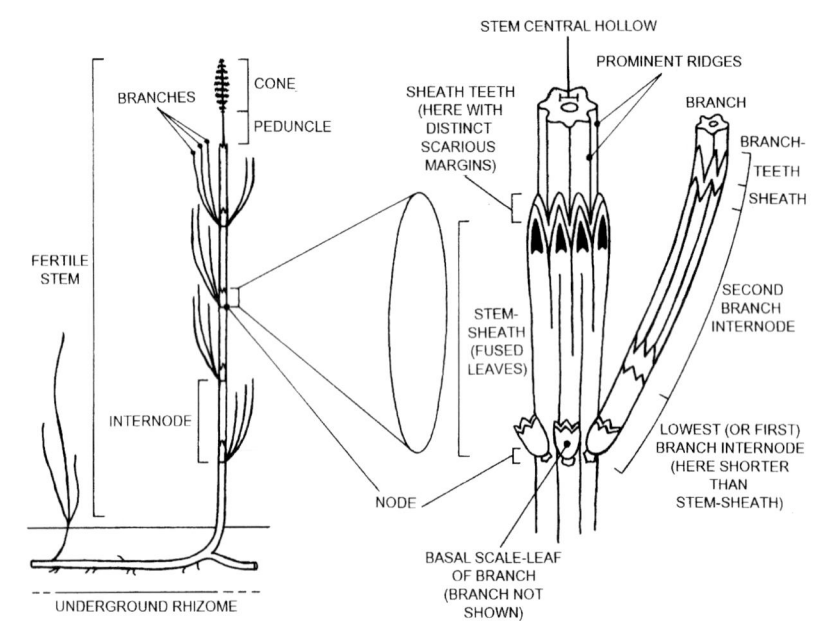

Fig. 23. Diagram of Horsetail (*Equisetum*) to illustrate certain descriptive terms.

becoming more widely spaced as the maturing cone elongates. Spores numerous, more or less spherical, with two strap-like elaters, attached centrally and spathulate at each end, forming four distinct appendages, spirally coiled when moist, and springing outwards when dry, assisting the dehiscence of the sporangium and possibly influencing wind dispersal by acting as wings. Prothalli are green, more or less cushion-shaped, lobed or branched, with numerous rhizoids on the underside. Male and female organs are often formed at different stages. About 23 species, almost cosmopolitan but absent from Australia and New Zealand apart from naturalized plants.

KEY TO SPECIES

1. Only fertile stems present, usually unbranched, not green, appearing in early spring　　　　　　　　　　　　　　　　　　　　　　　　　　**2**
   Sterile stems present (with or without fertile stems)　　　　　　　　**3**

2. Sheaths few (4-6), the uppermost less than 24mm long, with 6-12 teeth. Cone 1-4cm　　　　　　　　　　　　　　　　　　　　　　**E. arvense**
   Sheaths numerous (6-12), the uppermost 25-40mm long, with 20-30 teeth. Cone 4-10cm　　　　　　　　　　　　　　　　　　**E. telmateia**

3	Branches in whorls	**4**
	Sterile stems simple or with only a few branches near the base	**10**
4	Sterile stems whitish, robust (*c*.1cm diameter), smooth, with 20-40 inconspicuous ridges	**E. telmateia**
	Sterile stems green, usually less than 1cm diameter, rough or smooth	**5**
5	Stem-sheaths with 3-5 broad, obtuse or abruptly pointed lobes fewer than the 10-18 stem ridges; branches again branched	**E. sylvaticum**
	Stem-sheaths with subulate, lanceolate or triangular-ovate teeth as many as the ridges; branches usually simple (again branched in some forms of *E. arvense*)	**6**
6	Stem smooth, with 10-30 indistinct ridges, central hollow at least $4/5$ of stem's diameter	**E. fluviatile**
	Stem smooth, or rough, with 4-20 usually distinct ridges, central hollow less than $4/5$ of stem's diameter	**7**
7	Lowest internode of upper branches shorter than the adjacent stem-sheath	**8**
	Lowest internode of upper branches longer than the adjacent stem-sheath	**9**
8	Stem with 4-8 ridges, central hollow less than $1/2$ of stem's diameter. Cone obtuse. Rhizome smooth, glossy, dark purplish or black, glabrous (common)	**E. palustre**
	Stem with 10-20 ridges, central hollow $1/2$-$2/3$ of stem's diameter. Cone pointed. Rhizome rough, dull dark brown, hairy on the sheath (rare, suspected introduced - in only two localities, in Lincolnshire and Somerset)	**E. ramosissimum**
9	Stem-sheath 3-9mm, its teeth with a dark central line bordered by broader pale margins; stem central hollow $1/2$ of stem's diameter; lowest internode of lower branches shorter than adjacent stem-sheath. Rhizome dull dark brown, hairy on the sheath.	**E. pratense**
	Stem-sheath up to 10-22mm, its teeth with little or no pale margins; stem central hollow less than $1/2$ of stem's diameter; lowest internode of lower branches longer than adjacent stem-sheath. Rhizome glossy, dark brown, not hairy on the sheath	**E. arvense**
10	Stem slender, 1-2mm, simple or branched at the base, with 6-8 ridges, central hollow about $1/3$ of stem's diameter. Sheaths green with a dark apical band, or all black. Teeth with a narrow, dark mid-line and broad, pale scarious margin, their riangular-ovate or lingulate base persisting after the moderately persistent fine point is shed	**E. variegatum**
	Stem stout, 3-6mm, simple or occasionally sparsely branched where the apex has been damaged, with 10-30 ridges, central hollow $2/3$-$3/4$ of stem's diameter. Sheaths as broad as long; teeth subulate, soon shed	**E. hyemale**

EQUISETUM ARVENSE L.

Field Horsetail; Common Horsetail; Marchrhawn yr Ardir

Rhizomes shallowly ribbed, *c*.1.5-3mm in diameter near the stems, dark reddish-brown to almost black, rather glossy but this often concealed by dense rusty brown hairs. Deep underground rhizomes bear small ellipsoid tubers, singly or in strings. **Sterile stems** prostrate or erect, 10-90cm × 3-5mm, with numerous short internodes, green, with 8-20 prominent ridges, smooth or slightly rough. **Sheaths** 3-10mm, green, rather loose. **Teeth** 1-4mm, as many as the ridges, persistent, subulate, brown or blackish, especially towards their tips, with little or no scarious margin. **Branches**

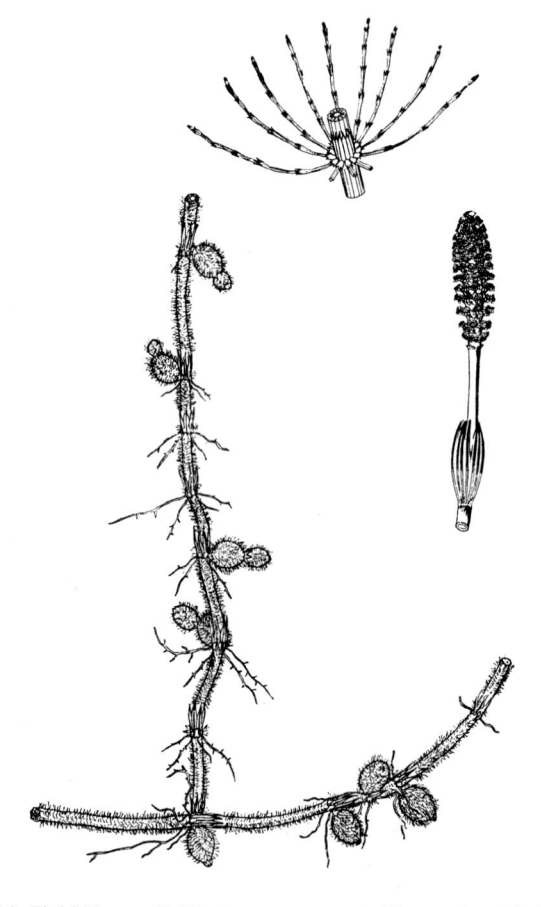

Fig. 24. Field Horsetail (*Equisetum arvense*). Top: node with branches. Left: rhizomes and tubers. Right: cone (all ×1/2). (*See also* Fig. 98).

numerous, usually simple, spreading or more or less fastigiate, usually diminishing in length and number per whorl towards the several unbranched basal and apical nodes, (3-)4-ridged, the lowest internode as long as or longer than the stem-sheath; the lowest branch-sheath 1-2mm long, with short triangular teeth, the other branch-sheaths longer, with 4 lanceolate, light brown teeth. **Stem central hollow** less than $1/2$ of stem's diameter. **Fertile stems** 10-25cm, unbranched, pale pinkish-brown, soft, fleshy, smooth and not conspicuously ridged; internodes 4-8, usually rather long, especially the upper ones. **Sheaths** 10-22mm, loose, light brown, with 6-12 darker teeth. **Cone** 10-40 × 4-9mm, cylindrical or conical, blunt-tipped, brown. Peduncle 20-60mm long. **Spores** ripening March-June. 2n=216.

E. arvense is a variable species and several forms or varieties have been named, most of them probably growth forms of no taxonomic significance. Unbranched dwarf forms in wet upland habitats in Wales have sometimes been confused with *E. variegatum*, but the latter may still be found in these areas. The apparent sterile stems of some *E. arvense* can on rare occasions produce cones, in some cases possibly due to damage to the plant earlier in the season. This feature has been recorded from ST18 (Radyr, Glam.) on small dumped mounds of earth and rubble.

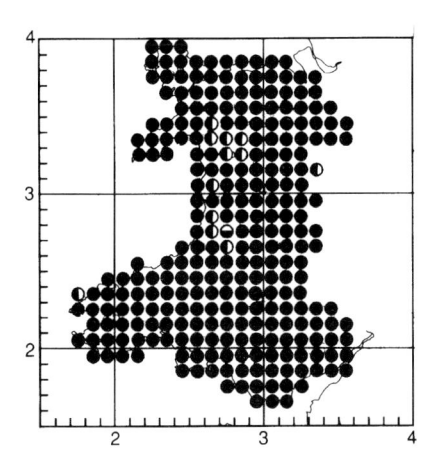

EQUISETUM ARVENSE

First recorded, 1813, Davies, *Welsh Botanology* I, 97 'Anglesey'.

Common in fields, waste places, hedge banks, dune slacks, etc. throughout the British Isles. Often a troublesome weed in fields and gardens.

Throughout Europe but more local in the south and east. Iceland and Faeroes. Northern Asia. Japan. Most of North America including Greenland. Introduced in New Zealand.

EQUISETUM FLUVIATILE L.

(*E. limosum* L.)

Water Horsetail; Marchrawn yr Afon

Rhizomes glossy, purplish dark brown, smooth, glabrous, long and stout, 3-6mm in diameter, bearing numerous wiry roots at the nodes. Tubers absent. **Stems** erect, simple or irregularly and often rather sparsely branched near the middle, 30-100(-140)cm × 2-8(-12)mm, the sterile often slightly taller than the fertile stems, dull green, glossy and often brownish near the base, smooth, with 10-30 indistinct, fine grooves. **Sheaths** 5-10mm, closely

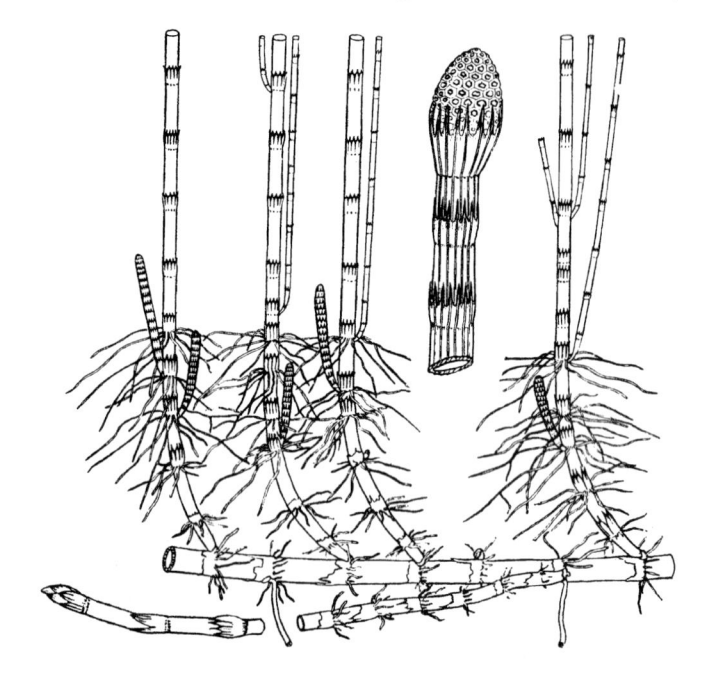

Fig. 25. Water Horsetail (*Equisetum fluviatile*). Rhizomes and stems (×1/4). Cone (natural size). (*See also* Fig. 98).

adpressed, usually the same colour as the stem, but the lower darker, cylindrical. **Teeth** persistent, as many as the grooves, subulate, 1-3mm long, usually dark brown (those of upper sheaths only brown-tipped), with little or no pale scarious margin. **Stem central hollow** at least $4/5$ of stem's diameter. **Branches** usually short, slender and ascending, 5-ridged; the sheaths with 4-5 teeth. **Cone** 10-20 × 6-10mm. Peduncle hidden by the apical sheath at first, ultimately 10-20mm long. **Spores** ripening May-July (-September). 2n=216.

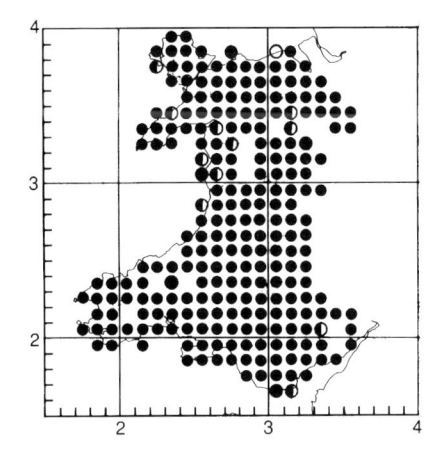

EQUISETUM FLUVIATILE

First recorded, 1813, Davies, *Welsh Botanology*, I, 97 'Anglesey'.

Locally common in suitable habitats, such as the margins of lakes, slow-moving rivers and canals, in ditches, fens, marshes, coastal dune slacks and other wet places, throughout the British Isles.

Europe: Throughout northern and central regions and southwards into central Spain, Italy, Macedonia, Rossia (N,B,C,W,E), Iceland and Faeroes. Caucasus, temperate Asia. North America: Canada and USA southwards to Virginia, Illinois, Idaho and Oregon.

EQUISETUM HYEMALE L.

(*Hippochaete hyemalis* (L.) Bruhin)

Rough Horsetail; Dutch-rush; Marchrawn y Gaeaf

Rhizomes dark, dull brown, roughly tuberculate, shallowly grooved, long and stout, 3-6mm in diameter. Tubers absent. **Stems** erect, 30-100cm × 3-6mm, persisting 3-4 years, simple, dull greyish or olive-green, with 10-30 distinct ridges each bearing a double row of tubercles, rough to the touch. **Sheaths** 3-10mm, about as broad as long, green at first with a narrow, black or dark brown apical band, becoming greyish-white with an additional, broader, dark band at the bottom or completely black. **Teeth** as many as the ridges, crinkly, slender pointed often with narrow, pale, scarious margins and up to 5mm long at first but usually shed at an early stage leaving their blunted bases as a dark crenulate apical margin to the sheath. Older sheaths sometimes splitting along the grooves, forming blunt pseudo-teeth. **Stem central hollow** $^{2}/_{3}$-$^{3}/_{4}$ of stem's diameter. **Cone** 7-15mm, sharply apiculate, the base and peduncle concealed at first by the uppermost sheath, which has larger and more persistent teeth, with broader scarious margins, than those lower down the stem. **Spores** ripening April-September. 2n=216.

EQUISETUM HYEMALE

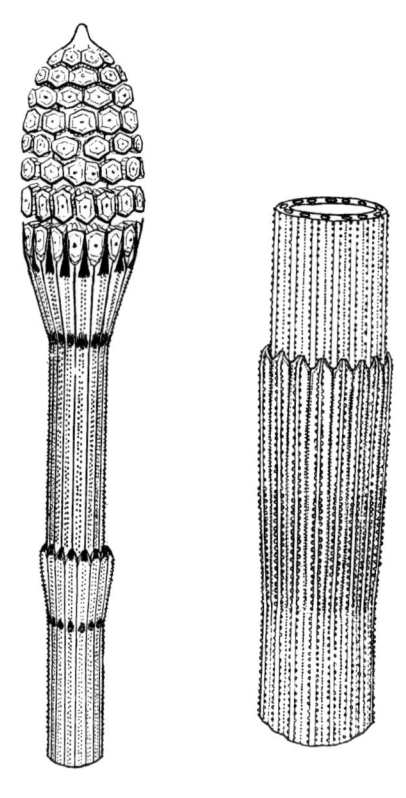

Fig. 26. Rough Horsetail (*Equisetum hyemale*). Cone (×2). Right: a single node (×4). (*See also* Fig. 98).

First recorded, 1805; J. W. Griffith in Turner and Dillwyn, *Bot. Guide*, I, 175. Denb. 'On the West side the brook that runs from Henllan Mill into the river Elwy, about 300 yards from Trap Bridge, less than a mile from Garn'.

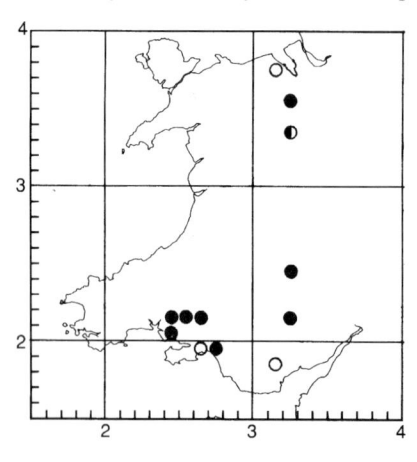

EQUISETUM HYEMALE

Local, although decreasing, in scattered localities on banks of streams and hedges, etc. throughout the British Isles; rare in southern England.

Throughout Europe and southwards to parts of the Mediterranean, Rossia (N,B,C,W,E), Iceland and Faeroes. Caucasus, northern and central Asia. Pacific coast south to Japan. North America: across the continent and southwards to Mexico and Central America.

EQUISETUM PALUSTRE L.

Marsh Horsetail; Marchrawn y Gors

Rhizomes glossy black or dark purplish, smooth, 1.5-3mm in diameter near the stems (stouter at greater depths), bearing numerous, wiry, brown-

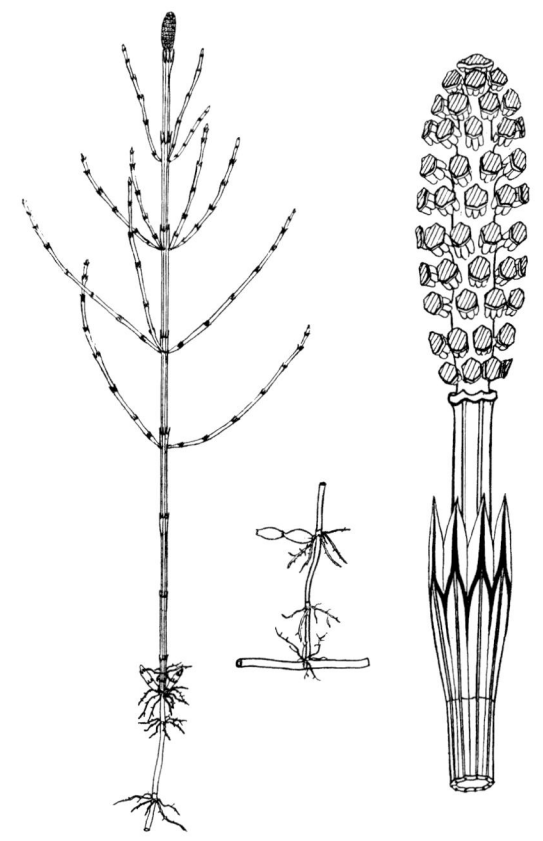

Fig. 27. Marsh Horsetail (*Equisetum palustre*). Left: fertile plant and portion of rhizome with tubers (*c.*×1/3). Right: cone (×2.5). (*See also* Fig. 98).

hairy roots at the nodes. At lower depths, numerous, ovoid or ellipsoid tubers are borne at the nodes, in strings, sausage-like. **Stems** arising in late spring or early summer, erect or sometimes decumbent, slender, 10-60cm × 1-3mm, pale greyish to dark green, with 4-8 prominent ridges, smooth or slightly verrucose. **Sheaths** 4-10mm, green, rather loose. **Teeth** 2-4mm, as many as the ridges, persistent, subulate blackish, with distinct pale scarious margins. **Branches** sometimes absent or sparse, but usually in rather irregular, more or less fastigiate whorls confined to the middle and upper parts of the stem, decreasing in length towards the apex, hollow, 4-6 ridged, the lowest internode about half the length of the stem sheath, the lowest branch sheath 1-2mm long, dark brown; the other branch sheaths are longer, green, with 5 or 6 more or less appressed, dark brown teeth. **Stem central hollow** usually less than $1/4$ of stem's diameter, surrounded by a ring of vallecular canals of similar size. **Cone** 10-30(-45) × 3-6mm, black, turning brown. Peduncle ultimately 20-40mm long. **Spores** ripening May-September. 2n=216.

E. palustre is a variable species and several forms or varieties have been named. One where a cone also appears at the apex of each branch, particularly in the upper part of the plant, has been named var. *polystachyum* Weigel.

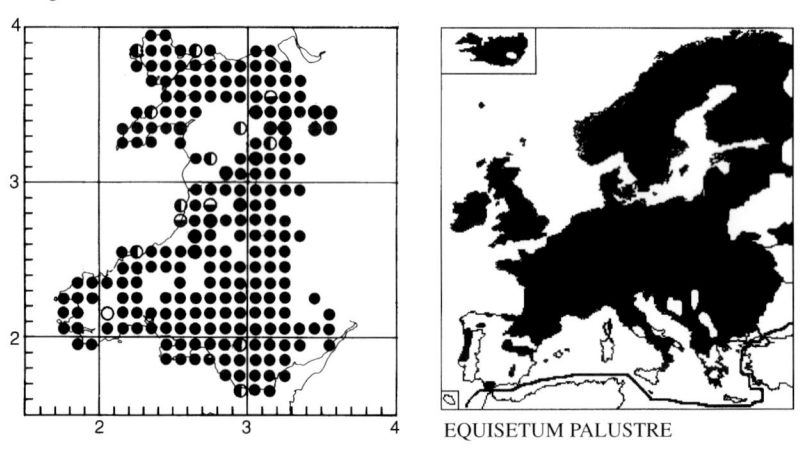

EQUISETUM PALUSTRE

First recorded, 1813; Davies, *Welsh Botanology* I, 97 'Anglesey'.

Common in most parts of the British Isles in bogs, marshes, lakesides, ditches, wet dune slacks and other wet places.

Throughout Europe apart from parts of the south. Iceland and Faeroes. Northern Asia and Japan. North America: across the continent, except northern Canada, and southwards to Vermont, Illinois, Montana, Idaho and California.

EQUISETUM PRATENSE Ehrh.

Shady Horsetail; Marchrawn y Cysgod

Rhizomes dark dull brown, noticeably grooved, glabrous except for their sheaths which are brown-hairy. Tubers absent. **Sterile stems** 20-60(-90)cm × 1-3mm, persisting only from spring until autumn, usually branched over the upper $^1/_2$ to $^2/_3$ of their length, pale green, with 8-20 prominent ridges, rough with tubercles. **Sheaths** 3-9mm, rather loose, pale green. **Teeth** as many as the ridges, ovate-lanceolate, sharply pointed, the subulate dark brown central line bordered by broader, pale brown, scarious margins. **Stem central hollow** $c.^1/_2$ of stem's diameter. **Branches** numerous, simple, spreading, slender, usually 4-12mm long, shorter in the lower than in the middle and upper whorls, 3-4 ribbed, the lowest internode shorter than the adjacent stem-sheath in the lower whorls but longer in the upper whorls. **Fertile stems** 10-30cm simple and withering soon after the spores are shed, or persisting longer and developing whorls of very short branches towards the apex. **Sheaths** larger, up to $c.13$mm long. **Cone** $c.15$-25mm, cylindrical, pale brown, darker towards the blunt apex. **Spores** ripening April. $2n=216$.

In the British Isles, rare, in a few scattered localities in N. England, Scotland and N.Ireland.

Throughout Europe apart from the south-west. Iceland and Faeroes. Northern and central Asia. Japan. North America: across the continent and southwards to British Columbia, South Dakota and New Jersey.

EQUISETUM RAMOSISSIMUM Desf.

(*Hippochaete ramosissima* (Desf.) Börner)

Branched Horsetail; Marchrawn Brigol

Rhizomes dark dull brown, slightly rough, prominently grooved, glabrous except for their sheaths, which are brown-hairy. Tubers absent. **Stems** 50-120cm × 3-9mm, persisting 1-2 years, usually branched about the middle and simple towards the apex and the base; greyish-green, with 10-20 distinct ridges, slightly rough with scattered tubercles. **Sheaths** 6-20mm green at first, then brown with a dark band at the bottom. **Teeth** as many as the ridges, dark brown, long pointed, with pale scarious margins. **Stem central hollow** $^1/_2-^2/_3$ of stem's diameter. **Branches** hollow, rarely more than six in a whorl, fastigiate, slender, 10-20cm or longer, with about 8 ridges; lowest internode shorter than the adjacent stem sheath. **Cone** $c.10$mm, its base concealed by the sheath. **Spores** ripening May-August. $2n=c.216$.

In Great Britain, established (suspected introduced) in two localities - near Boston, Lincs., on a raised artificial bank of a tidal estuary and N. Somerset in rough grassland near the sea.

Protected in Great Britain under the *Wildlife and Countryside Act, 1981*.

Widespread in central and southern Europe, Rossia (B,C,W,K,E), Azores. Central and southern Africa. Mascarene Islands. Madagascar. Throughout S. Asia. North America: introduced into North Carolina, Louisiana and Florida.

EQUISETUM SYLVATICUM L.

Wood Horsetail; Marchrawn y Coed

Rhizomes dull dark brown, reddish-brown hairy (glabrous at lower levels), 1.5-3mm in diameter near the stems, prominently 6-12 grooved;

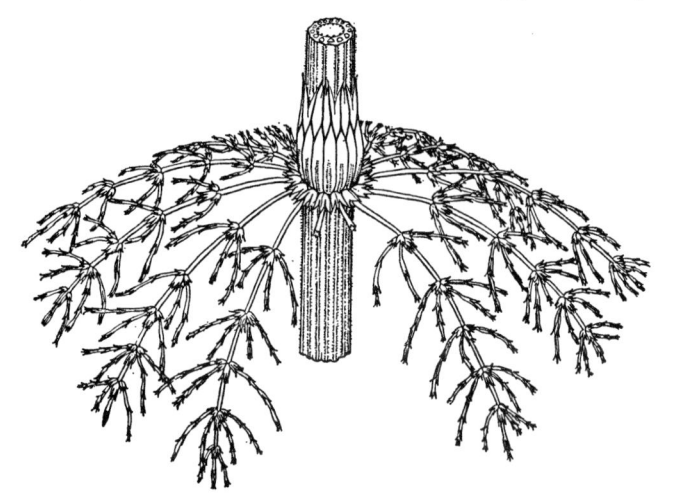

Fig. 28. Wood Horsetail (*Equisetum sylvaticum*). Above: a single node, with branches (natural size). Below right: cone (natural size). (*See also* Fig. 98).

bearing branching, wiry, reddish-brown hairy roots. **Tubers** subglobose or ellipsoid, *c*.7mm long, borne at the nodes, singly or in strings. **Sterile stems** 15-90cm×*c*.2-4mm, green, with 10-18 prominent ridges, variably tuberculate, smooth or fairly rough. **Sheaths** 10-15mm, loose, green below, pale reddish-brown in the upper half, where the teeth are united into 3-5 persistent, obtuse or abruptly pointed scarious lobes. **Branches** in nodal whorls numerous, throughout the upper 1/3-2/3 of the stem, pale or dark green, 3-4 ridged, very slender, themselves usually again branched, usually spreading and drooping at the tips. **Branch sheaths** *c*.2mm, inclusive of the

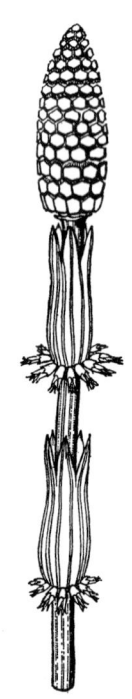

3-4 long subulate teeth. **Stem central hollow** $^{1}/_{4}$-$^{1}/_{2}$ of srem's diameter, with much smaller vallecular canals. **Fertile stems** usually shorter, with larger sheaths, up to 2cm long; branch whorls usually confined to the upper 2-5 whorls; branches and branchlets very short at first, lengthening as the cone matures. **Cone** 15-25 × 4-7mm, conical, pale brown, borne on a fleshy peduncle 20-50mm long. **Spores** ripening April-May. 2n=216.

A growth form with numerous, long, slender, dark green branches, occurs in moist, shady places. It is sometimes distinguished as 'var. *capillare* Hoffm.'

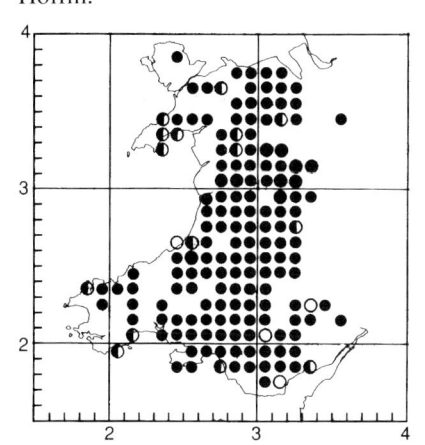

EQUISETUM SYLVATICUM

First recorded, 1804, Smith, *Flora Britannica*, III, 1102. 'By a wet dripping rock near the new walk, Hafod, Cardiganshire.'

Widespread in Wales, Scotland, N. England and N. Ireland, in woods and damp shady places, also mountain sides, moors, roadsides and fields, especially in upland areas. Less common in southern regions.

Most of Europe, although rare in the Mediterranean region. Rossia (N,B,C,W,E), Iceland. Temperate Asia including Japan. North America: across the continent, except for extreme northern Canada, southwards to Washington, West Virginia and Maryland. Greenland.

EQUISETUM TELMATEIA Ehrh.

(*E. maximum* auct.)

Great Horsetail; Marchrawn Mawr

Rhizomes dark brown, glossy, covered with dense reddish-brown hairs when young, 5-12mm in diameter; often bearing pyriform tubers up to 25 × 12mm. **Sterile stems** 30-180(-240)cm × *c*.1cm, whitish above, mottled dark

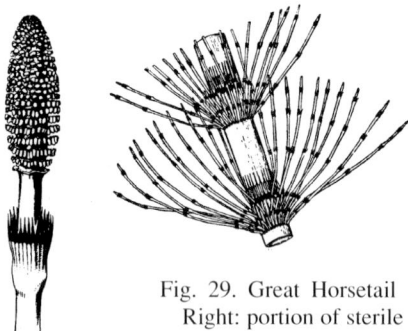

Fig. 29. Great Horsetail (*Equisetum telmateia*). Left: cone. Right: portion of sterile stem (×1/3).

brown or black towards base, smooth, with 20-40 inconspicuous ridges. **Sheaths** 10-40mm, rather loose, pale green, with long dark brown subulate teeth, as many as the ridges, with inconspicuous scarious margins. **Branches** on all but a few basal nodes, bright green, numerous, simple, long, spreading and often pendulous, 4-ridged and rough with tubercles; the lowest internode shorter than the stem-sheath; the lowest branch-sheath 1-2mm, glossy, dark chestnut-brown with lighter brown short triangular teeth; other branch-sheaths longer, green, with 4 linear-lanceolate, light brown teeth. **Stem central hollow** at least $2/3$ of stem's diameter. **Fertile stems** 15-40cm,

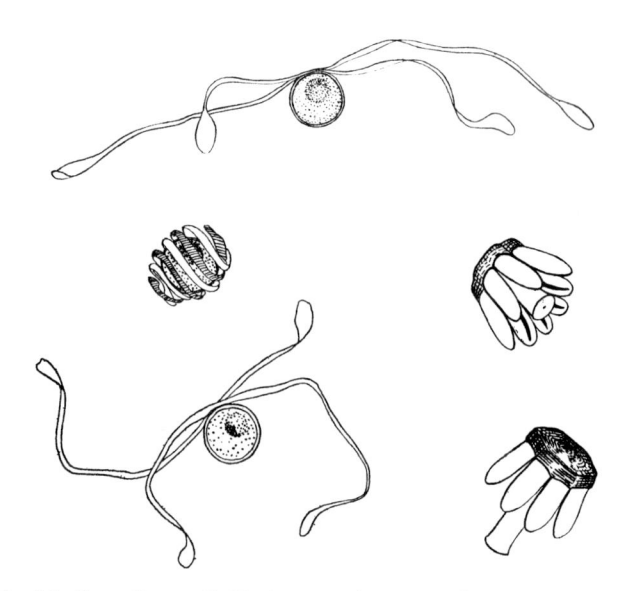

Fig. 30. Great Horsetail (*Equisetum telmateia*). Right: sporangiophores (×5). Left: spores, showing elaters, coiled and uncoiled (×135).

unbranched, pale pinkish-brown, soft, fleshy, smooth; concealed at first by the numerous (6-12) large, loose sheaths, the uppermost 25-40mm long, whitish in the lower half, brown above, their 20-30, long, lanceolate dark brown teeth with lighter margins. The 6-10 internodes elongate as the cone ripens. **Cone** 40-100 × 9-17mm, cylindrical or conical, narrowing to a blunt tip. Peduncle up to 85mm long. **Spores** ripening March-May. 2n=216.

A growth form bearing cones on a stem with green branches, similar to those of ordinary sterile stems, has been distinguished as var. *serotinum* A.Br., but it is of no taxonomic significance.

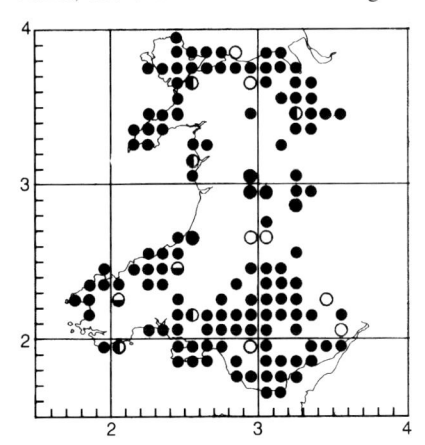

EQUISETUM TELMATEIA

First recorded, 1728, Brewer, Herb. Dillenius (Druce and Vines, *The Dillenian Herbaria*, 51, 1907). 'From ye meadows at Nant Francon [Caern.].'

Rather local, but widespread, in damp shady banks, roadsides, hedgerows and woods, sea-cliffs, etc. throughout the British Isles, but absent from much of Scotland.

Europe: southwards to the Black Sea region but absent from the extreme north of the continent and from Russia (N,C,E). Asia Minor and Caucasus. Azores. Madeira. Northern Africa. North America: west coast from British Columbia to California.

EQUISETUM VARIEGATUM Schleich.

(*Hippochaete variegata* (Schleicher ex Weber & Mohr) Bruhin)

Variegated Horsetail; Marchrawn Amlywiol

Rhizomes dark dull brown, glabrous, variably tuberculate, fairly prominently grooved, slender, 1.5-3mm diameter. **Stems** slender, decumbent to erect, variable in height, c.5-60cm × 1-2mm, persisting for several years, simple or branched at the base, dull green, with usually 6-8 prominent ridges, minutely tuberculate and slightly rough to the touch. **Sheaths** c.2-

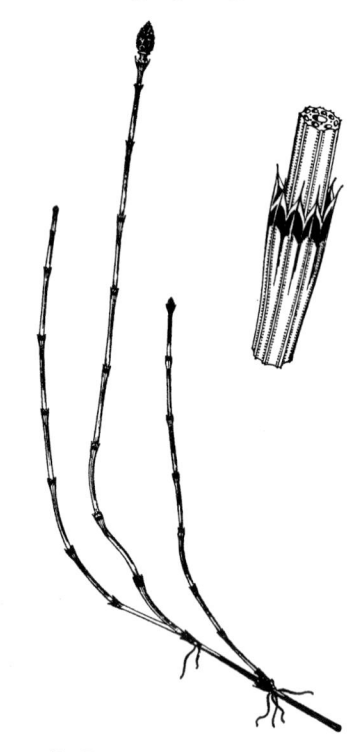

Fig. 31. Variegated Horsetail (*Equisetum variegatum*). Plant (×1/2). A single node (×5).

4mm, rather loose, green with a dark brown apical band. Teeth with a narrow, dark mid-line and broad white scarious margins, triangular-ovate or lingulate at the persistent base, tapering abruptly to a long, fine point, which is soon shed. **Stem central hollow** $c.^1/_3$ of stem's diameter. **Cone** 5-7 × 3-4mm, prominently apiculate, at first almost enclosed by the large apical sheath, with larger teeth than those lower down the stem. **Spores** ripening (March-) July-August (-October). 2n=216.

E. variegatum is a very variable species and has been divided into several varieties or even separate species on shoot size and growth habit (erect or decumbent). Many show good correlation with local ecology and maintain their growth form when brought into cultivation. Further investigation is needed to determine their distribution. Small growth forms of *E. arvense* have been mistaken for this species in wet upland areas in Wales, but

E. variegatum has been found on rare occasions in these areas and at other inland sites in Wales.

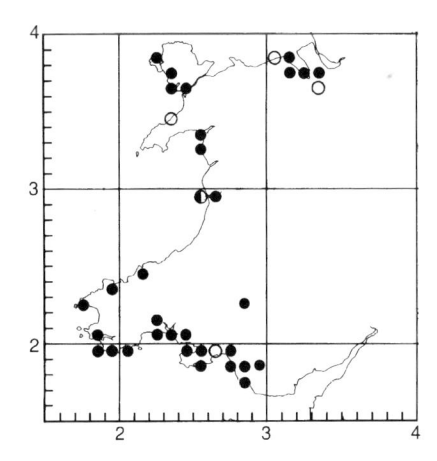

EQUISETUM VARIEGATUM

First recorded, 1874, Robinson cat. in Watson, *Topographical Botany*, edn 1, II, 511, '49 Carnarvon, 51 Flint'.

Damp dune slacks, lake shores, river and canal banks and can occur inland on wet stony mountain sites. Very local in the north and west of the British Isles, absent from most of central and southern England, widespread in central Ireland.

Europe: southwards into the Pyrenees and Apennines, Rossia (N,B,C,W), Iceland and Faeroes. Asia southwards to Mongolia and Japan. North America: across the continent southwards to Connecticut, Wisconsin and Oregon. Greenland.

HYBRIDS IN EQUISETUM

E. ARVENSE × *E. FLUVIATILE* = *E.* × *LITORALE* Kühlew ex Rupr.
Shore Horsetail; Marchrawn Croesryw

Rhizomes glabrous, glossy dark brown or blackish, smooth, 2-5mm diameter. At the nodes, bearing numerous dark brown wiry roots covered with brown hairs, and small, subglobose, glossy, black tubers, pointed at the tip, about 6mm long, in rows of 5 or 6, deep underground and therefore rarely seen. **Stems** erect or decumbent, resembling those of *E. fluviatile*, but usually more frequently and regularly branched, the length of the branches in each whorl decreasing progressively, from the middle region of the stem upwards, giving the plant a conical, tapering, 'Christmas Tree'-like appearance but with the apical region unbranched. **Grooves** 8-16, slightly more conspicuous than in *E. fluviatile*. **Sheaths** less closely adpressed especially towards the apex of the stem, often campanulate. **Teeth** similar to those of *E. fluviatile*. **Stem central hollow** 1/2-3/4 of stem's diameter. **Branches** slender, spreading more or less horizontally, or ascending, 4-ridged, with 4 slender black-tipped teeth. **Cone** 6-12 × 3-5mm yellowish, uncommon. Peduncle 5-10mm. **Spores** abortive. 2n=216 (showing irregular meiotic pairing of chromosomes).

Distinguished from *E. fluviatile* by the smaller stem central hollow, branches with fewer ribs, sheaths more campanulate than cylindrical, teeth often rather longer in proportion to the length of the sheath, cone yellowish. From *E. arvense*, by the larger stem central hollow and the glabrous rhizomes. The intermediate size of the stem hollow results in a characteristic springy, elastic feel to the internodes as each is repetatively squeezed gently between the thumb and a finger.

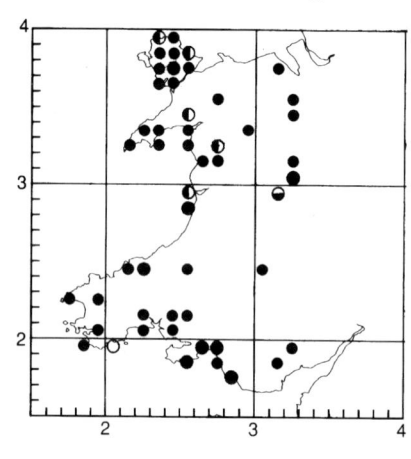

Wet dune slacks, ditches, lake and river margins, fens and damp commons, with or apart from the parents.

First recorded, 1948, P. Taylor, *B.S.B.I. Yearbook* 1950, p.50. Mer. 'Mynydd Gwerngraig'. Doubtless overlooked for many years, as the first British record was in 1886, from Bisley, Surrey.

EQUISETUM × LITORALE

Scattered throughout the British Isles, but probably more widespread than so far recorded, owing to confusion with *E. arvense*, *E. fluviatile* or *E. palustre*.

Many localities in N. and central Europe. Also known from North America.

E. ARVENSE × *E. PALUSTRE* = *E.* × *ROTHMALERI* C.N.Page

Ditch Horsetail; Marchrawn y Ffos

Stems erect, 25-50 cm, intermediate between the parents in general morphology but resembling *E. palustre* in being all alike except that the fertile stems are slightly smaller. Ridges 5-8. **Teeth** with little scarious margin, like E. arvense. **Stem central hollow** $1/4$-$1/2$ of stem's diameter. **Cone** 4-9mm long. **Spores** abortive.

Found on Skye (1971) and in Hertfordshire (1987).

E. FLUVIATILE × *E. PALUSTRE* = *E.* × *DYCEI* C.N.Page
Dyce's Horsetail; Marchrawn Dyce

Intermediate in character between the two parent species. Resembling a weak plant of *E. litorale* but with fewer whorled branches and a long branchless terminal segment.

Recorded from Wales in 1994, on peaty mud of drying-out reservoir near Ponterwyd (SN78, Cards.); with both parents. Outer Hebrides, west and south-west Ireland.

E. FLUVIATILE × *E. TELMATEIA* = *E.* × *WILLMOTII* C.N.Page
Willmot's Horsetail; Marchrawn Willmot

Shoots intermediate in character between the two parent species and resemble a taller and stouter *E.* × *litorale*. **Branches** numerous and unbranched. **Sheaths** long, green with long teeth. **Internodes** ivory-white and prominently ridged. **Stem central hollow** very large, $c.0.75$-0.9 of stem's diameter. **Fertile shoots** similar to vegetative shoots. **Cone** up to 15mm long. **Spores** abortive.

Found in County Cavan, Ireland (1984).

E. HYEMALE × *E. RAMOSISSIMUM* = *E.* × *MOOREI* Newman
Moore's Horsetail; Marchrawn Moore

Resembling *E. hyemale*, but stems more slender, rarely persisting more than one year, with 10-20 ridges; **sheaths** rather loose, longer than broad, remaining green longer; with dark bands at top and bottom; wholly dark only near the base of the stem; their **teeth** moderately persistent, dark brown, very slender, the pale scarious margins inconspicuous or absent. **Stem central hollow** $c.3/5$ of stem's diameter. **Spores** abortive. 2n=216 (showing irregular meiotic pairing of chromosomes).

Sand dunes and banks. Wicklow and Wexford.

W. and central Europe.

E. HYEMALE × *E. VARIEGATUM* = *E.* × *TRACHYODON* A.Braun
Mackay's Horsetail; Marchrawn Mackay

Stems simple or sparsely branched (not in whorls), 30-100cm × 2-4mm, greyish-green with 10-15 ridges, rough with tubercles, persisting for 2 or more years. **Sheaths** appressed, soon almost wholly black, with long subulate persistent dark brown **teeth**, usually with pale scarious margins. **Stem central hollow** $c.^1/_2$ of stem's diameter. $2n=c.216$ (showing irregular meiotic pairing of chromosomes).

Confined to a few localities, in Cheshire, S. Northumb. and Scotland; more widespread but still very local in Ireland. Also known from N. and central Europe, and North America including Greenland.

E. PALUSTRE × *E. TELMATEIA* = *E.* × *FONT-QUERI* Rothm.
Font-Quer's Horsetail; Marchrawn Font-Quer

Stems erect, 30-65cm, resembling *E. telmateia* in size and ivory-white internodes, more like *E. palustre* in general morphology and in the sterile and fertile stems being alike. **Ridges** 8-12, shallow. **Teeth** 2-ribbed, dark with scarious margins. **Stem central hollow** $c.^1/_4$ of stem's diameter. **Cone** 8-42mm long, abundant, spores only partly abortive.

Found in 1968 on Skye and since then in scattered localities from Worcs. to N. Ebudes, including on stabilized dune sand in Anglesey in 1989.

E. PRATENSE × *E. SYLVATICUM* = *E.* × *MILDEANUM* Rothm.
Milde's Horsetail; Marchrawn Milde

Intermediate in character between the two parent species. Not quite radially symmetrical in overall habit. **Stems** secondarily branched with shorter and fewer branchlets; **sheaths** tighter than in *E. sylvaticum*. Cone-bearing shoots not yet found.

Montane localities in Perthshire, and recently found in Outer Hebrides.

E. SYLVATICUM × *E. TELMATEIA* = *E.* × *BOWMANII* C.N.Page
Bowman's Horsetail; Marchrawn Bowman

Intermediate in character between the two parent species. **Fertile shoots** somewhat succulent, shorter and with looser leaf sheaths than those of *E. sylvaticum*. **Cones** similar to those of *E. telmateia* in shape and *E. sylvaticum* in size.

Found in the New Forest, S. Hants. in 1986.

FILICOPSIDA - FERNS

The Stem

One of the principal types of fern stem, i.e. the stock of the Male-fern, has been described on page 5. This type is found also in some other species of *Dryopteris*, and in Lady-fern (*Athyrium filix-femina*), Hard Shield-fern (*Polystichum aculeatum*) and many other ferns. The stock is short and thick and grows almost completely buried in the ground, and its growing point ascends at a steep angle to the horizontal. Although its leaves are apparently set in a circle or whorl round the growing point they are actually arranged in a very close spiral. The roots grow from the backs of the leaf-bases. In the Royal Fern (*Osmunda regalis*) there is a massive, branched and upright stock, while in the tree-ferns their upright stems may be several metres tall.

On the other hand, the other principal type of fern stem, the rhizome, is typically elongated and relatively narrow. It grows horizontally along, or beneath, the surface; its leaves are borne singly at intervals on its upper surface and usually in two rows. The growing point is naked except for hairs and scales and the roots all come from its lower surface. Such stems occur in Bracken (*Pteridium aquilinum*), Beech Fern (*Phegopteris connectilis*) and the Polypodies (*Polypodium* species).

There is no sharp line of distinction to be drawn between the two types: the stem in some ferns might with equal correctness be referred to either as a stock or as a rhizome. When the stem rises steeply, becoming almost erect, as in many of the stocky species, it is described as ascending. Other species, however, adopt a more or less horizontal, reclining attitude, with the growing point only ascending and the stem is then described as decumbent. Many ferns have stocky or rhizomatous stems which, by branching, produce several heads of leaves close together. Such stems are decribed as caespitose.

A few ferns possess stems which differ from any of the above types. Thus in the Adder's-tongues (*Ophioglossum*) and the Moonworts (*Botrychium*) the stem is short and upright, producing (in our native species) only one leaf a year.

Hairs and Scales

The surface of the fern stem, at least when young and commonly also later, is more or less thickly covered with hairs or scales. Usually the hairs (e.g. in Bracken (*Pteridium aquilinum*) and the Filmy-ferns (*Hymenophyllum*)) are coloured some shade of brown (from yellowish to almost black) and are jointed. The scales (Fig. 32) are coloured like the hairs, variously shaped (usually ovate to subulate) and more or less chaff-like. They are of two types (a) thin walled and consisting of a single layer of cells with their walls of uniform thickness and (b) latticed and consisting of

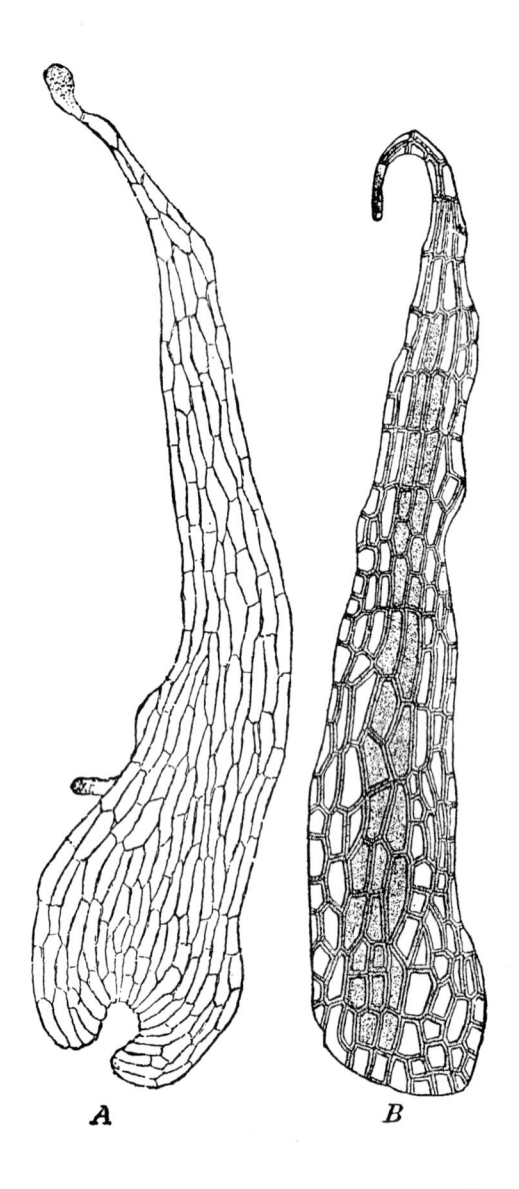

Fig. 32. *A*. Thin-walled scale of Brittle Bladder-fern (*Cystopteris fragilis*); the terminal cell is glandular. *B*. Latticed scale of Green Spleenwort (*Asplenium trichomanes-ramosum*) (*c*.×20). (After Sadebeck, from Engler and Prantl).

cells having their abutting walls strongly thickened and brown in colour, while the free walls are thin and colourless; scales of this type appear when magnified to be translucent with a lattice-like network of thick brownish-black lines. Thin-walled scales occur, e.g. in the Bladder-ferns (*Cystopteris*) and the Buckler-ferns (*Dryopteris*); latticed scales characterize the Spleenworts (*Asplenium*), the Rustyback (*Asplenium ceterach*) and the Hart's-tongue (*Asplenium scolopendrium*); a latticed scale may have a two- or more-layered midrib which then, on account of the thickening and colouration of the inside cell-walls, appears as a dark line, e.g. Maidenhair Spleenwort (*Asplenium trichomanes*). Both hairs and scales may be tipped with minute glands which exude some kind of secretion, e.g. Brittle Bladder-fern (*Cystopteris fragilis*) (Fig.32, A).

The Leaf

The leaves (or fronds) of ferns in general (Figs 33-35) are strongly developed relative to the stem which bears them. They may be either spirally placed or set in two ranks. They are very rarely provided with the paired leafy outgrowths at the base of the stipe (called stipules) which are so common in flowering plants. Among the native species described in this book stipules are confined to Royal Fern (*Osmunda regalis*), the Adder's-tongues (*Ophioglossum*) and Moonwort (*Botrychium lunaria*). The leaves of all native ferns are enfolded crozier-like in the bud except in the Ophioglossaceae and in the Water Fern, *Azolla*. Unless otherwise stated the leaves of all native species of fern unfold in spring and die down in autumn.

Fern leaves are commonly, and especially when young, more or less thickly covered with hairs or chaffy scales similar to those of the rhizome but when mature may be entirely without either (glabrous). Glands may occur either at the tips of hairs and scales (cf. Fig. 32, A) or as minute bodies closely attached to the surface of the leaf (e.g. Hay-scented Buckler-fern, *Dryopteris aemula*); glandular leaves are often fragrant and when fresh more or less sticky to the touch. In most ferns the leaf is constructed internally in a similar way to those of a flowering plant. The green assimilating tissue is protected on the outside by an upper and a lower epidermis and permeated by a system of air spaces which communicate with the outer atmosphere by means of adjustable pores called stomata. Leaves are supplied with water and nutrients by veins. A stipe (sometimes called the leaf stalk or petiole) is usually present. In most ferns when the leaf-blade dies, the base of the stipe remains attached to the stem until it decays. In others, however, the stipe becomes detached either wholly, leaving behind a distinct scar on the stem (as in Polypody, *Polypodium vulgare*) or partly, the surviving portion remaining attached to the stem (as in *Woodsia*).

The leaf-blade may be simple (undivided) as in Hart's-tongue (*Asplenium scolopendrium*), or it may be compound (divided), the characteristic form

72   FERNS (LEAF)

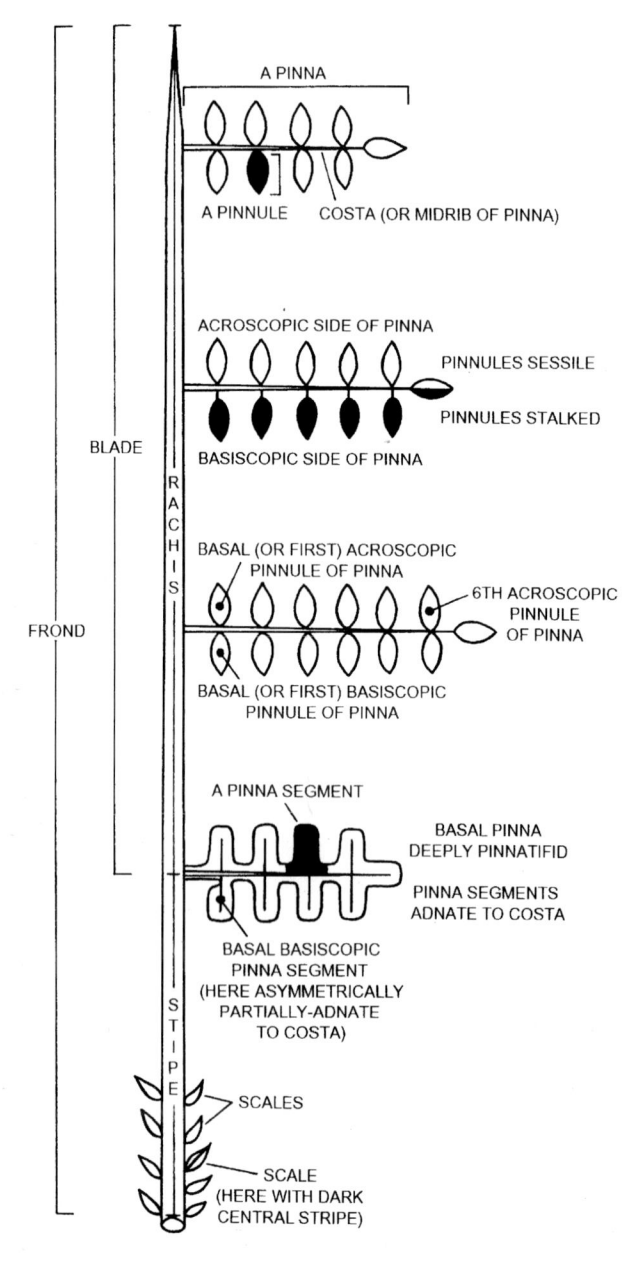

Fig. 33. Diagram of a fern (Filicopsida) to illustrate certain descriptive characters.

among ferns being that known as pinnate (*see* Fig. 34). In pinnate leaves a central axis or rachis bears right and left of it numerous distinct leaflets called pinnae. The primary pinnae may again branch bearing pinnae of the second order, or pinnules: the leaf as a whole is then said to be bipinnate. If the second order pinnae are themselves divided up to form pinnae of the third order, the whole leaf is then said to be tripinnate.

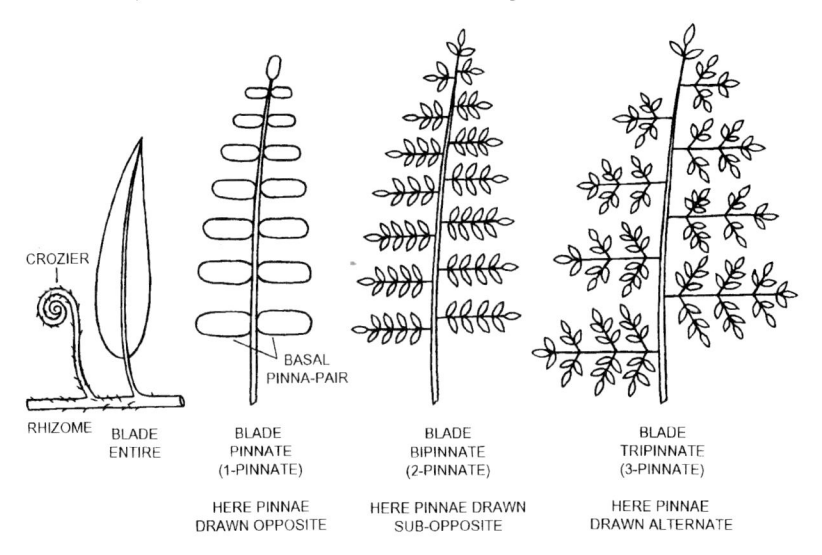

Fig. 34. Diagram of some fronds to illustrate different types of division (pinnation) of the leaf-blade.

A leaf or pinna may be built up on pinnate lines and yet not divided into distinct leaflets. The name segment is given to any division into which a leaf is cleft whether completely to the midrib or not; when the segments are shallow they are called lobes and the leaf (or pinna, etc.) is said to be pinnately lobed. When the segments are at all deeply cut the leaf (or pinna, etc.) is called pinnatifid and if cut to the midrib, pinnatisect.

In any leaf or part of a leaf the segments formed as the result of the last degree of division are called the ultimate segments. In general the degree of division is greatest in the basal part of the leaf and decreases progressively towards the tip or apex not only of the leaf as a whole but also of that of each of the pinnae; the ultimate segments at the base of the lowest pinnae may be of the third order while towards the leaf apex they may be of the first order. It is therefore convenient when describing a leaf completely to begin with the lowest pinnae and to work upwards. Thus a leaf may be bipinnate at the base 'becoming' pinnatifid towards the apex, or a stipe may be brown at the base 'becoming' straw-coloured half way up, and so on.

When it is necessary to distinguish between the two sides of a pinna the one facing towards the apex of the frond is called the acroscopic side, the other the basiscopic side. In the same way all pinnules which point towards the leaf apex are acroscopic pinnules, while those which point the other way are basiscopic and the pinnulets again may be similarly distinguished in relation to the apex of the pinna of which they form part. The pinnae of a compound leaf may be either stalked or stalkless (sessile). They may stand opposite to one another, or nearly so (subopposite), or they may be alternate.

The shape of the leaf-blade as a whole, or of a single pinna or pinnule of a compound leaf, may be described by the use of any of the following terms or combinations of them: ovate (egg-shaped in outline only, being broadest about one third up from the base), obovate (inversely egg-shaped in outline), lanceolate (tapering at both ends and somewhat broadened about one-third from the base), triangular (broadest at the base and tapering to the apex), deltoid (shaped like an equilateral triangle), linear (narrow and several times longer than wide), oblong (much longer than broad and with parallel sides). Frequently these terms are qualified by the use of an adverb such as 'broadly' or 'narrowly'.

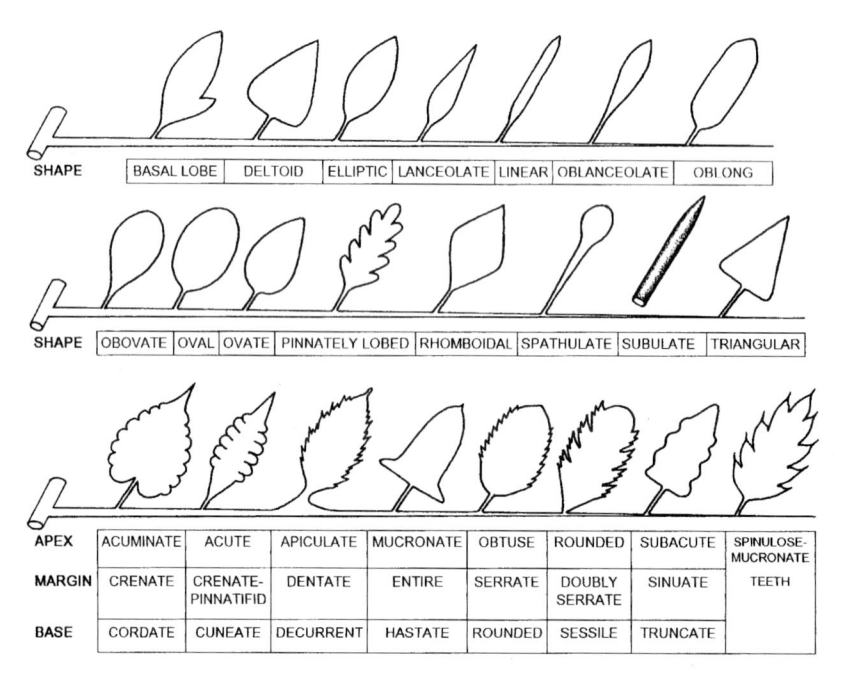

SHAPE	BASAL LOBE	DELTOID	ELLIPTIC	LANCEOLATE	LINEAR	OBLANCEOLATE	OBLONG

SHAPE	OBOVATE	OVAL	OVATE	PINNATELY LOBED	RHOMBOIDAL	SPATHULATE	SUBULATE	TRIANGULAR

APEX	ACUMINATE	ACUTE	APICULATE	MUCRONATE	OBTUSE	ROUNDED	SUBACUTE	SPINULOSE-MUCRONATE
MARGIN	CRENATE	CRENATE-PINNATIFID	DENTATE	ENTIRE	SERRATE	DOUBLY SERRATE	SINUATE	TEETH
BASE	CORDATE	CUNEATE	DECURRENT	HASTATE	ROUNDED	SESSILE	TRUNCATE	

Fig. 35. Diagram of some pinnae using pinnules to illustrate various shapes.

The apex of a leaf or pinna may be characterised according to its acuteness thus: acute (sharply pointed but not drawn out); acuminate (drawn out into a point), mucronate (ending in a short straight point), spinulose (ending in a small spine), obtuse (blunt or rounded), recurved (curved outwards away from the main apex), incurved (curved in towards the main apex).

The base of a leaf or segment may be described as: cuneate (tapering), truncate (cut off at right angles to the midrib), decurrent (running down the rachis, usually on one side only), unequal (asymmetrical). A segment is often so asymmetrical at the base that it is cuneate on one side and truncate on the other. A segment which is attached by the whole of its width to the rachis is said to be adnate.

The leaf margin may present a variety of shapes: entire (even, without toothing or division), serrate (toothed like a saw), crenate (with rounded teeth), sinuate (waved), incised (deeply cut), crisped (curled), inflexed (bent inwards and upwards), reflexed (bent outwards and downwards).

Throughout the ferns the shapes of the pinnae in one and the same plant vary very much according to their position on the frond.

The arrangement of the veins of a leaf is referred to as its venation. (The veins can often be seen more clearly if the leaf is held up to the light.) The primary or main vein of the leaf, or more usually of one of its segments, is called the midrib or, when less distinct, the mid-vein. Its branches are called secondary veins and their branches tertiaries. The tertiary veins are termed acroscopic when they branch off the secondary on the side facing towards the apex of the segment, basiscopic when on the opposite side. The type of veining in the pinnae varies very considerably from one kind of fern to another, but if attention is confined to mature leaves the venation is found to be in some respects at least fairly constant within any one species and sometimes throughout any one genus. Although the venation in ferns may generally be regarded as being made up of primary veins giving off secondaries, secondaries giving off tertiaries and so on, in actual fact the ultimate veins at least almost always branch by equal forking (true dichotomy). This is clearly seen, for example in Wilson's Filmy-fern (*Hymenophyllum wilsonii*) (Fig. 72) where the veins supplying the ultimate segments spring from the main vein which runs the whole length of a pinna.

Fertile Leaves

The fertile or spore-bearing leaves of ferns are in general hardly distinguishable from sterile or non-spore-bearing leaves. Among Welsh ferns the exceptions to this statement are: Hard Fern (*Blechnum spicant*) and Parsley Fern (*Cryptogramma crispa*), in which the fertile leaves are longer, have narrower pinnae and together form a distinct whorl; Royal Fern

(*Osmunda regalis*), in which the upper pinnae only of the fertile leaves bear spores; and the ferns belonging to the Adder's-tongue family, in which the leaves are divided into a sterile blade and a fertile spike. The most usual position of the sporangia is on the under surface of the leaf and away from the margin as in Male-fern (*Dryopteris filix-mas*), but they may be located near the margin as in Bracken (*Pteridium aquilinum*) and Maidenhair Fern (*Adiantum capillus-veneris*) or actually on the margin as in the Filmy-ferns. The sporangia usually arise above the veins and usually on a more or less strongly marked swelling called the receptacle. They form discrete clusters known as sori. In the Royal Fern (*Osmunda regalis*), however, the sporangia are not grouped into definite sori and in Moonwort (*Botrychium lunaria*) they are much larger and borne singly.

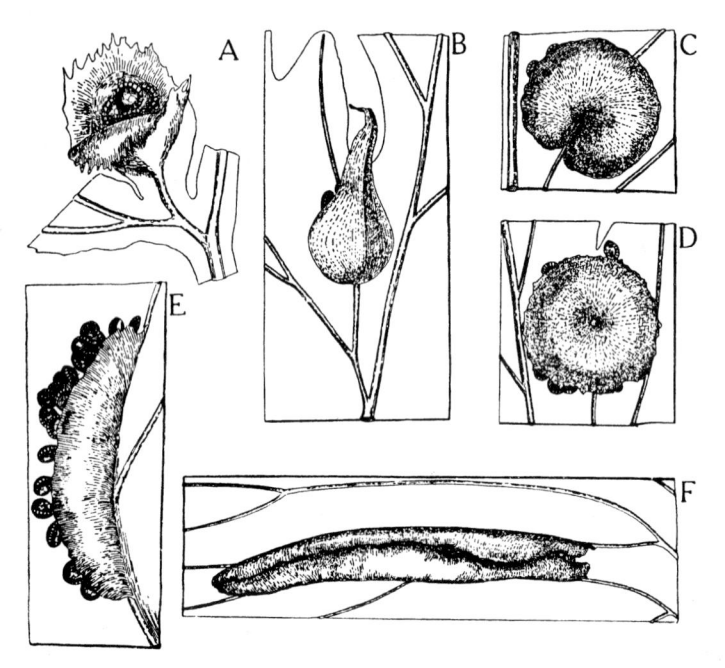

Fig. 36. Some types of sorus and indusium. A. Marginal sorus with two-lipped indusium (*Hymenophyllum tunbrigense*). All the remaining examples are superficial. B. Circular sorus with basal indusium (*Cystopteris fragilis*). C. Circular sorus with kidney-shaped indusium attached at the notch (*Dryopteris filix-mas*). D. Circular sorus with peltate indusium (*Polystichum setiferum*). E. Linear sorus and indusium (*Asplenium trichomanes*). F. Paired linear sori with indusia facing and overlapping (*Asplenium scolopendrium*).

The fern sorus may be entirely naked as in Polypody (*Polypodium vulgare*), where its form is approximately spherical or globose, or it may be protected by a usually membranous covering called an indusium. The indusia of some species remain attached even after the spores have been shed whereas in some other species the indusia are themselves shed. In the Spleenworts (*Asplenium*) (Fig. 36, E) the indusium is linear in shape and extends along a fertile vein; in the Filmy-ferns (Fig. 36, A) it encloses the base of the receptacle, forming a sort of bivalved cup; in Bladder-ferns (*Cystopteris*) (Fig. 36, B) and Woodsias (*Woodsia*) it is placed at the base of the sorus which it more or less covers from below; in the Buckler-ferns (*Dryopteris*) (Fig. 36, C) it surmounts the entire sorus as a kidney-shaped (reniform) shield and in the Shield-ferns (*Polystichum*) (Fig. 36, D) it forms a completely circular covering with a central stalk (i.e. it is peltate). In Bracken (*Pteridium aquilinum*) (Fig. 89) the principal (outer) indusium, formed or apparently formed by the leaf margin, faces an inner indusium arising from the leaf surface and the two together protect the continuous marginal sorus. A marginal sorus is also found in Hard Fern (*Blechnum spicant*).

In most ferns the capsule of the sporangium is set on a long stalk. However, sporangia may have a short stalk as in Royal Fern (*Osmunda regalis*) (Fig. 75) and the Woodsias (*Woodsia*), have no stalk as in the Filmy-ferns (*Hymenophyllum*), or be sunk in the receptacle as in Adder's-tongue (*Ophioglossum*) (Fig. 73). The process of opening of the sporangium, with consequent liberation of the spores, is called dehiscence. The most usual mode of dehiscence by the operation of a vertical annulus has already been described (p. 5). In the Filmy-ferns (Fig. 70), however, the annulus is oblique and the sporangium dehisces by a longitudinal split; in *Osmunda* (Fig. 75) the annulus is represented by a group of thick-walled cells at one side of the capsule and dehiscence follows the line of a longitudinal stomium running from the annulus over the top and down the opposite side; in the Ophioglossaceae (Fig. 73) there is no annulus and dehiscence takes place by a transverse slit; and finally in the Water Ferns the spores are liberated into the water by the decay of their various coverings.

In certain ferns all the sporangia in a sorus develop and mature simultaneously so sporangia of different ages are mixed up together. In others they are formed sequentially from the top of the receptacle downwards (thus the bristle of Killarney Ferns (*Trichomanes*) is the elongated receptacle bearing the developing sporangia).

Genus ADIANTUM L.

Maidenhair Ferns - Brigerau Gwener

Sori (Fig. 37) confined to special fertile lobes, or lappets, of the blade margin which are reflexed beneath the blade and membranous when mature, and each of which bears a group of parallel linear sori situated on a series of fine veinlets (or less often borne also on the blade surface between the veinlets); indusium none, the fertile lobes themselves protecting the sporangia. Blades undivided to several times pinnate; stipe usually black, shining and brittle; lamina of delicate texture and bright green in colour, its segments usually more or less triangular in outline; veins usually ending blindly.

A large genus, but a very natural one and clearly characterized by its foliage. Species about 200, mostly Tropical, especially in America.

ADIANTUM CAPILLUS-VENERIS L.

Maidenhair Fern; Briger Gwener

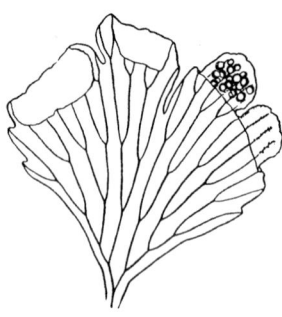

Fig. 37. Maidenhair Fern (*Adiantum capillus-veneris*): pinnule seen from below, showing the venation and a series of four fertile membranous lappets of the margin; the two lappets on the left are shown in the natural position, reflexed beneath the frond, the next has been straightened out to show the sori and the last has also had the sori removed to show the vein endings. (Enlarged rather more than 3 diameters). (*See also* Fig. 99A).

Rhizome creeping, up to 5mm thick, covered with narrow, chaffy brown scales. **Fronds** up to 30cm, erect when small, drooping when large, closely placed in two rows; **stipe** about as long as blade, beset with scales at the base, otherwise glabrous (like the rachis), dark brown to purplish-black. **Blade** up to 15 × 10cm, ovate or elongate-oval in outline, obtuse, bipinnate (simply pinnate towards the apex and in very small examples throughout); **pinnae** widely spaced and alternating on the rachis, stalked; **pinnules** up to 12mm broad, on very fine hair-like stalks, obovate or elongate-obovate and with a wedge-shaped base (less often practically rhomboidal or semicircular), the outer margin more or less deeply cut. **Fertile pinnules** more or less truncate at the apex; lappets as described above for the genus, half-moon shaped, each about 3mm wide by 1mm deep or smaller, 2-6 on each blade segment. **Sori** 2-10 per lobe. **Spores** ripening May-September. Diploid. 2n=60.

Very local in sheltered places on sea cliffs, usually on calcareous tufa.

First recorded, 1698; Edward Llwyd, in a letter to R. Richardson (*Phyt.*, 2nd ser., *1*, 268, 1855-6). '... growing very plentifully out of a marly incrustation both at Barry Island and Parth Kirig [Porthkerry], in Glamorganshire and out of no other matter.'

Confined to south-west England, Westmorland and the Isle of Man; in western Ireland and the Channel Islands. At most inland sites it has probably

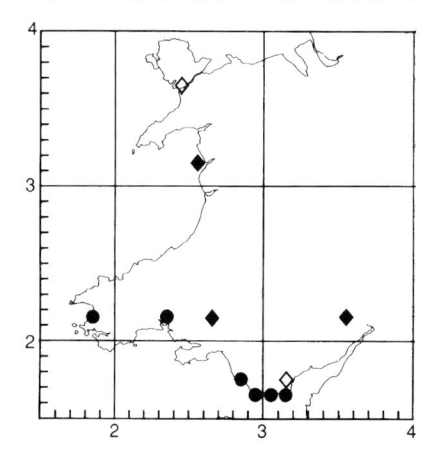

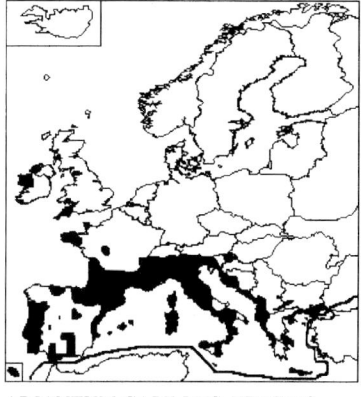

ADIANTUM CAPILLUS-VENERIS

become established from the spores of cultivated plants some of which may be the cultivated species *A. raddianum* or *A. venustum* but this has not been proved.

Cosmopolitan. Throughout the Tropical and warm Temperate zones especially in the Northern Hemisphere. Western and southern Europe, Rossia (K). Azores, Madeira, Canary Is. and the Cape Verde Is. Africa: north, central and South Africa, Mascarene Islands, Madagascar, Réunion. Scattered across the southern half of Asia. Southern USA, Mexico, Cuba southwards to the Amazon region. Australia, Polynesia.

ADIANTUM PEDATUM L.

American Maidenhair Fern; Briger Gwener America

Rhizomes short-creeping, branched, bearing fronds in close groups. **Fronds** to 30cm or more, ascending or arching, as wide as or wider than long, deciduous; **stipe** longer than blade. **Blade** first forking into two segments, then to 4-8 or more narrow, pinnately divided **pinnae**; **pinnules** almost rectangular, often deeply cleft along the upper margin, hairless somewhat flaccid, dull pale blue-green. 2n=58.

Garden escape established near Virginia Water, Berks. and wall of fern nursery, Dorset.

N. America and NE Asia.

Genus ANOGRAMMA Link

Jersey Ferns - Rhedyn Jersey

Rhizome very short with few scales. Sori elongated to linear, running the whole length of fertile veins of the second or higher order or only occupying the upper part of the veins; blade margin quite flat and indusium absent.

ANOGRAMMA LEPTOPHYLLA (L.) Link

Jersey Fern; Rhedynen Fach Jersey

A small winter annual (up to about 10cm) in general appearance somewhat reminiscent of *Cryptogramma* but the fertile fronds not so clearly

Fig. 38. Jersey Fern (*Anogramma leptophylla*). Erect fertile fronds and the earlier shorter sterile fronds (×0.6).

distinct from the sterile ones, in particular the margins of the ultimate segments not recurved; both kinds of fronds bipinnatifid. Rhizome not developed, the species perennating by means of a tuberous prothallus. **Spores** ripening March-May. Diploid. 2n=52.

Jersey and Guernsey.

Mediterranean region, Rossia (K). Azores, Madeira, Canary Is. and the Cape Verde Is. Africa. Caucasus, E. of Caspian Sea, southern India to New Zealand. Central and South America.

Genus ASPLENIUM L.

Spleenworts - Duegredyn

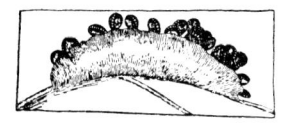

Fig. 39. Single linear sorus of Wall-rue (*Asplenium ruta-muraria*).

Sori usually oval to linear and single, or in twin pairs (*A. scolopendrium*), sometimes hooked at the end or rarely double back to back. Indusium conforming to the shape of the sorus, usually opening inwards, i.e. towards the midrib but occasionally outwards, i.e. towards the margin; those of a pair (*A. scolopendrium*) opening towards each other.

KEY TO SPECIES

1	Blade undivided	**A. scolopendrium**
	Blade divided	2

2 Blade segments attached to rachis across their whole base; dense scales covering underside, turning reddish-brown when mature **A. ceterach**
  Blade segments not attached to rachis across their whole base; no dense scales covering underside   3

3 Blades irregularly and sparsely divided; segments linear
                                                                       A. septentrionale
  Blades regularly divided, 1- to 3-pinnate; segments not linear   4

4	Fronds 1-pinnate	5
	Fronds 2- to 3-pinnate	7

5 Larger pinnae longer than 12mm, length usually two or more times width
                                                                       A. marinum
  Pinnae less than 12mm, length usually two or less times width   6

6	Rachis blackish or red-brown	**A. trichomanes**
	Rachis completely green	**A. trichomanes-ramosum**

7 Basal pair of pinnae slightly to considerably shorter than the adjacent pair **A. obovatum** subsp. **lanceolatum**
  Basal pair of pinnae the longest **8**

8 Pinna irregularly cut into ± equal segments; stipe green; indusia with fringed margins **A. ruta-muraria**
  Pinna regularly (1-)2-pinnate; stipe reddish-brown to blackish; indusia with entire margins **9**

9 Blade and pinnae acute to shortly acuminate **A. adiantum-nigrum**
  Blade and pinnae with finely drawn out (caudate) tips; (only in southern Ireland) **A. onopteris**

ASPLENIUM ADIANTUM-NIGRUM L.

Black Spleenwort; Duegredynen Ddu

Rhizome 10×0.5cm, creeping or decumbent, branched or caespitose, the younger parts clothed with brownish-black, subulate scales with long hair-like points. **Fronds** 10-45cm, tufted, winter-green; **stipe** not longer than the blade, thickened and scaly at the base, dark purplish-brown. **Blade** $5-25 \times 2.5-16$cm, triangular in outline, always widest at the base, often somewhat

Fig. 40. Black Spleenwort (*Asplenium adiantum-nigrum*). Median pinna from an average sized frond (×2). (*See also* Fig. 100A).

acuminate, firm in texture, dark shining green above, paler beneath, when young with scattered minute brownish-black hair-like scales, bi- or tri- or even almost quadri-pinnate at the base, becoming less divided towards the apex and ultimately only pinnatifid; **pinnae** straight, up to 15 on either side, alternate or sub-opposite, pinnate or bipinnate, rarely almost tripinnate at the base of the basal pinnae, the lowest pair larger than the succeeding, the lower and middle ones stalked, the upper ones becoming sessile, their rachises winged throughout; **pinnules** alternate, the basal ones of the lower and middle pinnae pinnate or pinnatifid, those on the side of the pinna towards the apex larger than the others; **ultimate segments** ovate to lanceolate, acute or rounded at the apex, more or less toothed in the upper half, cuneate at the base; **rachis** purplish-brown on the lower half of the under surface, otherwise green. **Sori** linear, situated towards the middle of the segment, on the secondaries or their acroscopic branches, the lowest ones sometimes paired back to back or occasionally hooked as in *Athyrium*; the mass of sporangia finally occupying the whole of the middle of the segment. **Indusium** conforming in shape to sorus, usually opening inwards, entire. **Spores** ripening June-October. Tetraploid. 2n=144.

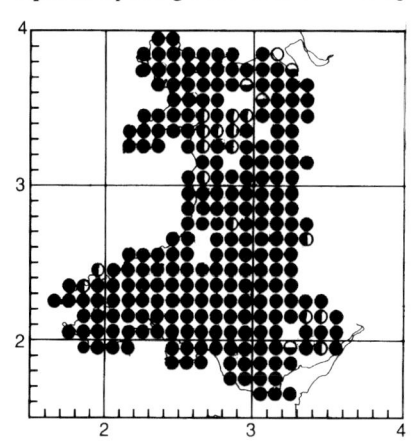

ASPLENIUM ADIANTUM-NIGRUM

Common in rock crevices and on walls and hedge banks.

First recorded, 1726; Littleton Brown in a letter to Dillenius. (Druce and Vines, *The Dillenian Herbaria*, lxxiii, 1907). Tenby.

Throughout the British Isles.

Cosmopolitan. Europe from southern Scandinavia southwards, excluding the Baltic regions. Russia (W,K). Faeroes. Azores, Madeira, Canary Is. and Cape Verde Is. North, west, central and southern Africa and the Mascarene Islands. Asia: Yemen, Asia Minor across the Caucasus to Himalaya; India and SE Asia. SW USA and Mexico. Australia, Micronesia and Hawaii.

ASPLENIUM CETERACH L.
(*Ceterach officinarum* Willd.)
Rustyback; Rhedynen Gefngoch

Stock short, upright or ascending, often caespitose, clothed with blackish-brown to blackish, lanceolate, acuminate scales. **Fronds** 3-20cm, tufted, wintergreen; stipe short, $1/6$-$1/4$ as long as the blade, at the base dark brown or blackish with scales similar to those of the stock, higher up with the same kind of scales intermixed with scales similar to those on the blade and with others intermediate between the two. **Blade** 2.5-17 × 0.5-3cm, thick and leathery, linear-lanceolate in outline deeply pinnatifid, deep or yellowish-green and with a few scattered scales especially on the midrib above, densely clothed below with tawny (at first silvery) ovate-acuminate, apparently peltate scales, which are attached near their bases set so as to overlap like tiles on a roof, and project beyond the margin; **segments** alternate, ovate or oblong, rounded at the apex, entire or crenate along the margin, decurrent at the base, the lowest much smaller and sometimes completely separated. **Sori** linear, situated about midway

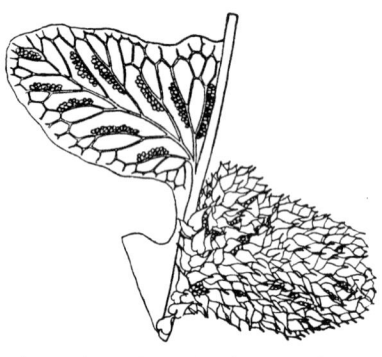

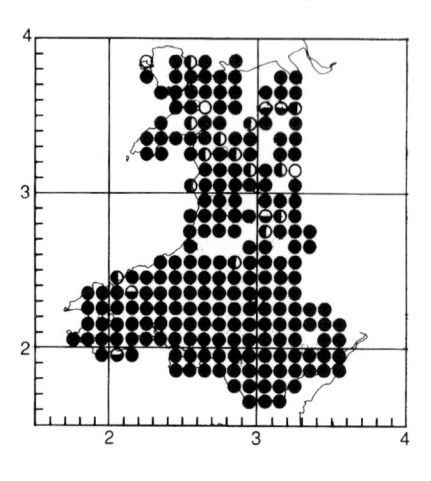

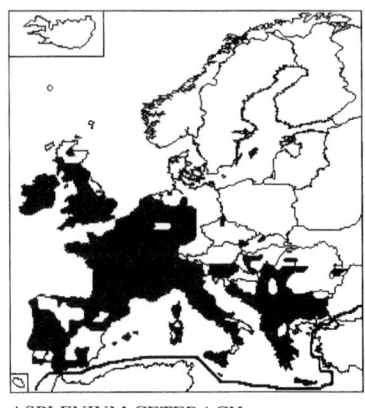

Fig. 41. Rustyback (*Asplenium ceterach*). Part of the upper third of a frond (×3) showing two segments, the lower one clothed with overlapping scales, the upper one with the scales removed, revealing the venation and the sori in their relation thereto. (*See also* Fig. 101).

ASPLENIUM CETERACH

between the midrib of the segment and the margin. **Indusium** rudimentary or absent. **Spores** ripening April-October. Tetraploid. 2n=144.

On old (lime-mortared) walls and limestone rocks.

First recorded, 1696, Edward Llwyd in a letter to Dr. R. Richardson. (*Phyt.*, 2nd ser., *I*, 268, 1855-6) 'In South Wales.'

Great Britain chiefly in the south and west, northwards to Easterness; scattered in eastern Britain; Channel Islands; throughout Ireland.

Europe: south-west of a line from Germany to Crimea (Rossia (K)). Madeira, Canary Is. and Cape Verde Is. North Africa. Asia Minor through the Caucasus to Afghanistan. Himalaya and southern part of central Siberia.

ASPLENIUM CUNEIFOLIUM Viv.

(*A. serpentini* Tausch)

Serpentine Black Spleenwort; Duegredynen Ddu Sarff-faen

Blades bipinnate or tripinnate near the base, softer, paler and less glossy than those of the other two species; **ultimate segments** fan-shaped, obovate, cuneate at the base, truncate at the apex, with broad, more or less mucronate teeth. Diploid. 2n=72.

Usually on serpentine (ultrabasic) rocks. Not confirmed for the British Isles, recorded in error for forms of *A. adiantum-nigrum* in Scotland and Ireland.

Usually on serpentine in central and southern Europe, Rossia (W).

ASPLENIUM FONTANUM (L.) Bernh.

Smooth Rock-spleenwort; Duegredynen Lefn y Creigiau

Essentially a mountain species on calcareous rocks in central and southern Europe.

As a rule much smaller than *A. obovatum* subsp. *lanceolatum*, from which it differs also in its entirely green rachis, much reduced lower pinnae, obovate, cuneate-based pinnules and mucronate teeth. Diploid. 2n=72.

Not confirmed for the British Isles. Records in a few English localities, as escapes growing on walls, have not been substantiated. All records are pre-1930. The specimen labelled 'A. fontanum' from 'betwixt Tan-y-bwlch and Tremaddoch' in Dr. J. Power's herbarium (Holmesdale, Natural History Club, Reigate) is not this species but probably an immature state of *Athyrium filix-femina*. The 1837 Jersey specimen in the Thompson collection at the University of Oxford, Botany School Herbarium has been considered doubtful due to the geographical distribution of the species.

ASPLENIUM MARINUM L.

Sea Spleenwort; Duegredynen Arfor

Stock caespitose, erect to decumbent, densely clothed with purplish- to brownish-black, linear-lanceolate scales. **Fronds** tufted 8-50cm, stipe about as long as the rachis, reddish to purplish-brown with a few scales at the base, otherwise glabrous, smooth and shining. **Blade** 4-40cm, narrowly to broadly lanceolate, somewhat leathery in texture, pinnate; **pinnae** bright green, oblong to broadly ovate, very unequal at the base (broadly truncate and often almost lobed on the acroscopic side (side facing towards the apex of the frond), cuneate on the basiscopic side (side facing towards the base of the frond)), and rounded or rarely subacute at the apex, the margin broadly crenate, the upper pinnae running together; **rachis** flanked by wings formed by the decurrent bases of successive pinnae. **Sori** linear, situated on the acroscopic forks of the secondary veins. **Indusium** also linear, opening inwards. **Spores** ripening June-September. Diploid. 2n=72.

Fig. 42. Sea Spleenwort (*Asplenium marinum*). Median pinna of a well-developed frond, twice natural size. (*See also* Fig. 100*B*).

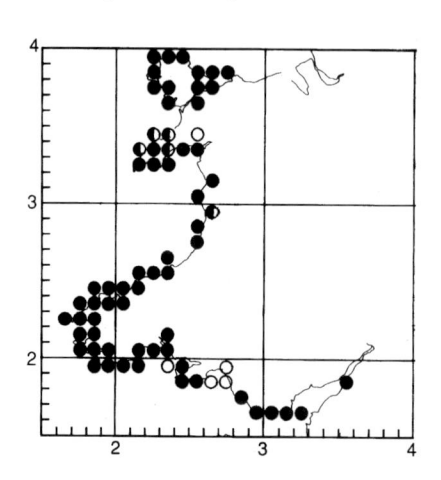

ASPLENIUM MARINUM

Clefts of sea-cliffs, locally frequent or common; less frequently on walls near the coast.

First recorded, 1639; Johnson, *Merc. Bot.*, pars alt., 9, 1641. Llanddwyn, Anglesey.

All round the coasts of the British Isles, except east and south-east England from south Yorkshire to Hants.; formerly in E. Sussex.

Europe: from the coast of southern Norway and Atlantic coasts southwards to Spain and Portugal, and the Mediterranean coast east to Italy. Azores, Madeira, Canary Is., Cape Verde Is. and St Helena. Coast of N. Africa.

ASPLENIUM OBOVATUM Viv. subsp. *LANCEOLATUM* (Fiori) P.Silva

(*A. billotii* F.W.Schultz; *A. lanceolatum* Hudson, non Forssk.)

Lanceolate Spleenwort; Duegredynen Reiniolaidd

Stock short, 1.5-3 × *c*.0.5cm, erect to decumbent, caespitose, densely clothed with brownish-black, shining, subulate scales, with long hair-like points. **Fronds** 10-30(-40)cm tufted; **stipe** $1/3$-$2/3$ as long as the rachis, dark reddish- to purplish-brown (the colour extending upwards on the under side), semicircular in section, with a few subulate or hair-like scales especially at the base, otherwise glabrous. **Blade** 6-20(-30) × 3-7(-10)cm, lanceolate in outline (slightly narrowed at the base), rigid, bipinnate; **pinnae** up to about 18 or 20 on either side (the lowest wide apart), lanceolate, ovate-lanceolate or ovate, very shortly stalked, pinnate for most of their length, becoming pinnatifid and ultimately simple towards the apex of the frond; basal pair of pinnae slightly to considerably shorter than the second pair; **pinnules** obliquely oblong, or obovate, coarsely toothed with acute mucronate teeth, the lower ones cuneate at the base, the upper ones running together, the basal acroscopic pinnule usually larger than the rest and frequently lobed or even pinnatifid; **rachis** usually reddish- to purplish-brown on the under side except towards the apex, often with scattered jointed hair-like scales, which occur on the rachises of the pinnules also. **Sori** shortly oblong, situated nearer to the margin of the pinnule than to the midrib, the sporangia sometimes almost covering the whole frond beneath. **Indusium** oblong, entire, opening inwards. **Spores** ripening June-September. Tetraploid. 2n=144.

Fig. 43. Lanceolate Spleenwort (*Asplenium obovatum* subsp. *lanceolatum*). Median pinna from an average-sized frond (×2). (*See also* Fig. 100C).

Rare (or in a few places frequent), in rock crevices and on old walls, the sides of wells and those of old mine shafts.

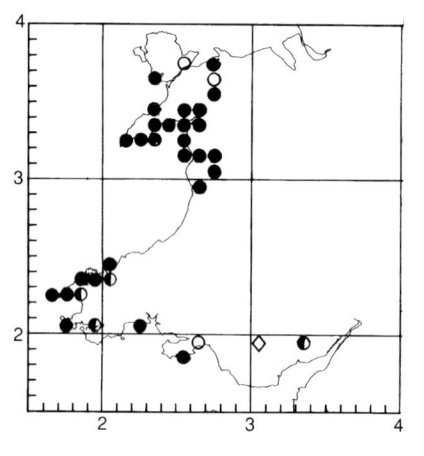

ASPLENIUM OBOVATUM ssp. LANCEOLATUM

First recorded, 1804, Rev. J. Evans, *Tour through North Wales*, p. 229. Borth, Anglesey.

Chiefly near the sea: SW England and Channel Islands. Very local in southern Ireland; scattered north to Sutherland, formerly in SE England.

Protected in the Republic of Ireland under the *Flora Protection Order, 1987*.

Coasts of western Europe from the north of England to the Mediterranean region, with small outliers in NE France, Switzerland, S. Italy and Germany. Azores, Madeira, Canary Is. and St. Helena. Algeria.

ASPLENIUM OBOVATUM Viv. subsp. *OBOVATUM*

A diploid subspecies (2n=72), with a Mediterranean distribution. It has been confused with subsp. *lanceolatum*, from which it differs in having shorter more triangular pinnae, with fewer, less distinct pinnules bearing obtuse mucronate teeth. Diploid. 2n=72.

Not recorded from the British Isles.

ASPLENIUM ONOPTERIS L.

(*A. adiantum-nigrum* subsp. *onopteris* (L.) Luerss.)

Irish Spleenwort; Duegredynen Wyddelig

Blades always tripinnate; **stipe** often longer than the rachis; **pinnae** and **pinnules** curved with tapering apices; **segments** narrowly lanceolate to linear, with long acuminate teeth. Diploid. 2n=72.

Very local in the south of Ireland. Specimens outside Ireland have been redetermined as *A. adiantum-nigrum.*

Mediterranean region and parts of central Europe. Azores, Madeira and Canary Is. Turkey.

ASPLENIUM RUTA-MURARIA L.

Wall-rue; Duegredynen y Muriau

Stock short, creeping, branched, the younger parts clothed with blackish-brown subulate scales with hair-like points. **Fronds** 2-12(-15)cm, tufted, winter-green; stipes as long or twice as long as the blade, dark purplish-brown at the base only, beset when young with numerous minute, globose, deciduous glands and a few hair-like scales. **Blade** 1-6 × 1-4.5cm, triangular-ovate or ovate-lanceolate in outline, leathery in texture, bipinnate or almost tripinnate at the base; **pinnae** 4 or 5 on either side, alternate or the basal ones subopposite, widely spaced, variable in their degree of subdivision, the lowest two pinnae in robust plants pinnate with the basal pinnules trifid, the third and fourth pinnae pinnate or ternate, the rest less and less divided towards the apex of the frond, the terminal one being entire or slightly lobed; **ultimate segments** very variable, obovate, oblanceolate or rhomboidal, cuneate at the base, obtuse or rounded at the apex, serrate or crenate in the upper half. **Sori** linear, borne on the veins, 1-3 on either side of the middle line, occasionally paired back to back, the sporangia ultimately covering the under surface. **Indusium** similar in shape to the sorus, finely crenate, usually opening inwards. **Spores** ripening June-October. Tetraploid. 2n=144.

Fig. 44. Wall-rue (*Asplenium ruta-muraria*). Single frond from a plant of average development, natural size. On the left: a portion of the frond bearing an ultimate segment magnified (×3) to show the venation and the sori with their crenate-edged indusia.

Common on rocks and walls especially limestone (including mortar).

First recorded, 1695; Llwyd in Camden's *Britannia* (ed. Edmund Gibson), 699. 'On Snowdon hill.'

Throughout the British Isles but more local in northern Scotland.

Circumpolar. Throughout Europe. North Africa. Western Asia. Caucasus through Iran to Afghanistan then splitting to Himalaya and north-east through central and eastern Siberia. Japan. Eastern USA and in Canada just north of the Great Lakes.

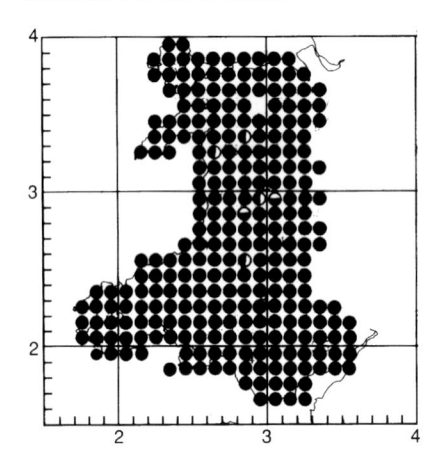

ASPLENIUM RUTA-MURARIA

ASPLENIUM SCOLOPENDRIUM L.

(*Phyllitis scolopendrium* (L.) Newman; *Scolopendrium vulgare* Sm.)

Hart's-tongue; Tafod yr Hydd

Stock up to 6 × 0.5cm, ascending to almost erect, the brown surface of the younger parts very thickly clothed with brown (tinged with violet), linear-lanceolate to lanceolate, acuminate scales, having cordate bases. **Fronds** 10-60cm, tufted, strongly curved outwards and downwards, winter-green; **stipe** very variable in length, up to half as long as the rachis, usually less, green to dull brown merging into purple-brown at the base, practically semicylindrical with a flat or feebly convex upper surface and usually with shallow grooves along the flanks, the swollen base clothed with scales like those of the rhizome, the upper part of the stipe with gradually smaller and

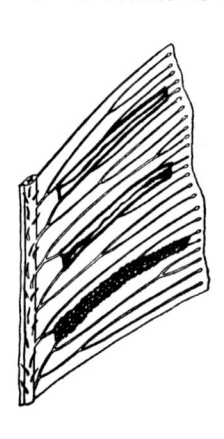

Fig. 45. Hart's-tongue (*Asplenium scolopendrium*). Portion of blade (about natural size) showing venation and three twinned sori, the top two each covered by two indusia. (*See also* Fig. 101).

narrower scales, becoming more or less glabrous when old. **Blade** 5-40 × 1-6cm, clear green, fleshy or leathery in texture, more or less strap-shaped (lanceolate, linear-lanceolate or oblong-lanceolate), cordate at the base and often narrowed slightly to a little bay just above, narrowed towards the apex, obtuse or subacute, the margin more or less wavy; midrib stout, beset, when young at least, with scattered subulate or hair-like scales. **Sori** linear, usually more abundant in the upper part of the frond, in vigorous plants occupying almost the whole width from midrib to margin, but usually less, the sporangia of a pair of sori running together to form a dense brown cigar-shaped mass. **Indusium** membranous, its margin entire, at first colourless, finally brown and reflexed. **Spores** ripening July-August. Diploid. 2n=72.

A large number of variants of Hart's-tongue Ferns (many of them having the blade curled and cleft into lobes and segments) may be found wild and many more have arisen under cultivation. Whatever their origin, when cultivated they are classed as cultivars.

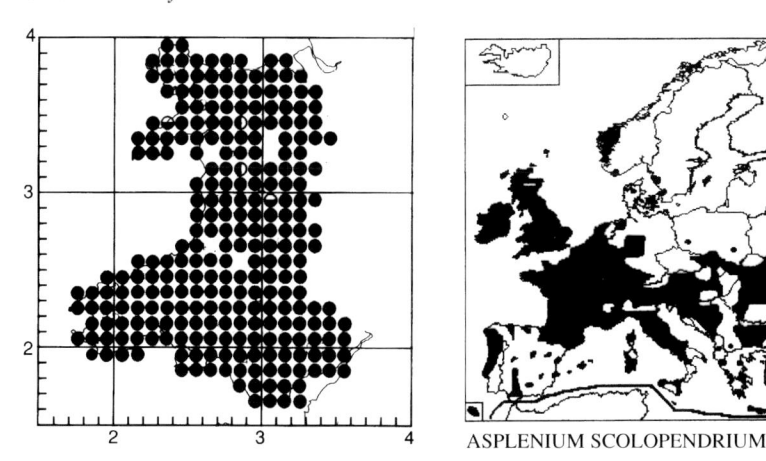

ASPLENIUM SCOLOPENDRIUM

Woods, hedge banks, walls, and rock ledges; common on limestone, less so elsewhere.

First recorded before 1597 by William Salesbury in his manuscript herbal, 'yn eymyl tal acre [Talacre, Flints.].'

Throughout the British Isles but scattered in northern Scotland.

Europe: south-west of a line from southern Scandinavia to Rossia (W). Azores, Madeira and Canary Is. N. Africa. Asia Minor through the Caucasus to Iran. Japan. Local in eastern North America and Mexico.

ASPLENIUM SEPTENTRIONALE (L.) Hoffm.

Forked Spleenwort; Duegredynen Fforchog

Rhizome creeping, branched, the younger parts densely clothed with brownish-black subulate scales with hair-like points and ciliate margins. **Fronds** 5-15cm, tufted; **stipe** as long or up to several times as long as the rachis, dark shining chestnut- to blackish-brown at the base only and with scattered very minute brownish hairs. **Blade** stiff and leathery in texture, irregularly forked rather than pinnate or bipinnate; **pinnae** only one or two in addition to the terminal one, the lowest pinna (and less often also the next) sometimes divided into a larger terminal pinnule and a smaller lateral one standing on the side towards the frond apex; **pinnae** and **pinnules** 5 in all at the most, linear-lanceolate, acute or acuminate at the apex, with a few long irregular subulate teeth in the upper half, sometimes narrowly cuneate at the base. **Sori** linear, 1-5 on each segment, one above the other or more or less parallel, practically covering the whole under surface. **Indusium** also linear, entire, opening inwards. **Spores** ripening June-October. Tetraploid. 2n=144.

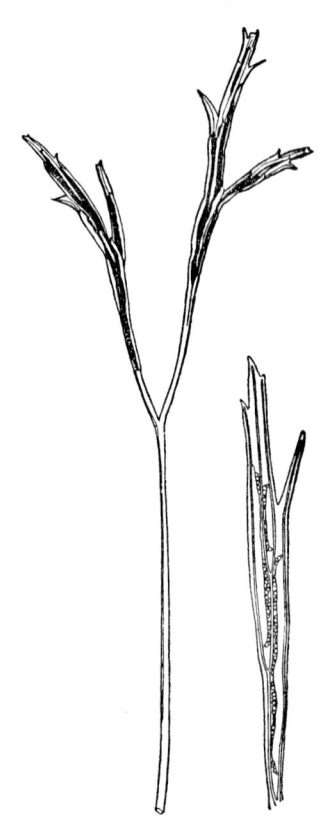

Fig. 46. Forked Spleenwort (*Asplenium septentrionale*). Single frond (natural size). On the right a single pinna magnified (×3) to show venation and sori.

Crevices of rocks and walls (not limestone).

First recorded, 1695; Llwyd in Camden's *Britannia* (ed. Edmund Gibson), 700. 'On top of Carnedh Llewelyn, near Llan Lhechyd [Caern.].'

Very scattered on east side of Scotland, much reduced on west side; Co. Galway; rare in N. England, and very rare in SW England.

Protected in the Republic of Ireland under the *Flora Protection Order, 1987.*

Circumpolar. Europe, generally in the mountains, from Norway to the Mediterranean region. Iceland. Madeira and Canary Is. Morocco. Caucasus through Afghanistan to Himalaya; western Siberia. W. Virginia and SW USA. Mexico.

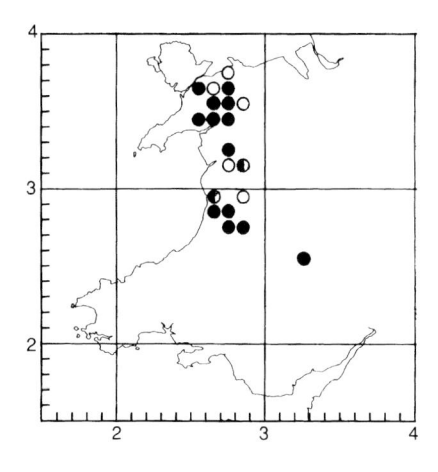

ASPLENIUM SEPTENTRIONALE

ASPLENIUM TRICHOMANES L.

Maidenhair Spleenwort; Duegredynen Gwallt y Forwyn

Stock up to 5 × 0.2cm, creeping or decumbent, densely caespitose, the younger parts clothed with dark brown, linear-lanceolate, acuminate scales, usually with a black central stripe. **Fronds** 5-35cm, tufted, winter-green; **stipe** $1/6$-$1/4$ as long as the rachis, wiry, brownish to purplish-black, rounded behind and flat in front with a very narrow pale brown (at first greenish), membranous, brittle, winged border on either angle, glabrous and shiny. **Blade** 4-30 × 0.5-2cm, linear in outline, pinnate; **pinnae** up to about 30 or even 40 pairs, dark green, thick, glabrous, roundish oblong (except the lowest which are usually broadly ovate and also smaller), unequal at the base (truncate on the acroscopic side, cuneate on the basiscopic side), rounded at the apex, crenate along the margin except on the basiscopic side, becoming detached from the rachis during the second season leaving behind a small brown tooth-like projection; **rachis** brownish to purplish-black, winged, eventually becoming bare of pinnae. **Sori** linear, situated midway between the margin and the midrib, on the acroscopic branches of the secondary veins or in part on the secondaries themselves. **Indusium** also linear, entire or slightly crenate, opening inwards. **Spores** ripening May-October.

Common on old walls and in rock crevices, more rarely on hedge banks. The diploid, subsp. *trichomanes*, is much rarer than the tetraploid, subsp. *quadrivalens*. The tetraploid, subsp. *pachyrachis*, has recently been identified for the British Isles as has its hybrid with subsp. *quadrivalens* (Rickard, 1989).

94 ASPLENIUM TRICHOMANES subspecies

ASPLENIUM TRICHOMANES

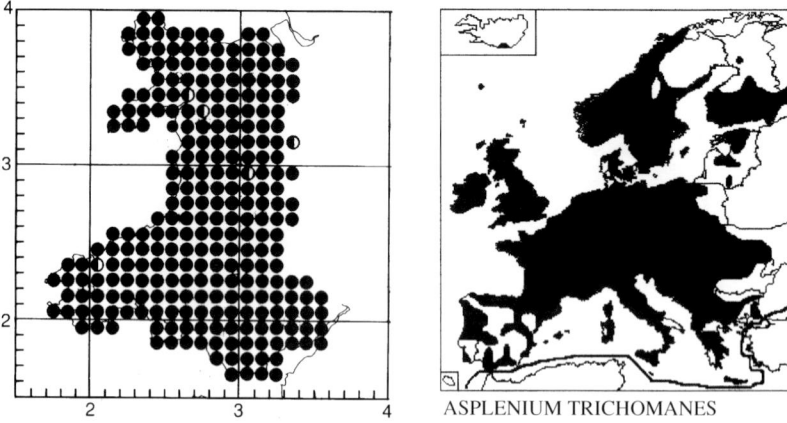

The Welsh distribution map of ssp. *quadrivalens* is identical with this.

First recorded by William Salesbury in his 1597 manuscript herbal, 'ar y bont vaen ar Lugwy [Betws-y-coed].'

Throughout the British Isles.

Cosmopolitan. In the Temperate and Sub-Arctic zones of both hemispheres and in the high mountains of the Tropics. Europe except north Lapland, but including Iceland, Faeroes and Azores. Madeira and Canary Is. North Africa, tropical east Africa, Zimbabwe and S. Africa. Asia: Yemen. Turkey through Caucasus to northern Iran. Scattered across the western half of the middle of Siberia. Himalaya, China, Japan, Korea and eastern Indonesia. America: Southern Canada, USA, Cuba and Central America to northern and western South America. SE Australia, Tasmania, New Zealand and Hawaii.

Four subspecies of *Asplenium trichomanes* have been recognized in Europe: subspecies *trichomanes*, *quadrivalens*, *pachyrachis* and *inexpectans*. The first three subspecies are native to Wales. They are often difficult to determine on macroscopic characters, which tend to overlap. It is usually necessary to consider a whole range of characters rather than any single one. For instance, pinnae shape can vary greatly, even on a single frond, and by itself would often prove to be an unreliable character.

KEY TO THE SUBSPECIES OF ASPLENIUM TRICHOMANES

Not all plants can be identified to subspecies even after microscopic examination of their spores (mean exospore length in air, Fig. 47).
1 Pinnae often hastate (with projecting basal lobe(s)); oblong to subtriangular   a) subsp. **pachyrachis**
Pinnae not hastate; suborbicular to oblong   **2**

2   Calcicole. Pinnae to 11mm, oblong, symmetric, almost sessile and more crowded looking and wider than (c) (compare illustrations). Margins sometimes rolled under but the apex remaining flat or slightly rolled under. Generally, fronds with a more robust appearance than (c)
    b) subsp. **quadrivalens**
Calcifuge. Pinnae to 8mm, oval/rhombic and asymmetric, to almost orbicular, and with distinct stalk. Pinnae looking more widely spaced and narrower than (b) (compare illustrations) due to them often curling under longitudinally and sometimes turning up at the apex. Generally, fronds with a more delicate appearance than (b)   c) subsp. **trichomanes**

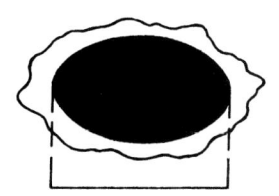

Fig. 47. Spore of Maidenhair Spleenwort (*Asplenium trichomanes*). Bar indicates where measurements are taken of exospore.

ASPLENIUM TRICHOMANES L. subsp. *PACHYRACHIS* (H.Christ) Lovis & Reichst.

Lobed Maidenhair Spleenwort; Duegredynen Gwallt y Forwyn Labedog

Rhizome scales up to 5mm long, lanceolate to linear-lanceolate, with a dark brown central stripe. **Stipe** black-brown or dark brown and remaining so throughout the season eventually becoming dull, often persistent but less so than in subsp. *trichomanes*, the dead fronds tending to break off just above the stipe. **Pinnae** up to 11mm long, usually subtriangular or even hastate, usually with a distinct auricle in the acroscopic margin, symmetrical, almost sessile or sometimes shortly stalked, opposite or alternate above, alternate below, with a square insertion. Margin often deeply crenately toothed and wavy, generally smaller and more delicate than in subsp. *quadrivalens*. **Mean exospore length in air** (26-)32-38(-50)µm similar to subsp. *quadrivalens*. **Stomatal guard-cell length** 40-49µm. Tetraploid. 2n=144.

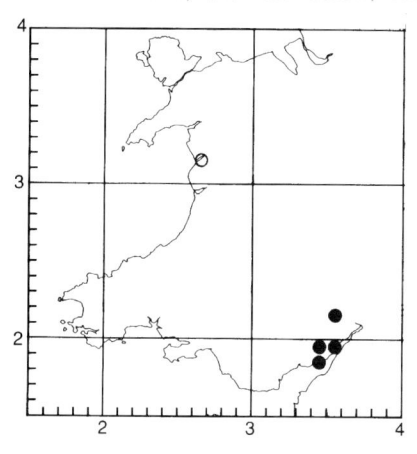

ASPLENIUM TRICHOMANES subsp. PACHYRACHIS

On limestone rocks and walls especially old castles and often under overhangs.

Recently found in Mons. (1988); W. Gloucs. and Herefs. 19th century specimens could well be this subspecies in the Moore herbarium at Kew, as named varieties of the species, from Merioneth, Mid-West Yorks. and Co. Clare (Rickard, M. H., 1989).

S. and W. Europe.

ASPLENIUM TRICHOMANES L. subsp. *QUADRIVALENS* D.E.Mey. emend. Lovis

Common Maidenhair Spleenwort; Duegredynen Gwallt y Forwyn Gyffredin

Rhizome scales up to 5mm long, lanceolate to linear-lanceolate, with a dark brown central stripe. **Stipe** dark brown. **Pinnae** 16-30 pairs, robust in texture (except in shaded habitats), up to 12mm long, usually oblong and

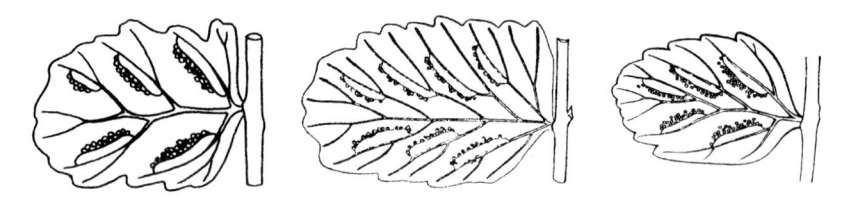

Fig. 48. Maidenhair Spleenwort (*Asplenium trichomanes*). Left: single pinna from subsp. *pachyrachis*. Centre: from subsp. *quadrivalens*. Right: from subsp. *trichomanes*. All show the sori on the acroscopic branches of the secondary veins (c.×5). (*See also* Fig. 102).

parallel-sided; the upper pinnae usually relatively crowded, their insertion transverse or somewhat oblique. **Sori** (4-)5-8(-10), on middle pinnae. **Mean exospore length in air** (26-)32-39(-47)μm similar to subsp. *pachyrachis*. **Stomatal guard-cell length** 40-48μm. Tetraploid. 2n=144.

On basic (sometimes neutral) rocks and walls.

Throughout the British Isles. Common in the wetter areas of the west and north, less so in the drier east.

In Europe, generally less frequent in the north. Widely distributed in both northern and southern hemispheres.

The Welsh distribution map of subsp. *quadrivalens* is identical to the one for *A. trichomanes* (q.v.).

ASPLENIUM TRICHOMANES L. subsp. *TRICHOMANES*

Delicate Maidenhair Spleenwort; Duegredynen Gwallt y Forwyn Gain

Rhizome scales up to 3.5mm, lanceolate, with a reddish-brown central stripe. **Stipe** reddish-brown. **Pinnae** (10-)14-28 pairs, delicate in texture, 2.5-7.5mm long, usually sub-orbicular, auriculate when luxuriant, the upper pinnae distant and obliquely inserted. **Sori** 4-6(-8), on middle pinnae. **Mean exospore length in air** (23-)28-32(-34)μm being an important character for separating the subspecies from the other two British subspecies over most of the range. Intermediate mean figures in the low 30's forbid determination using this technique. **Stomatal guard-cell length** 32-40μm. Diploid. 2n=72.

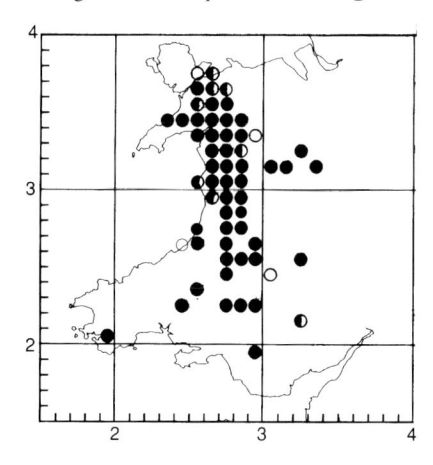

On non-calcareous rock, mostly in upland regions.

Scattered in Scotland, local in the Lake District, also Co. Down; under-recorded. Throughout Europe. Also in Asia, Africa, N. America and Australia.

ASPLENIUM TRICHOMANES subsp. TRICHOMANES

ASPLENIUM TRICHOMANES-RAMOSUM L.

(*A. viride* L.)

Green Spleenwort; Duegredynen Werdd

Stock up to 10 × 0.25cm, decumbent or ascending, caespitose, the younger parts clothed with dark brown, linear-lanceolate, acuminate scales, usually without a central stripe. **Fronds** 20cm, tufted, not usually wintergreen; **stipe** 1/5-1/2 as long as the rachis, reddish-brown to dark purplish-brown at the base only, green higher up, with a few scattered hair-like scales, becoming glabrous, approximately semicircular in section, not winged. **Blade** 3-15 × 0.5-1.5cm, linear in outline, pinnate, **pinnae** up to about 30 pairs, bright green, delicate, glabrous, ovate (the lowest broadly ovate), somewhat unequal and more or less cuneate at the base, obtuse at the apex, crenate-margined (often deeply so), not falling from the rachis but dying down with the frond as a whole; **rachis** green, slightly grooved on the upper side, with a few hair-like scales in the lower part when young. **Sori** linear,

Fig. 49. Green Spleenwort (*Asplenium trichomanes-ramosum*). Median pinna from a well developed plant showing the sori on the secondary veins (×3). (*See also* Fig. 102*D*).

situated towards the midrib and at a distance from the margin, on the secondary veins and extending up their acroscopic forks. **Indusium** also linear, entire or slightly toothed. **Spores** ripening June-September. Diploid. 2n=72.

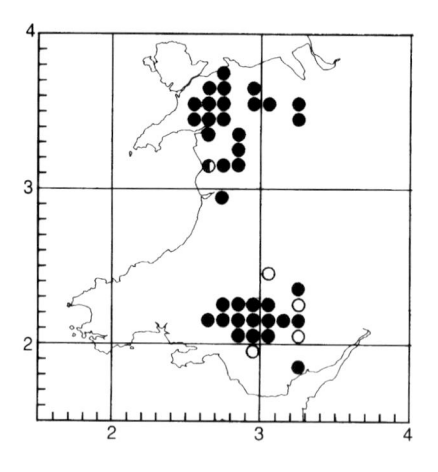

ASPLENIUM TRICHOMANES-RAMOSUM

Locally frequent in crevices of shady rocks (especially limestone in mountainous districts).

First recorded, 1695; Llwyd in Camden's *Britannia* (ed. Edmund Gibson), 702. 'Snowdon.'

Local in Scotland, northern England south to Derbys., also very local on west side of Ireland.

Cosmopolitan, mainly arctic-montane. Europe: from the Arctic to southern Spain. Iceland and Faeroes. Morocco. Asia: in Asia Minor and Caucasus, scattered in western and southern central Siberia; Afghanistan and Himalaya, Japan. N. America: mostly in NW USA and western Canada, and NE USA and SE Canada. Greenland.

HYBRIDS IN ASPLENIUM

Several *Asplenium* hybrids have been recorded. Their parentage can sometimes be deduced from genetical and cytological as well as morphological evidence.

A. ADIANTUM-NIGRUM × *A. OBOVATUM* = *A.* × *SARNIENSE* Sleep

Guernsey Spleenwort; Duegredynen Guernsey

In some ways intermediate between the parents but resembling *A. adiantum-nigrum* in its triangular, bipinnate or tripinnate fronds with well-developed basal pinnae. The influence of *A. obovatum* subsp. *lanceolatum* shows in the middle third of the frond where the pinnules are oval in shape, rather crowded, distinctly short-stalked, and with broad, mucronate teeth. Spores abortive. Tetraploid hybrid. 2n=144. Guernsey. Also, recently recorded from NW France.

The English name should not be confused with Guernsey Fern (*A. obovatum A. scolopendrium*).

A. ADIANTUM-NIGRUM × *A. ONOPTERIS* = *A.* × *TICINENSE* D.E.Mey.

Hybrid Black Spleenwort; Duegredynen Ddu Groesryw

Fronds intermediate between the parents in many characters, tripinnate towards the base. Pinnae segments less narrowly lanceolate than in *A. onopteris*, with acute to acuminate teeth. Spores usually abortive. Triploid hybrid. 2n=108. Found in Co. Cork and Co. Kilkenny. Switzerland, Italy, France.

A. ADIANTUM-NIGRUM × *A. SCOLOPENDRIUM* = *A.* × *JACKSONII* (Alston) Lawalrée

(× *Asplenophyllitis jacksonii* Alston)

Jackson's Fern; Rhedynen Jackson

Blade 9-20cm, triangular-lanceolate, pinnate; pinnae more or less hastate, tapering to an acute point, the basal pair usually longest, *c*.3-4cm, shortly stalked and with large rounded basal lobes running half way to the midrib, the rest more or less sessile and less markedly lobed, and in the upper half confluent into an irregularly pinnatifid apex. Sori elongate, on each side of the midrib, frequently twinned as in *Asplenium scolopendrium*. Triploid hybrid. 2n=108. Found in Devon, Cornwall and the Channel Islands in the 19th Century. Recently found in NW France.

A. ADIANTUM-NIGRUM × *A. SEPTENTRIONALE* = *A.* × *CONTREI*
Callé, Lovis & Reichst.

Caernarvonshire Fern; Rhedynen Sir Gaernarfon

Somewhat resembling *A.* × *alternifolium*, but with proportionately broader fronds, more triangular in outline, like *A. adiantum-nigrum*; pinnate (bipinnate at the base); pinnae and ultimate segments slender oblanceolate, tapering, irregularly slender-toothed. Sporangia abortive in the specimens examined. Tetraploid hybrid. 2n=144. Specimens collected by the Rev. T. Butler in 1870, from the Pass of Llanberis (SH65, Caerns.), have been determined as this hybrid (Anon., 1968; Callé, Lovis & Reichstein, 1975). Also recorded from France.

A. CETERACH × *A. RUTA-MURARIA* = *A.* × *BADENSE* (D.E.Mey.)
Jermy

(× *Asplenoceterach badense* D.E.Mey.)

Its fronds are scaly, like those of the Rustyback, but they are very irregularly and distantly lobed, their segments being irregularly oblong, 5-10mm, shorter towards the apex. This curious hybrid was found at Baden, Kaiserstuhl, Germany, in 1956. It has not yet been reported in the British Isles, but should be looked for, as both parents are native here.

A. OBOVATUM × *A. SCOLOPENDRIUM* = *A.* × *MICRODON* (T.Moore)
Lovis & Vida

(× *Asplenophyllitis microdon* (T.Moore) Alston)

Guernsey Fern; Rhedynen Guernsey

Blade 15-30cm, lanceolate, widest about or below the middle, pinnate; pinnae obliquely triangular or cordate sub-hastate, shallowly lobed, toothed, the longest *c*.2-4cm, the lowest shorter than the succeeding pair, the upper narrower and confluent; sori small, short, dispersed. Triploid hybrid. 2n=108. Very rare, formerly West Cornwall and Guernsey. Rediscovered in Guernsey in 1965, about 80 years since it was last reported there.

The English name should not be confused with Guernsey Spleenwort (*A. adiantum-nigrum* × *A. obovatum*).

A. OBOVATUM subsp. *LANCEOLATUM* × *A. TRICHOMANES* = *A.* × *REFRACTUM* (T.Moore) Lowe

Described from plants in cultivation at Peperharrow Park, Surrey, since 1851, but which were reputed to have originated in Scotland.

Fronds up to 20cm high, 2cm wide. Stipe dark brown, not winged, bulbil-bearing. Blade linear-lanceolate, partially bipinnate. Pinnae refracted; the upper pinnatifid; the lower with roundish, more or less confluent, coarsely toothed pinnules. Sori short, oblique oblong.

A. RUTA-MURARIA × *A. SEPTENTRIONALE* = *A.* × *MURBECKII* Dörfl.

Murbeck's Spleenwort; Duegredynen Murbeck

Fronds 3-5cm; similar to small forms of *A.* × *alternifolium*, from which it may be distinguished by the more conspicuous, narrowly acute or even lanceolate teeth at the tips of its pinnae; basal pinnae sometimes deeply lobed, but not bipinnate in the specimens examined. Tetraploid hybrid. 2n=144. Recorded in Cumberland; formerly in E. Scotland. Rare. Continental Europe.

A. RUTA-MURARIA × *A. TRICHOMANES* subsp. *QUADRIVALENS* = *A.* × *CLERMONTIAE* Syme

Lady Clermont's Spleenwort; Duegredynen y Fonesig

Distinguished from *A. trichomanes* as follows: stipes green above and lacking the membranous wings on either angle; pinnae distinctly stalked, the lower ones three-lobed and longer than the succeeding pair. Indusium conspicuously fine-toothed. Tetraploid hybrid. 2n=144. Extinct, formerly Co. Down and possibly Westmorland. Continental Europe. USA.

A. SCOLOPENDRIUM × *A. TRICHOMANES* subsp. *QUADRIVALENS* = *A.* × *CONFLUENS* (T. Moore ex Lowe) Lawalrée

(× *Asplenophyllitis confluens* (T.Moore ex Lowe) Alston)

Confluent Maidenhair Spleenwort; Duegredynen Gwallt y Forwyn Gydlifol

Blade 10-15(-30) cm, linear-lanceolate; pinnae irregularly orbicular or triangular, crowded, in the upper third overlapping and becoming confluent. Presumed to be a triploid hybrid. [2n=108]. Very rare. S. Kerry, formerly North-east Yorkshire, Westmorland, N. Kerry.

A. SEPTENTRIONALE × *A. TRICHOMANES* subsp. *TRICHOMANES* = *A.* × *ALTERNIFOLIUM* Wulfen

(*A. breynii* auct.; *A. germanicum* auct.)

Alternate-leaved Spleenwort; Duegredynen Dail Bob yn Ail

Stock short, creeping to ascending, caespitose, the younger parts thickly clothed with blackish-brown to black subulate scales having no central stripe. **Fronds** densely tufted, 2.5-17cm, winter-green; stipe about as long as the rachis or longer, channelled above (as is the rachis) up to the middle,

Fig. 50. Alternate-leaved Spleenwort (*Asplenium* × *alternifolium*). Single frond (natural size). On the left: a median pinna magnified (×3) to show venation and sori.

shining chestnut- to blackish-brown (less often brown up to and including the rachis, but beneath only). **Blade** 1.5-7.5 × 0.8-2.8cm, linear-lanceolate in outline, fairly thick and herbaceous in texture, pinnate (at the base bipinnate); **pinnae** alternate to sub-opposite, set widely apart, the lowest trifid or ternate (with the lowest (acroscopic) pinnule only more or less distinctly stalked), the middle pinnae unequally lobed (the lowest (acroscopic) lobe the largest), the upper pinnae undivided and incurved towards the rachis and the terminal one pinnatifid; **ultimate segments** lanceolate or oblanceolate, cuneate at the base, irregularly toothed towards the apex, the teeth obtuse or subacute. **Sori** linear, at most 2 in two rows on each segment, occasionally paired back to back, the sporangia when ripe covering the middle part of the segment. **Indusium** also linear, its margin entire or at most slightly waved, usually opening inwards. Triploid hybrid. 2n=108.

On rocks, very rare.

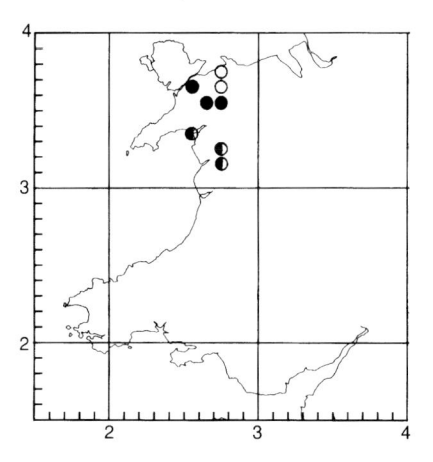

First recorded, 1847; E. Newman, *The Phytologist*, 2, 975. 'Near Llanrwst'.

Very rare in Lake District; formerly Somerset, Northumberland and the south-eastern part of Scotland.

Mountainous districts of Europe from Norway to the southern Alps, Kashmir and Hong Kong.

ASPLENIUM × ALTERNIFOLIUM

A. TRICHOMANES subsp. *PACHYRACHIS* × subsp. *QUADRIVALENS* = *A. TRICHOMANES* L. nothossp. *STAUFFERI* Lovis & Reichst.

A vigorous plant with a more upright habit than subsp. *pachyrachis*. It is distinctly intermediate between the parents with some slightly triangular pinnae, most of which are attached to the rachis by the centre as in subsp. *pachyrachis*. Fronds 15-30 × 1.5-2.0cm. It is usually a darker green than subsp. *pachyrachis*, more like subsp. *quadrivalens*. Spores mainly abortive, the most reliable character, but sometimes has a few good spores. Tetraploid hybrid. 2n=144, with partial meiotic pairing of chromosomes. Discovered within recent years in Mons. (1989) and Herefs. (1988). Known from central Europe.

A. TRICHOMANES subsp. QUADRIVALENS × subsp. TRICHOMANES
= A. TRICHOMANES L. nothossp. LUSATICUM (D.E.Mey.) Lawalrée
(A. × lusaticum D.E.Mey.)

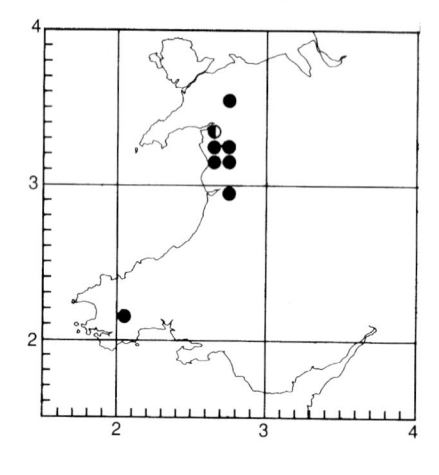

Hybrid Maidenhair Spleenwort; Duegredynen Gwallt y Forwyn Groesryw

Said to be intermediate between the parents but in practice only identifiable by its chromosome number, abortive spores and hybrid vigour. Triploid hybrid. $2n=108$. Recorded from Scotland and Ireland. Parts of N. and E. Europe, and USA.

ASPLENIUM × LUSATICUM

Genus ATHYRIUM Roth
Lady-ferns - Rhedyn Mair

Sori variously shaped from round to linear, the indusia in typical species hooked (i.e. consisting of two unequal arms placed back to back, the longer arm opening inwards and joined at the outer end to the shorter arm which opens outwards), in other species horseshoe-shaped or kidney-shaped (cf. *Dryopteris*) or straight (cf. *Asplenium*). Terrestrial, usually erect ferns with somewhat limp, soft-textured pinnate to tripinnate leaves, with free veins, without hairs but with some scales. About 180 species, mostly eastern Asiatic, few in the tropics.

KEY TO SPECIES

1 Fronds having sori, frequent. Sori oblong to curved with noticeably curved indusia   **A. filix-femina**
Fronds having sori, infrequent. Sori orbicular with indusia absent or imperfectly formed. (Scottish Highlands)   **2**

2 Suberect fronds (to 75cm) arising as in a shuttlecock **A. distentifolium** var. **distentifolium**
Smaller, spreading fronds (to 35cm) arising as in a flattened shuttlecock with the very short stipes mostly bent rather sharply backwards at a distinct elbow   **A. distentifolium** var. **flexile**

ATHYRIUM DISTENTIFOLIUM Tausch ex Opiz var. *DISTENTIFOLIUM*

(*A. alpestre* auct., non Clairv.)

Alpine Lady-fern; Rhedynen Fair Alpaidd

Resembling *A. filix-femina* in habit but usually with the pinnules less acutely toothed and distinguished by the very rudimentary indusium (invisible or missing when the sorus ripens), the small number of sporangia in each sorus, and the reticulate (not warty) markings on the spores. Spores winged, ripening July-August. Diploid. 2n=80.

Central and northern Scotland.

The species is circumpolar, of the alpine and sub-alpine regions of N. Europe including Iceland and Faeroes. The mountains of central and S. Europe, Russia (N,C,W), Turkey, Caucasus, western and eastern Siberia, Japan. Western N. America, Eastern Canada around the Gulf of St Lawrence. Greenland.

ATHYRIUM DISTENTIFOLIUM Tausch ex Opiz var. *FLEXILE* (Newman) Jermy

(*A. flexile* (Newman) Druce)

Newman's Lady-fern; Rhedynen Fair Newman

Fronds 7.5-30cm, deflexed, with very small sori, often of only 3-5 sporangia; stipe very short, 1-3cm. Spores ripening July-August. Diploid. 2n=80.

High alpine corries in central Scotland where it is endemic.

ATHYRIUM FILIX-FEMINA (L.) Roth

Lady-fern; Rhedynen Fair

Stock rather stout, up to 10 × 0.5cm, erect or ascending, simple or sometimes branched, beset with the bases of old fronds, the younger parts clothed with dark brown or sometimes blackish, lanceolate or ovate-lanceolate scales. **Fronds** 20-90cm, erect or gracefully spreading or drooping; **stipe** $^1/_4$-$^1/_3$ as long as the rachis, its base purplish-black, swollen and broadened, tapering above and below and beset with brownish, lanceolate scales, the upper part pale green to dull purplish-red with fewer scales. **Blade** 15-70 × 5-25cm, lanceolate in outline, bi- or tri-pinnate, usually rather flaccid and thin in texture, bright clear or yellowish-green; **pinnae** numerous (up to about 30 on either side), alternating or less often sub-opposite, sessile or shortly stalked, mostly close set (but the lowest wider apart and sometimes bent downwards), linear-lanceolate or linear-oblong, acuminate; **pinnules** alternate (except the lowest), oblong or oblong-lanceolate, often a little curved towards the apex of the pinna, sessile, unequal and more or less decurrent in the direction of the base of the pinna, pinnately lobed, the lowest segments pinnately toothed, the upper ones with 2-3 obtuse or subacute teeth or simple; **rachis** with minute hairs or scattered subulate scales or glabrous. **Sori** situated on the lowest acroscopic veinlets, forming a single row on either side near the midrib of the pinnule, in tripinnate leaves two short rows also on the lowest segments of the third order, the lowest sori of each row kidney-shaped, horseshoe-shaped or more or less hooked, the upper ones nearly or quite straight, and oblong or linear. **Indusium** similar in shape to the sorus, attached on one side, more or less inflated, membranous, the free margin usually ciliate or irregularly toothed. **Spores** not winged, ripening July-August. Diploid. 2n=80.

Common in woods and on hedge banks and in other damp, shady and sheltered places, and in rocky situations in mountainous districts.

First recorded, 1726; Herb. Dillenius. (Druce and Vines, *The Dillenian Herbaria*, 47, 1907.) 'Anglesey and Llanberrys [Llanberis, Caern.].'

Fig. 51. Lady-fern (*Athyrium filix-femina*). Median pinnule from fifth pinna on one side of an average-sized frond (×5). (*See also* Fig. 103*A*).

Throughout the British Isles.

Cosmopolitan. Throughout Europe. Azores, Madeira, Canary Is. and the Cape Verde Is. N. Africa and southern Africa. Asia: in a broad band across western Asia and Siberia to Japan and the Kamchatka peninsula. Asia Minor through Iran to Himalaya and India. China and Java. N. America: mostly south of a line from southern Alaska to Newfoundland, being more frequent in the eastern half. Greenland. Haiti and Mexico southwards to southern Brazil.

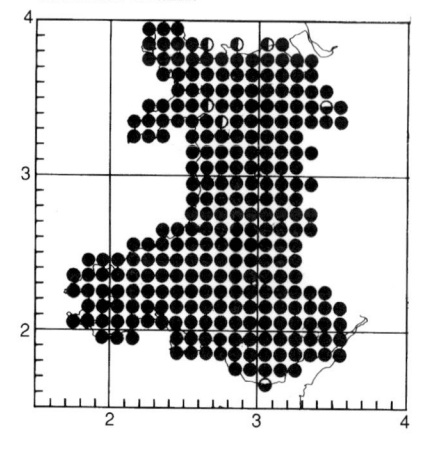

ATHYRIUM FILIX-FEMINA

Genus AZOLLA Lam.
Water Ferns - Rhedyn y Dŵr

Very small floating aquatic plants, being moss-like in general appearance and recalling *Lemna* (Duckweed) from a distance. Stem much branched, bearing roots and leaves. Leaves borne on the upper side of the stem in two rows alternating right and left; each blade deeply divided into two lobes, the upper one a green floating assimilating organ (containing also a chamber inhabited by threads of its symbiont the blue-green alga *Anabaena azollae*), the lower one thinner and submerged, the upper lobes of neighbouring leaves covering one another like tiles on a roof, so that the upper surface of the stem is hardly visible, the lower lobes only covering one another at their bases so that the under surface of the stem is more or less uncovered. Roots numerous, solitary, arising from the under surface of the stem at the points of origin of the branches. Sori borne in one or two pairs on the lower lobe of the first leaf of a lateral branch, the entire group of sori being covered by a cowl-like (hooded) flange of the upper lobe of the same leaf; each sorus wholly surrounded by its own indusium and, when mature, either globose, consisting of numerous microsporangia, or smaller and acorn-shaped, consisting of a single megasporangium. Microsporangium long-stalked, containing a number of little clumps of microspores, called massulae, each of which (in the two species naturalized in Europe) is beset with numerous barbed hairs (glochidia). The massulae are liberated by the decay of the indusium and the sporangium wall, and some become attached by the glochidia to the megaspores; fertilization follows. 6 species in the Tropical and Warm Temperate zones.

AZOLLA FILICULOIDES Lam.
(*A. caroliniana* Willd., non auct.)
Water Fern; Azolla; Rhedynen y Dŵr

Plants growing in dense tufted masses, 1-2cm across, the ends of the shoots directed forward and outward and often protruding, not lying flat on the surface. Upper lobes of the **leaves** about 1mm across, glaucous green becoming reddish in colour in the autumn, having a broad, distinct, clear margin and numerous one-celled hairs on the upper surface. **Sori** in pairs, either assorted or two **megasori** together; **microsori** up to 1.5mm, **megasori** up to 0.5mm in diameter. **Indusium** of the microsorus translucent, so that the microsporangia are clearly visible under a lens. Stalks of the glochidia 75µm long, not divided up into cells (i.e. non-septate). **Spores** ripening June-September. Diploid. 2n=44.

Naturalized, and fruiting freely in ponds, drainage ditches and slowly-flowing streams; becoming more widespread in Wales.

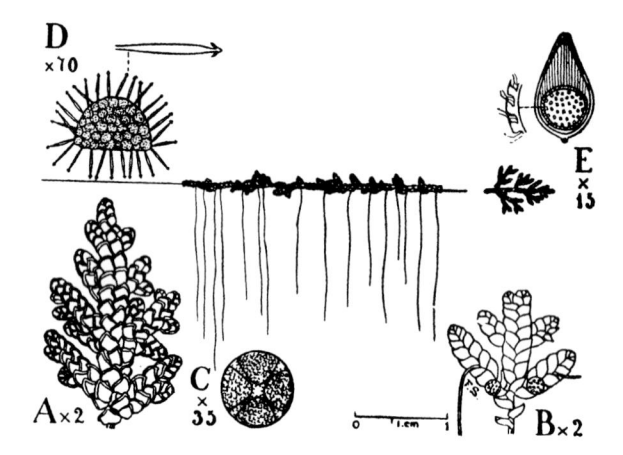

Fig. 52. Water Fern (*Azolla filiculoides*). A. Portion of plant. B. Plant with sporangia. C. Microsporangium. D. Massula (with single glochidium more highly magnified). E. Megasporangium in vertical section. (From Butcher and Strudwick, *Further Illustrations of British Plants*).

First recorded, 1922; J. D. Dean, *Botl Soc. Exch. Club Br. Isl. Rep.* for 1922, 6, 755 (1923), Goldcliff, near Newport, Mon.

Naturalized in England and Ireland, the Channel Islands, fourteen other countries in western, central and southern Europe, southern Africa, NE Asia, Australia, New Zealand and Hawaii. Native to America where it is also spreading as an introduced species. Found as a fossil in Interglacial deposits in Suffolk as well as in Holland, Germany and the former USSR.

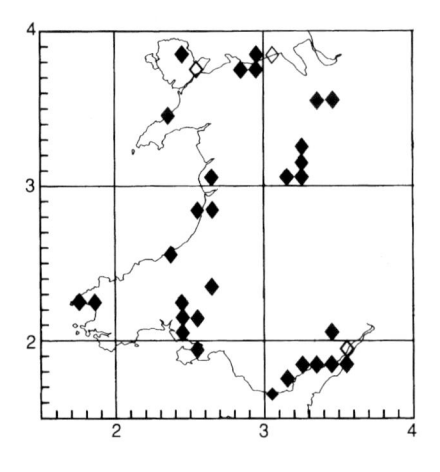

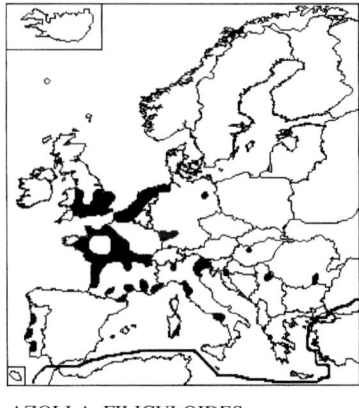

AZOLLA FILICULOIDES

The English name should not be confused with Alpine Water Fern, one of the English names for *Blechnum penna-marina*.

Azolla mexicana C.Presl (*A. caroliniana* auct., non Willd.), native to N., central and S. America, has been recorded as an alien in western Europe and needs to be looked for in the warmer parts of the British Isles. It differs from *A. filiculoides* in having leaves less glaucous-green, becoming brownish-red in autumn, having 2-celled hairs on the upper surface rather than 1-celled hairs. The stalks of the glochidia are divided up into cells (i.e. septate), although this is not constant.

Genus BLECHNUM L.

Hard-ferns - Gwibredyn

Sori (coenosori) linear and, as a rule, occupying practically the entire length of the segments of the fertile fronds, seldom interrupted. Indusium linear. Scales thin-walled (not latticed). About 220 species, only one of which occurs in the North Temperate regions.

BLECHNUM CORDATUM (Desv.) Hieron.

(*B. chilense* (Kaulf.) Mett.)

Chilean Hard-fern; Gwibredynen Chile

Fronds to 150 × 30cm, the fertile scattered among the sterile. **Pinnae** subentire, the sterile up to 15 × 2.5cm, the fertile as long but much narrower, both only narrowly attached to the rachis. *See* Fig. 113, A.

Naturalized in shady places and by ditches and streams.

First recorded 1995; Ellis, Hutchinson & Stokes, Clyne Woods, Swansea (10-km square (hectad) SS69). Welsh Plant Records. *BSBI Welsh Bulletin* (in press).

SW England, W. Scotland, SW Ireland. The identity of British naturalized material needs clarification as more than one species may be involved, in fact material from botanic gardens near the Swansea locality has been labelled *B. magellanicum* Mett.

Native to the southern tip of S. America.

BLECHNUM PENNA-MARINA (Poir.) Kuhn

Antarctic Hard-fern; Alpine Water Fern; Gwibredynen Antarctig

Resembles a very small *B. spicant*, with fronds not larger than 20 × 1.5cm in the British Isles (to 35(-45)cm where native) and often reddish when young; slender creeping rhizome. **Pinnae** on sterile fronds with indented venation on upperside. 2n=66. *See* Fig. 114, A.

Has been naturalized for a short time in a few places. Native to the southern tip of South America, some of the subantarctic islands, SE Australia and New Zealand.

One of the English names should not be confused with Water Fern (*Azolla filiculoides*).

BLECHNUM SPICANT (L.) Roth

Hard-fern; Gwibredynen

Stock upright or decumbent, usually 4-5cm × 6mm, in old plants caespitose, the young parts clothed with dark tawny-brown, lanceolate, acuminate scales. **Fronds** tufted, of two kinds, sterile and fertile, the stipes

Fig. 53. Hard-fern (*Blechnum spicant*). Left: portion of sterile frond; right: portion of fertile frond showing two pinnae, one with two coenosori intact, the other with parts of the coenosori cut away in order to show the commissures which link up to the secondary veins. (*c.*×3). (*See also* Fig. 104).

of both dark to purple-brown, slightly swollen behind and grooved in front, beset at the base with scales like those of the rhizome. **Sterile leaves** 10-15 × 2-7.5cm, lying spread out on the ground, winter-green; **stipe** $^1/_6$-$^1/_3$ as long as rachis. **Blade** lanceolate, leathery in texture, glabrous, dark green, shiny above, paler and dull below, pinnate; **pinnae** 30-60 on either side, set close like the teeth of a comb, linear-oblong with a slight curve towards the apex of the frond, obtuse or acute (sometimes mucronate), entire-margined or slightly crenate, decreasing in size towards the apex and especially towards the base, the lowest segments becoming small and rounded and often separated from one another; **rachis** deeply furrowed on the upper side and darker in colour in its lower half. **Fertile fronds** taller, 15-75 × 2-5cm, placed in the middle of the tuft of leaves and marking the end of the year's growth, stiffly erect, not persisting over the winter; **stipe** up to about as long as blade. **Blade** linear-lanceolate in outline, pinnate; **pinnae** linear, suddenly narrowing from a broad base, acute, the higher more and more closely set, the lower increasingly shorter and roundish, wider apart and without sori; occasionally, plants are found which have fertile pinnae much broader than usual sometimes about as broad as normal sterile pinnae; **rachis** almost to the apex dark like the stipe. **Sori** linear, situated midway between the margin and midrib of the pinna and parallel with the latter, running almost the entire

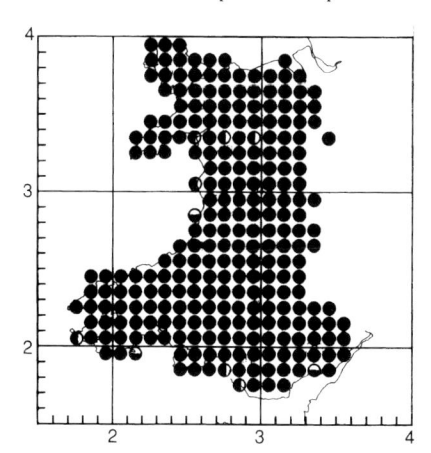

BLECHNUM SPICANT

length of the pinna. **Indusium** whitish at first becoming brown, swollen and somewhat blistered in appearance. **Spores** ripening June-August. Diploid. 2n=68.

Common on acidic soils in woods, heaths and hedge banks and on rock ledges.

First recorded, 1798, Bingley, *A Tour round North Wales...*, vol. 2, 430. 'Moist heaths between Caernarvon and Llanberis.'

Throughout the British Isles, but absent from much of central eastern Scotland.

A disjunct distribution world-wide. Most of Europe from northern Scandinavia to Spain except the Hungarian Plain and local in the Mediterranean. Russia (B,W). Iceland and Faeroes. Azores and Madeira. North Africa. Asia Minor. Caucasus. Iran. China. Japan. Western N. America from Alaska to California.

Genus BOTRYCHIUM Sw.

Moonworts - Lloerlysiau

Sterile blade entire or pinnately lobed or up to 3-4 times pinnate; fertile spike simply pinnate or more highly branched. Sporangia free, sessile, arranged in two rows one on either side of the spike or segment and opening by a transverse slit.

Small terrestrial herbs with a short, brown, upright, fleshy, usually unbranched, subterranean stock, which is sheathed in the dry bases of old fronds and produces each year as a rule a single frond, the base of which is surrounded by a conspicuous membranous and stipule-like sheath derived from the frond of the previous year, and which is borne on a thick, fleshy stipe. About 40 species, chiefly in the Arctic and North Temperate zones (especially North America), a few in the Tropics and Antarctic.

BOTRYCHIUM LUNARIA (L.) Sw.

Moonwort; Lloerlys

Stock subterranean, upright, usually unbranched, cylindrical, brown, beset below with thick fleshy, radiating roots. **Frond** 5-15(-25)cm overall, taking three years to develop; stipe about half or as long as the blade, surrounded at the base by long, brown sheaths formed from the persistent bases of old fronds, hollow and fleshy. **Blade** 3-10(-12)cm, oblong, pinnate; pinnae 4-7 pairs, fan-shaped to crescent-shaped, their margins entire or slightly and irregularly crenate or (in the variety *subincisa* Roep.) deeply crenate. **Spike** 2-10(-16)cm, its stipe in well-developed specimens often overtopping the sterile blade, pinnate to bipinnate. **Sporangia** borne on the margins of the ultimate segments, golden brown when mature. **Spores** ripening June-August. 2n=90.

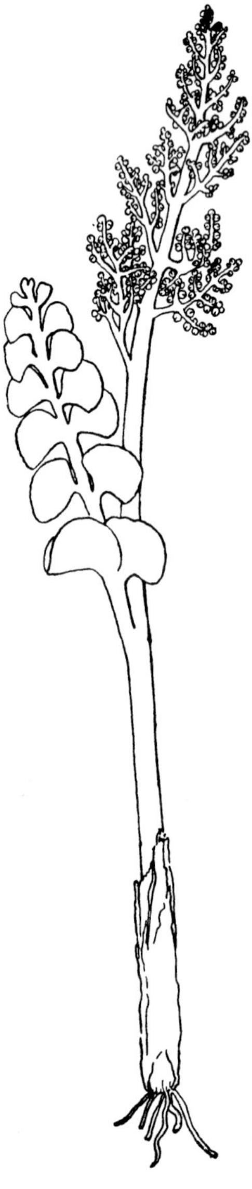

Fig. 54. Moonwort (*Botrychium lunaria*), a large plant, one-half natural size.

Abnormal forms with two fertile branches or with capsules on the edges of the sterile fronds are occasionally met with.

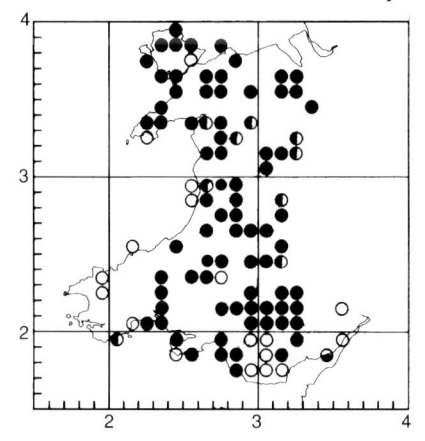

BOTRYCHIUM LUNARIA

Pastures, hillsides, heaths, and fixed sand dunes. Easily overlooked and probably more common than records suggest.

First recorded in 1606-08 by Sir John Salusbury: MS. record in his copy of Gerard's *Herbal* now in the Library at Christ Church, Oxford. (*See* Gunther, R. W. T.: *Early British Botanists*, Oxford, 1922). '*Lunaria minor* is found in Cunnygree of the Right Hon. Sir John Salusburys, Knight, lying between Botuarry [Bodfari] and Carewis and great plenty of them are found in Place y Chambers fielde lying hard by Snodioge parke neare Denbigh.'

Throughout the British Isles but scarce in many areas.

Bipolar. In the Arctic, North Temperate, Antarctic, and South Temperate regions. Europe, except in the coastal parts of the Mediterranean region and the Hungarian Plain. Iceland and Faeroes. Azores and Madeira. Asia: across much of the middle of Siberia. Himalaya, China, Japan. N. America: Canada and northern USA and in its western third down to California. Greenland. S. America: in Chile and Patagonia. SE Australia and Tasmania. New Zealand.

CETERACH Willd., *see under* ASPLENIUM L.

Ceterach officinarum Willd., see ***Asplenium ceterach*** L.

Genus CRYPTOGRAMMA R.Br.

Parsley Ferns - Rhedyn Persli

Fronds of two kinds (a) with narrow divisions and bearing sori (fertile frond), (b) with broader divisions and without sori (sterile frond). Sori situated on the vein endings and rounded or more or less elliptical in outline. Indusium absent, but sori covered when young by the recurved margin of the frond. Sporangium pear-shaped and shortly stalked; the annulus sometimes obliquely placed. A small genus of four species; occurring chiefly in the North Temperate zone, only one species in Wales.

CRYPTOGRAMMA CRISPA (L.) R.Br.

Parsley Fern; Rhedynen Bersli

Rhizome 2-3mm thick, creeping or ascending, caespitose, scaly covered with dead frond-bases. **Fronds** tufted in a dense spiral and very numerous. **Sterile** (outer) **fronds** 7-15cm; **stipe** as long or twice as long as blade. **Blade**

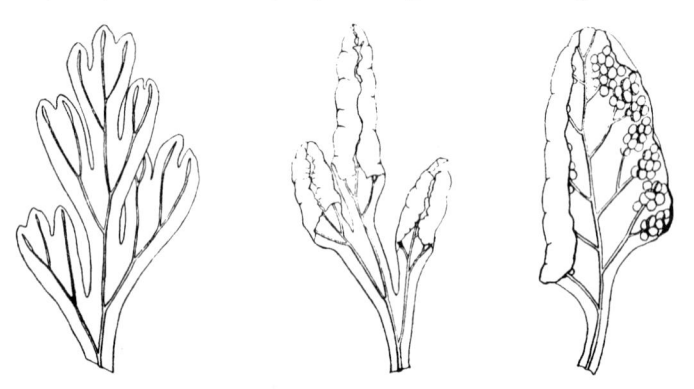

Fig. 55. Parsley Fern (*Cryptogramma crispa*). Left: single pinna from sterile frond (×5). Middle: single pinna from fertile frond (×5). Right: fertile segment with its margin flattened out on one side to display the sori (×9). (*See also* Fig. 99B).

4-7 × 3-5cm, ovate to triangular-ovate, obtuse, pinnate, pinnae 7-15, stalked to sessile, ovate, obtuse, deeply bi- or tri-pinnatifid; **segments** alternate, normally obovate-cuneate, obtuse to truncate at the apex, with 2 to 4 teeth or lobes in the upper half. **Fertile** (inner) **fronds** longer stalked, 11-30cm overall, broadly ovate to lanceolate, obtuse, pinnate; **pinnae** 7-11, stalked, ovate, once or twice pinnate, the ultimate pinnules stalked, mostly linear (less often more or less ovate), obtuse, their margins membranous and strongly inrolled beneath. **Sori** elliptical in outline, borne on the vein endings, at first distinct, but eventually appearing to form a continuous band running parallel with the margin of the pinnule. **Sporangia** sometimes with a slightly oblique annulus. **Spores** ripening June-August. Tetraploid. 2n=120.

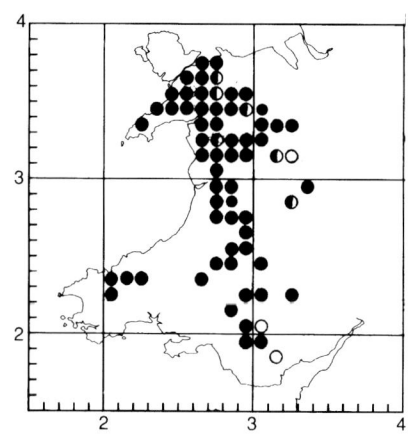

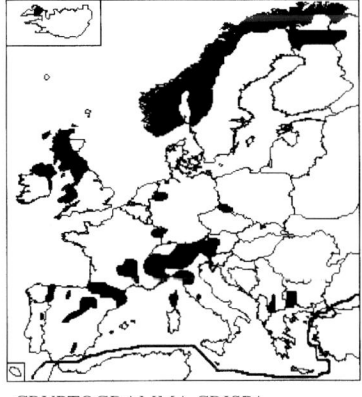

CRYPTOGRAMMA CRISPA

Mountainous districts only: a pioneer plant on screes formed from non-calcareous rocks, seldom if ever on a stable substratum.

First recorded, 1662; Ray, Itinerary III (*Memorials*, ed. E. Lankester, 171, 1846). 'On Snowdon hill we found that species of *Adianthum floridum* which we had before observed in Westmorland.'

Western Britain with much of Scotland but very local in SW England (Devon) and Ireland.

Protected in the Republic of Ireland under the *Flora Protection Order, 1987*.

Arctic-alpine. Europe scattered in most countries up to Lapland. Iceland. Commoner in the Pyrennes, the Alps and Norway. Rossia (N). Asia: Asia Minor, Caucasus, scattered populations of varieties from Afghanistan to W. China, southern central Siberia and Kamchatka peninsula. Western side of N. America and central Canada.

Genus CYATHEA Sm.

Silver Tree-ferns - Coedredyn Arian

Differing from *Dicksonia* in possessing dense scales on the rhizomes (trunk) and stipes rather than hairs. Large, tree-like, terrestial plants. Rhizomes erect, sometimes very tall and trunk-forming, bearing fronds in large, spreading crowns from the apex. Rhizome apex and stipe bases usually densely scaly. Fronds all of one kind, very large and several times pinnately divided to produce a very open, lace-like effect; blades ovate. Sori rounded, with or without indusia, these when present, very variable usually simple and scale-like. A large genus of about 600 species.

CYATHEA DEALBATA (G.Forst.) Sw.

(*Alsophila tricolor* (Colenso) Tryon)

Silver(y) Tree-fern; Ponga; Coedredynen Arian

Rhizome forming a trunk to 20cm in diameter. **Fronds** to almost 4m with glaucous stipe bases. **Blades** 3-pinnate, bright green above but distinctly white or bright silvery-grey beneath. **Sori** large, covered by cup-shaped indusia. The silver colour on the lower side of the fronds clearly distinguishes it from *Dicksonia antarctica*. 2n=138.

A garden relic reproducing in S. Kerry, Ireland.

Tree-fern native to New Zealand.

Genus CYRTOMIUM C.Presl

(PHANEROPHLEBIA C.Presl)

House Holly-ferns - Rhedyn Celynnog

Tropical Asia, Hawaii, tropical America and South Africa. Twenty species, of which only one is at all likely to be found as an escape in the British Isles.

CYRTOMIUM FALCATUM (L.f.) C.Presl

(*Phanerophlebia falcata* (L.f.) Copel.; *Polystichum falcatum* (L.f.) Diels)

House Holly-fern; Rhedynen Celynnog

Stock short, erect, bearing large brown scales. **Blades** pinnate, 15-40 cm, leathery; **pinnae** up to 27, 5-10 × (1.5-)2-4cm, ovate-acuminate, short-stalked, entire or undulate, often with a single deep lobe near the base. **Sori** round, scattered. Apogamous. 2n=123. *See* Fig. 113, B.

An eastern Asiatic species, often grown in cold greenhouses and as an ornamental house plant, sometimes reported as an escape. The English name should not be confused with Holly-fern (*Polystichum lonchitis*).

Recorded from near the Afon Llynfi, Llanllynfi (SH45, Caerns.) by M. Wood in 1929.

Scattered in western Britain from Scilly to west central Scotland; Channel Islands and W. Cork. Recently found in north London.

Native to E. Asia.

Genus CYSTOPTERIS Bernh.

Bladder-ferns - Ffiolredyn

Sori round, situated on a vein which extends beyond the receptacle. Indusium sub-globose, acuminate, attached by its broad base below the sorus on its inner side only, at first covering the sorus like a hood, later reflexed. 18 species various in habit, in both hemispheres but chiefly north temperate.

KEY TO SPECIES

1 Blade triangular-ovate with lowest pinna-pair the longest; rhizome elongated bearing fronds singly **C. montana**
   Blade narrowly oblong to lanceolate with longest pinna-pair near the middle of the blade; rhizome short, bearing terminal tuft of fronds 2

2 Most adjacent pinnae and pinnules partly overlapping, making the blade look congested. Spores rugose (microscope) **C. dickieana**
   Adjacent pinnae and pinnules not or scarcely overlapping. Spores spinose (microscope) **C. fragilis**

CYSTOPTERIS DICKIEANA R.Sim*

Dickie's Bladder-fern; Ffiolredynen Arfor

Frond pinnate; **pinnae** deeply pinnatifid; margins of segments crenate. **Spores** 41-48µm, rugose not spinose, ripening July-August. Tetraploid. 2n=168. The combination of these features has been maintained in cultivation from material originating from the sea-caves of Kincardineshire, distributed in the 19th century by George Dickie.

Very rare on basic rocks in sea-caves; Kincardineshire.

Protected in Great Britain under the *Wildlife and Countryside Act, 1981*.

* Pteridologists have differing views about *C. dickieana*. Either the species as a whole is conspecific with *C. fragilis* or the Kincardineshire plants from sea-caves are considered to be a distinct species (as described above) or are a variety of the commoner *C. dickieana* of mainland Europe, Turkey and Siberia, which has been recorded recently inland in the British Isles (Tennant 1995,1996). This shows a morphology closer to *C. fragilis* with less overlap, if at all, of adjoining pinnae or of pinnules or pinna segments, but retaining the rugose surface to the spores. See under *C. fragilis* for discussion of the *C. fragilis* complex world-wide.

CYSTOPTERIS FRAGILIS (L.) Bernh.

Brittle Bladder-fern; Rhedynen Frau

Rhizome usually short 4-5 × 0.5cm, decumbent or horizontal, often branched, the older parts covered with spirally arranged dead frond bases, the younger parts, including the outermost frond bases and the young leaves in the bud, covered with thin yellow-brown lanceolate to ovate-lanceolate acuminate scales. **Fronds** 6-35(-45)cm, in a small tuft of 5-6, dying down in winter; **stipe** $^1/_3$-$^2/_3$ as long as rachis, brittle, dark brown and scaly at the base, straw-yellow or green, and glabrous or with a few scales above. **Blade** 4.5-25(-30) × 2-10(-12.5)cm, oblong-lanceolate to ovate-lanceolate in outline, acute or acuminate, thin, flaccid and delicate, bipinnate or almost tripinnate; **pinnae** (primary segments) up to about 15 on either side, opposite or nearly so, becoming alternate towards the apex, shortly stalked, ovate- to oblong-lanceolate, pinnate or almost bipinnate, the lowest pair often somewhat distant from the others and distinctly shorter; **pinnules** (secondary segments) ovate at the base of the pinna, oblong towards the apex, all more or less decurrent, the larger (basal) ones deeply pinnatifid, the others pinnately toothed, the teeth obtuse or acute; **tertiary segments** pinnately toothed. **Sori** in two rows one on either side of the midribs of the pinnules or of the tertiary segments where these are strongly developed. **Indusium** pale-coloured, membranous, ovate-lanceolate, acuminate. **Spores** relatively large (*c*.40-55μm long), coarsely spiny, ripening July-August. Tetraploid. 2n=168; and hexaploid. 2n=252; with one pentaploid (2n=210) record from Derbys.

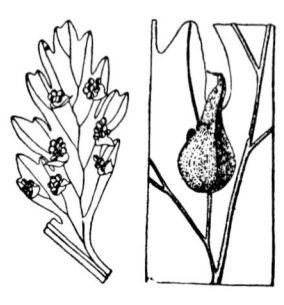

Fig. 56. Brittle Bladder-fern (*Cystopteris fragilis*). Left: one of the lower pinnules of the fourth pinna (out of 15 on one side); the lowest (acroscopic) tertiary segment bears two sori; in all the sori the indusium has become reflexed (×3). Right: small portion of a fertile frond more highly magnified to show the shape of the indusium before it becomes reflexed (×16). (*See also* Fig. 103B).

The fronds vary remarkably, according to habitat, in size and degree of dissection and in relative width and acuteness of ultimate segments.

Fissures of moist, shady rocks and walls.

First recorded, 1696; Llwyd in Ray, *Syn*, edn 2, 50. 'On Snowdon.'

Throughout Great Britain, thinning out towards the south-east; similarly in Ireland.

Throughout Europe. Azores and the Canary Is. North, central and southern Africa, Réunion. Kerguelen. Most of Siberia to the Arctic coast.

Asia Minor through Himalaya, Burma and China to Japan. From northern Greenland to Mexico and Chile. S. Orkney Is., S. Australia, New Zealand and Hawaii.

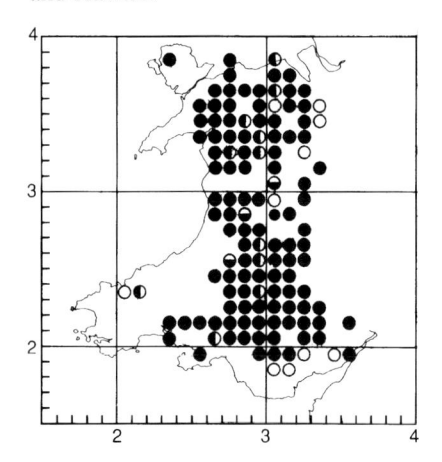

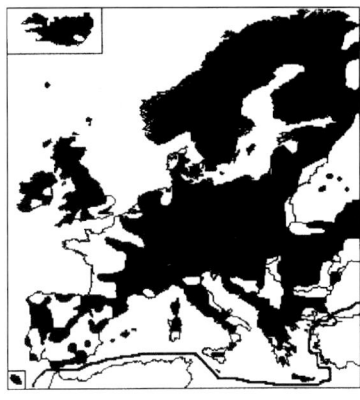

CYSTOPTERIS FRAGILIS

World-wide, *C. fragilis* forms a widespread, polymorphic, polyploid complex with diploids to octoploids reported. Fronds vary remarkably, according to habitat, in size and degree of dissection, in relative width and acuteness of ultimate segments, and in the overlap, if any, of adjoining pinnae and of pinnules or pinna segments. Spores vary in size and surface structure from spiny and smooth to rugose and granular. Various combinations of morphology and spore type are found and sterile hybrids occur between the cytotypes. Because of this, it is often not possible to distinguish morphologically between the variously described species, subspecies and varieties so that some pteridologists treat these all as one species - *Cystopteris fragilis*.

CYSTOPTERIS MONTANA (Lam.) Desv.

Mountain Bladder-fern; Ffiolredynen y Mynydd

Rhizome creeping, about 2mm thick; **fronds** at irregular intervals; stipe twice as long as rachis; **blade** tripinnate, triangular, the lowest pinnae much the longest, the **pinnulets** pinnatifid. **Spores** small (c.35µm long), coarsely spiny, ripening July-August. Tetraploid n=84, 2n=168.

Central Scotland, formerly Lake District. Recorded in error from Caernarvonshire.

N. and NE Europe. Mountains of central and S. Europe from the Pyrennes to the E. Carpathians, Rossia (N,C,W,E). Asia. North America: mostly in the NW and NE Greenland.

Genus DAVALLIA Sm.

Hare's-foot Ferns - Rhedyn Troed-yr-Ysgafarnog

Small to moderately large, mostly epiphytic plants. Rhizomes long-creeping, branching occasionally, mostly densely scaly bearing fronds individually at regular intervals. Fronds of one kind semi-evergreen or deciduous, mostly broadly triangular to narrowly ovate, of a soft flexible texture. Blade typically rather finely dissected, with ultimate segments decurrent. Stipes jointed to the rhizome. Sori terminal on veinlets or margins of ultimate segments, each covered by an elongate, vase-shaped indusium which opens near the margin. A genus of about 40 species.

DAVALLIA BULLATA Wall. group

Squirrel's-foot Ferns; Rhedyn Troed-y-Wiwer

Rhizome less than 5mm thick, often very long creeping and climbing. **Fronds** 15-45cm, **stipe** less than half to about equalling the rachis. **Blade** length about equalling to slightly longer than width. 3-pinnate with the ultimate segments sometimes deeply notched.

A former greenhouse escape in Surrey.

E. Asia.

DAVALLIA CANARIENSIS (L.) Sm.

Hare's-foot Fern; Rhedynen Troed-yr-Ysgafarnog

Rhizome long, densely silky-scaly producing 3(-4)-pinnate. **Fronds** to 70cm or more. **Stipe** about as long as to slightly longer than rachis. **Blade** about as broad as long, mostly 4-pinnate. **Sori** on lowerside, each covered by an **indusium** attached at the base and sides and opening towards blade-margin. 2n=80.

It was naturalized on a wall in Guernsey.

Native to SW Europe, Azores, Madeira, Canary Is. and Cape Verde Is.

Genus DICKSONIA L'Hér.

Australian Tree-Ferns - Coedredyn Awstralia

Medium to large or tree-like terrestial plants (cf. *Cyathea*). Rhizome thick or thin, usually unbranched, covered with matted growths of old stipe bases, bristly hairs and interwoven roots, which build up the bulk of the girth, and with dense masses of rust-brown hairs amongst the crown, erect, bearing fronds in regular, spreading crowns. Fronds dark green, 3-pinnate, of one kind, evergreen, ovate, harsh, rather leathery and often slightly prickly. Sori marginal, on pinnules with deep lobed margins, each protected by a 2-valved cup, formed on the inner side by an indusium and on the outer by a portion of the reflexed margin of the pinnule, both hard when mature. About 25 species in Australasia, Polynesia and Mexico.

DICKSONIA ANTARCTICA Labill.

Australian Tree-fern; Soft Tree-fern; Coedredynen Awstralia

Rhizome (trunk) eventually to 15m but often less than 1m. **Fronds** up to 200 × 60cm. **Stipe** and rachis sparsely hairy, with long, brown shaggy hairs at the base which do not stand out perpendicularly. *Dicksonia fibrosa* Colenso, native to New Zealand, is also grown in gardens and may be confused with this species. Separating the two species is difficult. Generally *D. antarctica* is larger. When mature the trunk is larger (to 15m × 60cm compared to 7m × 30cm in *D. fibrosa*) and the length to breadth ratio of mature fronds of the same length has been provisionally quoted as larger (6:1 compared to 4.5:1) although further work is needed on these two species in the British Isles. *See* Fig. 114, B.

Naturalized in woods and shady places in Scilly, W. and E. Cornwall and S. Kerry.

Native to southern and eastern Australia.

Genus DRYOPTERIS Adans.

Buckler-ferns - Marchredyn

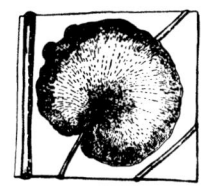

Fig. 57. Male-fern (*Dryopteris filix-mas*). Part of a fertile segment (×13) showing a single sorus at an early stage, still almost entirely covered by the kidney-shaped indusium.

Sori generally large with large reniform indusia. Rhizome usually oblique or erect, and densely clothed with broad soft often lacerated scales. Leaves tufted, mostly of thick somewhat fleshy texture and light green colour, lanceolate and bipinnatifid, or triangular bi- to tri-pinnate and with unequal-sided pinnae having their largest basal pinnules on the side away from the frond apex. Veins free, as a rule forked, the midribs decurrent. About 150 species mostly in the North Temperate region, many in eastern Asia and Africa, few in Tropics.

KEY TO SPECIES

The use of the term 'pinnules' here includes the deeply divided pinna segments of some species.

1. Pinnules divided transversely at least half way to their midrib, most narrowed at base or shortly stalked — **2**

   Most pinnules divided transversely to less than half way to their midrib, most attached by their whole base to the pinna midrib (costa) — **7**

2. Basal pair of pinnules on lowest pair of pinnae almost equal; pinnules without spinulose-mucronate teeth — **D. submontana**

   Basal pair of pinnules on lowest pair of pinnae usually very unequal in length; pinnule teeth spinulose-mucronate (×10 lens; often very short and few in *D. aemula*) — **3**

3. A few to most scales on stipe with a darker central stripe — **4**

   Scales on stipe with a pale uniform tint — **5**

4. Rachis about twice the length of the stipe or more; pinnules convex with down-curved edges. Scales mostly with dark brown to black central stripes. Spores dark brown to black with dense tubercles (microscope) — **D. dilatata**

   Rachis 1 - *c*.1.5 × length of stipe; pinnules flat but sometimes with teeth turned upwards; pinnule segments generally deeper cut and more obviously toothed than *D. dilatata* (cf. Figs. 62, 63, 105, 106). Scales with some darker central stripes, but not as intense as in *D. dilatata*. Spores paler than in *D. dilatata* with sparse tubercles (microscope) — **D. expansa**

5. Fronds many arising as a broad shuttlecock with drooping tips and sometimes hay-scented; spinulose-mucronate pinnule-teeth often few and short (×10 lens) — **D. aemula**

   Fronds few, not forming a complete shuttlecock, frequent spinulose-mucronate teeth — **6**

6. Fronds usually less than 30cm; blade with a more or less triangular outline; sori development poor giving impression of an immature plant; rachis about twice the length of stipe — Forms of **D. dilatata**

   Fronds 30-80cm; blade linear-lanceolate sometimes parallel-sided; rachis equalling or slightly longer than stipe — **D. carthusiana**

7. Fertile fronds longer than vegetative fronds; some pinnule teeth clearly spinulose-mucronate — **D. cristata**

   Fertile and sterile fronds similar; pinnules not spinulose-mucronate — **8**

8. Base of pinna midrib with blackish spot on upper side and sometimes a more extensiv e one on underside (often absent on dried specimens of *D. affinis* subsp. *borreri*) — **9**

   No blackish spot (sometimes seemingly present late in the season on old coriaceous fronds of *D. filix-mas* exposed to the sun) — **10**

9   Most pinnule side-margins unlobed, or with a few small acute teeth or lobes especially on large shaded plants   **D. affinis**
Most pinnule side-margins lobed along their entire length or acutely toothed (Microscopic examination of mature spores is essential)
D. affinis × D. filix-mas (D. × complexa)

10 Most pinnule side-margins toothed along their entire length, apical teeth acute; indusia without a continuous concentric outer rim   **D. filix-mas**
Pinnule side-margins entire or with a few shallow lobes, apical teeth broad, blunt-tipped and sometimes divergent; some indusia with a continuous concentric outer rim   **D. oreades**

DRYOPTERIS AEMULA (Aiton) Kuntze

(*Lastrea aemula* (Aiton) Brack.; *L. foenisecii* (Lowe) H.C.Watson)

Hay-scented Buckler-fern; Marchredynen Aroglus

Stock stout, erect or ascending, the older parts clothed with the bases of dead fronds, the younger with narrow-lanceolate, uniformly pale brown, sometimes lacerated, erect scales. **Fronds** 15-60cm, numerous, tufted, fragrant with the smell of new-mown hay; **stipe** as long as rachis, rigid, brownish-purple, clad (densely below, sparingly above) with narrowly lanceolate to ovate, uniformly rusty-brown, sometimes lacerated scales. **Blade** 8-30 × 4-20cm, triangular to ovate-lanceolate in outline, bright green above, paler beneath, both surfaces sprinkled with minute sessile glands, bipinnate (to tripinnate at the base); **pinnae** about 15-20 on either side, stalked, the lowest pairs opposite or subopposite, the uppermost only alternate, the lowest pair unequally triangular and distinctly the longest, the rest lanceolate to linear-lanceolate, all acute; **pinnules** alternate, oblong, the basal ones stalked, the remainder sessile and becoming decurrent, pinnate becoming pinnatifid and ultimately only toothed, the lowest basiscopic pinnule of the lowest pinna on either side twice as

Fig. 58. Hay-scented Buckler-fern (*Dryopteris aemula*). Median pinnule from second or third pinna of an average-sized frond (×6). (*See also* Fig. 105*A*).

long as the lowest acroscopic pinnule on the same pinna and unlike the rest usually broadly lanceolate in outline; **ultimate segments** (whether pinnulets or pinnules) oblong, the margins upturned (producing a crisped appearance) and pinnately toothed, the teeth not or only slightly incurved towards the apex of the segment, somewhat coarse and shortly mucronate; **rachis** bearing scattered lanceolate scales with hair-like points. **Sori** in two rows on the ultimate segments, one on either side of the midrib and somewhat nearer to it than to the margin, each individual sorus on the acroscopic fork of a secondary vein. **Indusium** reniform, the margin jagged and beset with sessile glands. **Spores** ripening July-September. Diploid. 2n=82.

Shady, rocky places, especially about streams and in rocky woods; rare.

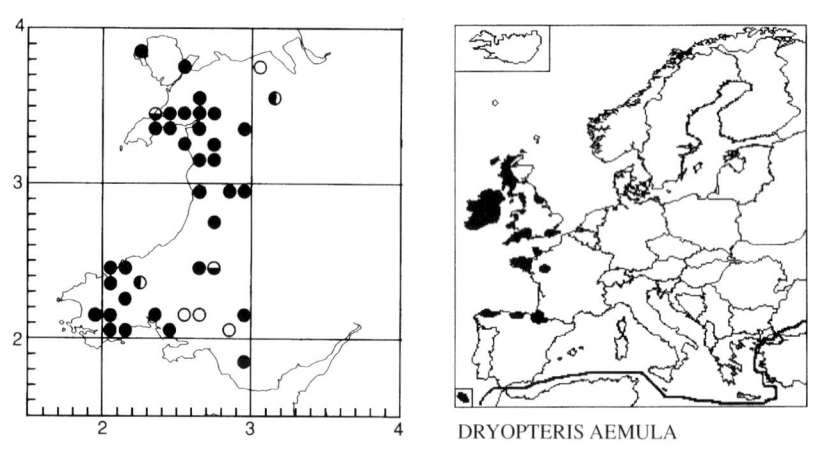

DRYOPTERIS AEMULA

First recorded, 1690; Ray, *Syn.*, 27, 1690, as *Filix montana ramosa minor argute denticulata*, from the Glyder, Caerns.

In Great Britain chiefly in the west, SW England, western Scotland to Orkney, with an outlier on the Weald of SE England. Mostly in the west of Ireland; scattered records in N. England and central highlands of Scotland.

Western Europe from N. Scotland to NW Spain. Azores, Madeira and the Canary Is., Turkey and the Caucasus.

DRYOPTERIS AFFINIS (Lowe) Fraser-Jenk.

(*Dryopteris paleacea* auct.; *D.filix-mas* auct., non (L.) Schott)

Scaly Male-fern; Marchredynen Euraid

As *D. filix-mas* except: **stipe** and **rachis** shaggy with orange-yellow to golden or dark-brown, linear-lanceolate to ovate-lanceolate, acuminate scales; stipe as long as blade. **Fronds** somewhat leathery, yellowish-green,

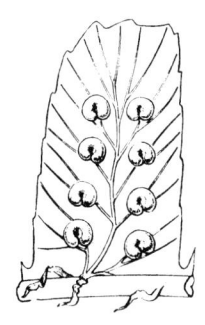

Fig. 59. Scaly Male-fern (*Dryopteris affinis*). Single fertile segment from a median pinna (×6). (*See also* Fig. 107A).

glossy, persisting until late in season. **Pinnae** (when living) with a dark brown or blackish patch near their junction with the rachis; **pinna segments** variable and sometimes similar to those of *D. filix-mas* in shape but often straight- and almost parallel-sided, usually more or less subtruncate and with a few obtuse or acute triangular teeth at the apex, the rest of the margin subentire. **Sori** (3-)4-5 on each side of the midrib. **Indusium** with margin tucked under the sorus (Fig. 60). **Spores** ripening July-October. Reproducing apogamously only. Apogamous diploid. 2n=82; and apogamous triploid. 2n=123. All its hybrids are apogamous.

Common in hedge banks, woods and woodland clearings.

Throughout the British Isles but less common than *D. filix-mas*.

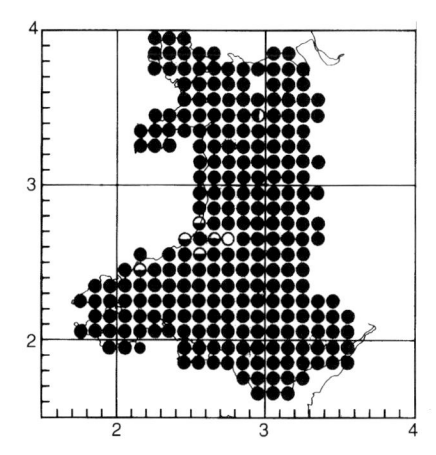

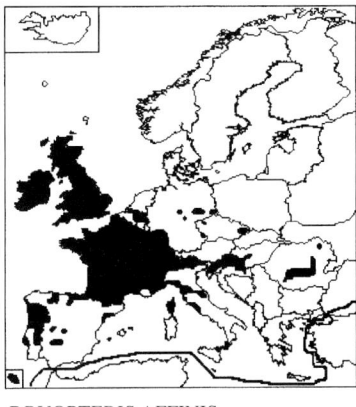

DRYOPTERIS AFFINIS

Norway and most of Europe south of the Baltic, Russia (W,K). Azores and Madeira. Morocco. Asia Minor, Caucasus and Iran. Closely related taxa are reported from S. and SE Asia, Central and S. America.

Current view of *Dryopteris affinis* (Scaly Male-fern)

D. affinis is very variable and what have been referred to as species, subspecies, varieties, forms and morphotypes have been described over the years. Some pteridologists disagree with the interpretation of *Dryopteris affinis* as described below. It is anticipated that future work will clarify the situation.

D. affinis consists of apogamous diploids and triploids considered to be derived in various combinations from hybridization between *D. oreades*, at least one other ancestral diploid, and *D. caucasica* (A. Braun) Fraser-Jenk. & Corley, a species found in some countries which border the Black Sea.

In order to keep a balance between the artificial and the biological aspects of a species concept, the **subspecies of *D. affinis*** are retained here, a treatment suggested for any similar apogamous complexes where the taxa are artificially close. Several subspecies have been recognised, but their characters and delimitations are still under study. The use of the term hybrid, and the associated 'x' sign, are confined to sterile hybrids arising *de novo* from the parental species and not for hybrid-derived species (Fraser-Jenkins, 1988, 1996).

The following three subspecies are the best understood in the British Isles at present and all have been recorded in Wales. Recognition is possible only after some experience and there are many plants which cannot be assigned to these three subspecies even by specialists on the subject. *D. affinis* subsp. *affinis* and subsp. *cambrensis* are the most clearly defined and plants which do not resemble these are sometimes placed in subsp. *borreri* by recorders, the last being the most variable with a few 'varieties' or morphotypes having been described informally in the literature.

KEY TO THE SUBSPECIES OF DRYOPTERIS AFFINIS

The use of the term 'pinnule-teeth' here refers to the teeth of the deeply divided pinna segments.

1 Pinnule-teeth acute; indusium thin and not splitting radially at maturity but lifting round the edges to become reflexed radially then falling early (deciduous)   subsp. **borreri**

   Pinnule-teeth obtuse; indusium thick, occasionally splitting radially at maturity and more or less persistent   2

2 Indusium well tucked under as sporangia mature, occasionally splitting radially at maturity; fronds persisting in winter sometimes with many indusia   subsp. **affinis**

   Indusium lifting slightly round the edges, occasionally splitting radially at maturity; fronds not persisting in winter   subsp. **cambrensis**

DRYOPTERIS AFFINIS (Lowe) Fraser-Jenk. subsp. *AFFINIS*

(*Lastrea pseudomas* Woll.; *Dryopteris pseudomas* sensu (Woll.) Holub & Pouzar pro parte)

Yellow Scaly Male-fern; Marchredynen Euraid

Fronds persisting in winter. **Lamina** thicker and more coriaceous and more glossy on upper surface than in other subspecies and flat, i.e. with no reflexed pinnae or pinna segments. **Pinna segments** rounded to rounded-truncate at their apex usually with a few obtuse teeth, these sometimes prominent but occasionally not obvious. The **first basiscopic pinna segment** of the lowest pinna, on either side of the rachis, partially adnate to the costa, not symmetrically but mostly on one side of the pinna segment midrib. **Indusium** inflexed, thick and persistent (resembling a bun pressed down in the middle when viewed from above), some eventually splitting radially to their centres (resembling a sponge cake with a slice taken out) (Fig. 60). Apogamous diploid. 2n=82.

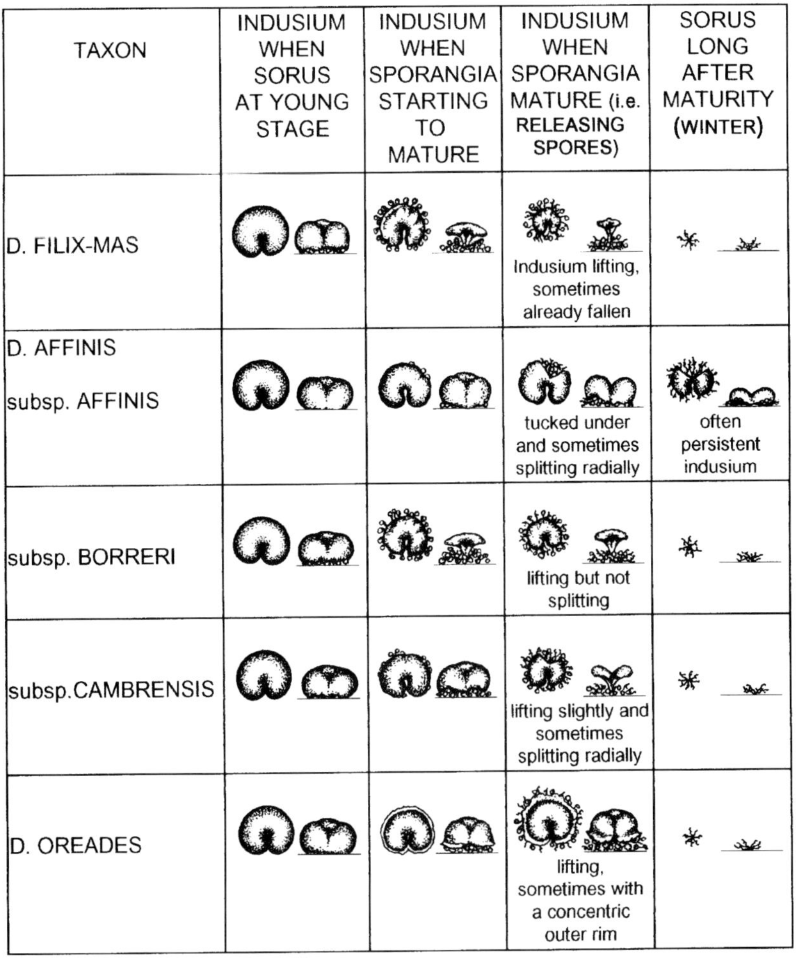

TAXON	INDUSIUM WHEN SORUS AT YOUNG STAGE	INDUSIUM WHEN SPORANGIA STARTING TO MATURE	INDUSIUM WHEN SPORANGIA MATURE (i.e. RELEASING SPORES)	SORUS LONG AFTER MATURITY (WINTER)
D. FILIX-MAS			Indusium lifting, sometimes already fallen	
D. AFFINIS subsp. AFFINIS			tucked under and sometimes splitting radially	often persistent indusium
subsp. BORRERI			lifting but not splitting	
subsp. CAMBRENSIS			lifting slightly and sometimes splitting radially	
D. OREADES			lifting, sometimes with a concentric outer rim	

Fig. 60. Scaly Male-fern (*Dryopteris affinis*) and associated species showing sequential changes to the indusium.

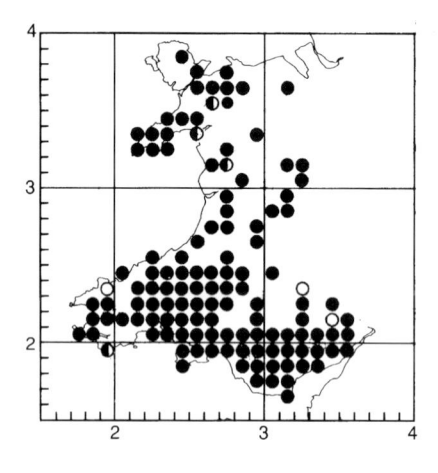

This subspecies is widespread in Wales but tends to be absent or scarce in base-rich areas, being less tolerant of such conditions than subsp. *borreri*.

SW Europe and W. central Europe.

DRYOPTERIS AFFINIS subsp. AFFINIS

DRYOPTERIS AFFINIS (Lowe) Fraser-Jenk. subsp. ***BORRERI*** (Newman) Fraser-Jenk.

(*Dryopteris borreri* (Newman) ex Oberh. & Tavel; *D. tavelii* Rothm.; *D. affinis* subsp. *stilluppensis* (Sabr.) Fraser-Jenk.; *D. affinis* subsp. *robusta* Oberh. & Tavel ex Fraser-Jenk.; *D. woynarii* auct., non Rothm.; *D. mediterranea* Fomin; *D. paleacea* (D.Don) Hand.-Mazz. pro parte, non (Sw.) C.Chr.)

Common Scaly Male-fern; Marchredynen Feddal

Fronds mostly not persisting in winter. **Lamina** less shiny, paler green and less coriaceous than in the other subspecies, and flat in plane of frond. **Pinnae**: the dark patch at their junction with the rachis tends to disappear on dried herbarium specimens. **Pinna segments** truncate to subacute with sometimes prominent acute teeth round their apex (resembling cat's ears when prominent in the two corners). Pinna segments often with two sharp corners at their apex, the corner on the basiscopic side of the pinna obtuse while that on the acroscopic side acute, the pinna segment thus resembling a parallelogram. The **first basiscopic pinna segment** of

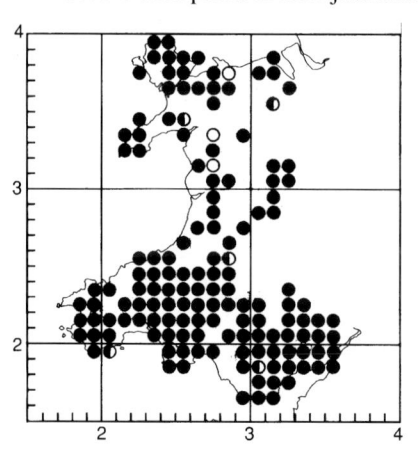

DRYOPTERIS AFFINIS subsp. BORRERI

the lowest pinna, on either side of the rachis, attached only round its midrib to the costa. **Indusium** inflexed but soon becoming reflexed (shuttlecock-shaped) and falling early (Fig. 60). Apogamous triploid. 2n=123.

This subspecies is the most variable and the most common and widespread in Wales and is tolerant of more base-rich areas than subsp. *affinis*.

In Europe, throughout the range of the species.

DRYOPTERIS AFFINIS (Lowe) Fraser-Jenk. subsp. ***CAMBRENSIS*** Fraser-Jenk.

(incl. *Dryopteris affinis* subsp. *stilluppensis* sensu Fraser-Jenk., non Sabr.)

Narrow Scaly Male-fern; Marchredynen Gulddail

Fronds not persisting in winter and generally narrow in outline relative to the other two subspecies. **Lamina** glossy on upper surface and somewhat coriaceous with pinna segments often reflexed from the costa of the pinna (pinna in transverse cross-section resembling the hull of a rowing boat) and pinnae often swept upwards giving the frond the narrowest outline of the three subspecies; glandular on the axes, at least when young. **Pinna segments** rounded with obtuse teeth round their apex, with the two basal pinna segments of each pinna often overlapping the rachis, the basiscopic of these often bent outwards above the plane of the frond (resembling projecting steps up the rachis). The **first basiscopic pinna segment** of the lowest pinna, on either side of the lamina, attached only round its midrib to the costa. **Indusium** inflexed as in subsp. *affinis* but shrivelling later like subsp. *borreri* (Fig. 60). Apogamous triploid. 2n=123.

This subspecies is the least common in Wales of the three and its full distribution is not known because it is only now becoming recognised by recorders. It is a subspecies mostly of northern Britain being one of the commonest subspecies encountered in Scotland. In South Wales the sparse locations are mostly confined to the higher ground despite extensive field work in some vice-counties.

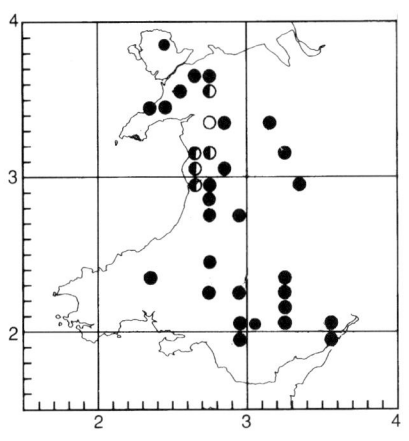

In Europe, throughout the range of the species except in parts of central and S. Europe.

DRYOPTERIS AFFINIS
subsp. CAMBRENSIS

DRYOPTERIS CARTHUSIANA (Vill.) H.P.Fuchs

(*D. spinulosa* (O.F.Müll.) Kuntze; *D. lanceolato-cristata* (Hoffm.) Alston)

Narrow Buckler-fern; Marchredynen Gul

Stock short, up to 10 × 3-4cm, decumbent or horizontal, very scaly, the older parts clothed with dead frond bases. **Fronds** 30-120cm, few, not forming a complete shuttlecock-like tuft; **stipe** about equal to the rachis in length, relatively thin, dark brown below and pale above, clothed at the base and bestrewn here and there above with pale brown ovate to ovate-lanceolate, cordate, hair-pointed scales having no dark central portion. **Blade** 15-60 × 10-25cm, oblong-lanceolate to lanceolate (but not narrowed at the base) or rarely ovate-oblong in outline, light to yellowish-green glabrous except rarely for minute glandular hairs on the under surface, bipinnate; **pinnae** about 15-20 on either side, shortly stalked, the lowest

Fig. 61. Narrow Buckler-fern (*Dryopteris carthusiana*). Median pinnule from third or fourth pinna of an average-sized frond (×6). (*See also* Fig. 105B).

pairs opposite or sub-opposite, the uppermost only alternate, the lowest pair equally triangular in outline and usually a little shorter than the next pair above, the upper pinnae becoming more nearly lanceolate; **pinnules** oblong-ovate decurrent, the basal ones deeply pinnatifid, the rest becoming less and less deeply cut towards the apex of each pinna, the basiscopic pinnules of the lowest pinnae larger than the acroscopic ones, **ultimate segments** ovate-oblong, pinnately toothed, the teeth spinulose-mucronate and incurved; **rachis** with scattered subulate scales. **Sori** numerous, mostly on the acroscopic forks of the secondary veins of the pinnules and rather nearer to the midrib than to the margin of the pinnule, or, in the larger tertiary segments, forming two rows on the lower secondary veins of those segments. **Indusium** rounded-reniform, its margin entire, wavy or slightly toothed, rarely with a few stalked glands. **Spores** ripening July-September. Tetraploid. 2n=164.

D. carthusiana may be distinguished from *D. dilatata* (a) by the colouring of the scales on the stipes: dark in the centre and pale at the margins in *D. dilatata* and uniformly pale throughout in *D. carthusiana*. This does not apply to immature looking forms of D. dilatata (q.v.) (b) by the margin of the indusium: usually fringed with glands in *D. dilatata*, very rarely so in *D. carthusiana*. (c) The relative length of the stipe to the rachis is also useful: $1/4$-$2/3$ in *D. dilatata*, stipe nearly as long as rachis in *D.carthusiana*. (d) In the field *D. carthusiana* can sometimes be recognised from a distance as it usually has fewer, more erect, paler green fronds than *D. dilatata*.

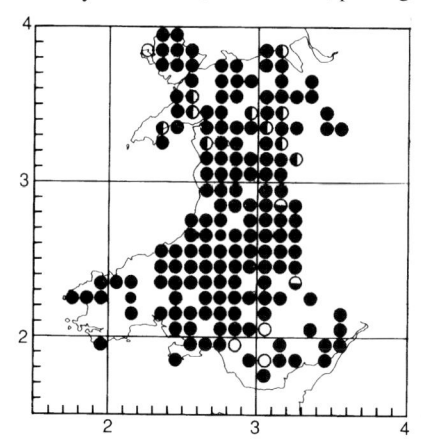

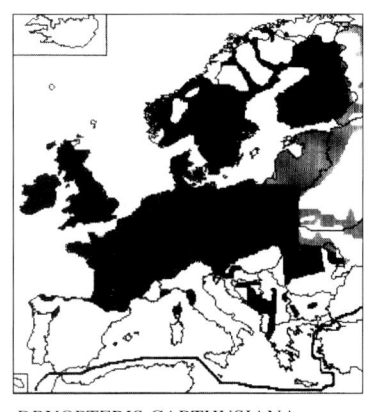

DRYOPTERIS CARTHUSIANA

The 'grey' area refers to records that may strictly be *D. carthusiana, D. dilatata* or *D. expansa*

Locally frequent in damp or marshy woods and damp heathland but often overlooked for *D. dilatata*. See under that species for immature forms confused with *D. carthusiana*.

First recorded, 1813; Davies, *Welsh Bot.*, pt 1, 98. Bodafon Mountain, Anglesey.

Throughout most of the British Isles, but much less common than *D. dilatata* with which it has often been confused.

Most of Europe south to N. Spain but absent from much of the Mediterranean. Western and central Siberia. Caucasus. N. America: across southern Canada to Newfoundland and NE USA.

DRYOPTERIS CRISTATA (L.) A.Gray

(*Lastrea cristata* (L.) C.Presl)

Crested Buckler-fern; Marchredynen Gribog

Rhizome decumbent or creeping. **Fronds** in an open tuft, sterile (shorter - up to 45cm - somewhat spreading) distinct from fertile (taller - up to 100cm - erect); **stipe** $1/2$ as long as sterile blade, $2/3$ to equally as long as fertile blade, when mature with broad pale scales at the base only. **Blade** linear-lanceolate, glabrous, pinnate; **pinnae** short triangular-oblong, shortly stalked, pinnatifid, the lowest pair not shorter than the next; **pinna segments** oblong, usually rounded at the apex, serrate-mucronate, attached by their whole width and connected at the base, the lowest pinnately lobed, the acroscopic and basiscopic segments of the lower pinnae nearly equal. **Fertile pinnae** with their stipes twisted so that their upper surfaces are directed towards the apex of the frond. **Sori** forming two rows about midway between the mid-vein and the margin of each segment. **Spores** ripening July-August. Tetraploid. 2n=164. *See* Fig. 108.

Bogs and wet heaths.

Very local and decreasing, East Anglia; rare elsewhere in SE England and Scotland.

Most of Europe from Finland and southern Scandinavia to Spain but excluding most of the Mediterranean. Commoner from Germany and the Alps eastwards to Rossia. Western Siberia. Scattered across southern Canada. Great Lakes to Newfoundland and southwards through the Appalachians.

DRYOPTERIS DILATATA (Hoffm.) A.Gray

(*Dryopteris austriaca* (Jacq.) Woyn.; *D. aristata* (Vill.) Druce; *Lastrea dilatata* (Hoffm.) C.Presl)

Broad Buckler-fern; Marchredynen Lydan

Stock stout, ascending or upright, almost stem-like, sometimes branched, the older parts thickly clothed with brownish-black dead frond bases, the younger parts with light brown lanceolate scales having a darker brown or almost black central stripe. **Fronds** (15-)30-150cm, tufted (shuttlecock-like when the stock is erect); **stipe** $^{1}/_{4}$-$^{2}/_{3}$ as long as rachis, dark purplish-brown below to straw-coloured or pale green above, densely clothed at the base, and somewhat less so higher up, with lanceolate light brown scales with a dark central stripe. **Blade** (10-)20-90×(7-)15-40cm, triangular- to lanceolate-ovate in outline (not narrowed at the base), dark green above and glandular on the veins, paler beneath, bi- or tri-pinnate; **pinnae** up to 25 on either side, stalked, the lower ones opposite or sub-opposite, the uppermost ones only becoming alternate, the lowest pair unequally triangular in outline and as long as or slightly longer than the succeeding pair, the remaining pinnae becoming less unequal and more nearly lanceolate; **pinnules** alternate, oblong-ovate to oblong-lanceolate, pinnate or pinnatifid, the basal ones stalked, the remainder sessile or decurrent, the lowest acroscopic pinnule of the lowest pinna usually somewhat shorter than the next succeeding pinnule and often only about half as long as the lowest basiscopic pinnule of the same pinna; **pinnulets** (and other segments of the third order) ovate to oblong, coarsely toothed, the teeth mucronate and incurved towards the apex of the pinnule; **rachis** with scattered scales and either with stalked glands or otherwise glabrous. **Sori** numerous, round, in two rows one on either side of the midrib, and somewhat nearer it than the margin of the pinnulets or pinnules, each individual sorus on the acroscopic fork of a secondary vein. **Indusium** reniform, prominent, convex, its margin usually dentate and sometimes glandular. **Spores** dark brown, with dense, obtuse tubercles, ripening July-September. Tetraploid. 2n=164. For points of distinction from *D. carthusiana* and *D. expansa* see under those species.

Fig. 62. Broad Buckler-fern (*Dryopteris dilatata*). Median pinnule from a pinna in the upper part of an average frond (×6). (*See also* Fig. 105C).

Immature looking forms of *D. dilatata* exist particularly in sheep-grazed upland areas round the bases of outcropping rocks, usually less than 30cm and with scales without any dark central venation. They have often been wrongly recorded as *D. carthusiana*. The broad triangular outline to the frond indicates *D. dilatata*.

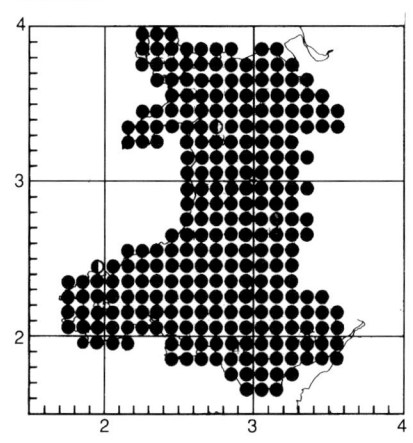

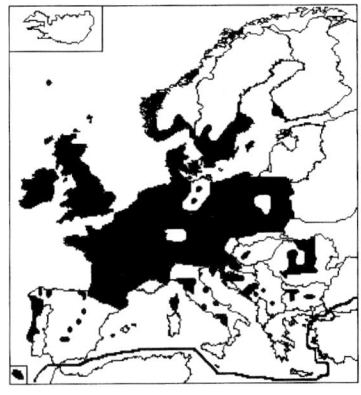

DRYOPTERIS DILATATA

Common in woods and thickets, and on hedge banks and shady mountain rock ledges.

First recorded, 1726; Herb. Dillenius. 'Cumglas [Cwm Glâs] in monte Snodon [Snowdon, Caerns.].' (Druce and Vines, *The Dillenian Herbaria*, 49, 1907.)

Throughout the British Isles but sometimes confused with the less common *D. carthusiana* and *D. expansa*.

Most of Europe from southern Scandinavia to Portugal but excluding much of the Mediterranean and south-eastern Europe. Rossia (N,E). Iceland and Faeroes. Turkey, Caucasus, western and central Siberia.

DRYOPTERIS EXPANSA (C.Presl) Fraser-Jenk. & Jermy

(*D. dilatata* var. *alpina* T.Moore; *D. assimilis* S.Walker)

Northern Buckler-fern; Marchredynen y Gogledd

As *D. dilatata* except: **Fronds** generally smaller, 7-60cm; **scales** on the stipe either uniformly light brown or with a dark central stripe, often more abruptly acuminate. **Blade** light green, thin. **Sori** usually relatively large, *c.*0.5-1.5mm across. **Spores** pale brown, with small, widely spaced, acute tubercles. Diploid. 2n=82.

The lowest acroscopic pinnule of the lowest pinna is usually about half as long as the lowest basiscopic pinnule of the same pinna, but this is not a reliable diagnostic character as it is too variable in both species.

Rock crevices in mountainous areas but much less common than *D. dilatata* with which it has sometimes been confused.

Fig. 63. Northern Buckler-fern (*Dryopteris expansa*). Median pinnule from third or fourth pinna of an average-sized frond (×6). (*See also* Fig. 106).

First recorded, 1895 (as *Lastraea dilatata* var. *alpina*); J. E. Griffith, *Flora of Angl. and Caern.*, 169, 'Cwm Glâs! Cwm Cywion. J. Ll. Williams. West Cliff of Carnedd Dafydd. Rev. A. Ley.'

Northern England, central and northern Scotland.

Northern Europe to the mountains of southern Europe, southwards to Portugal. Recently identified for the Netherlands. Iceland and Faeroes. Turkey. Possibly across the middle of Siberia. E. China, Korea, Japan. Kamchatka peninsula across through the Aleutian Is. to Alaska and down the western side of Canada and USA. Lake Superior north-eastwards to Labrador and Greenland.

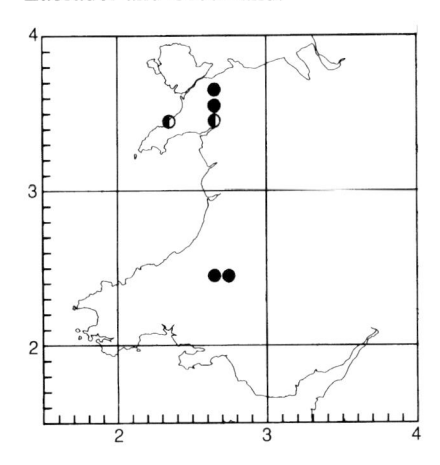

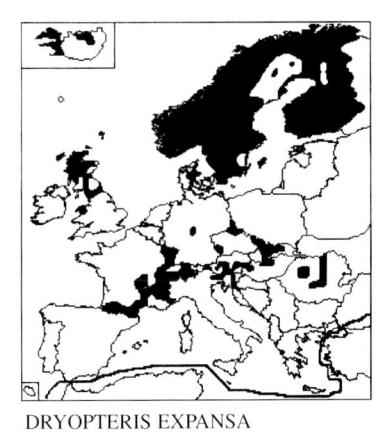

DRYOPTERIS EXPANSA

DRYOPTERIS FILIX-MAS (L.) Schott

(*Lastrea filix-mas* (L.) C.Presl)

Male-fern; Marchredynen

Stock stout, up to 20×2-2.5cm, erect or ascending, the older part densely covered with spirally set, blackish-brown dead frond bases, the younger parts thickly clothed with chaffy scales similar to those of the full-grown stipe. **Fronds** 35-150cm, arranged in shuttlecock-like tufts, dying in autumn; **stipe** $1/5$-$1/3$ as long as rachis sparsely or moderately clothed with pale-brown, linear-lanceolate to ovate-lanceolate, cordate, hair-pointed scales of various sizes up to 1.5×0.5cm. **Blade** 30-125×10-30cm, soft, dark green above, paler below, lanceolate, pinnate; **pinnae** about 20-35 on either side, sometimes sub-opposite at the base, otherwise alternate, sessile or nearly so, linear-lanceolate, truncate at the base, usually acuminate, deeply pinnatifid (almost to the rachis) the lowest parts of the lowest pinnae rarely pinnate; **pinna segments** alternate (except the lowest pair on each pinna), touching by their broadened bases, oblong, rounded at the apex, serrate (rarely deeply so or even lobed) along the margin (the teeth not spinulose); **rachis** and lower sides of midribs sparsely scaly, or sometimes without scales. Usually, only the upper third of the frond fertile. **Sori** on the acroscopic branches of the secondary veins and forming two lines near and parallel with the midrib of the segment. **Indusium** large, *c*.1.5mm diameter, convex, reniform, with a deep notch, usually not glandular, falling early often before the spores are shed, the margin lying flat on the surface of the blade when young (Fig. 60). **Spores** ripening July-September. Tetraploid. 2n=164.

Common in woods, woodland clearings and hedge banks.

First recorded 1726; Herb. Dillenius, 'prope Dolegelle [Dolgellau, Merioneth]'. (Druce and Vines, *The Dillenian Herbaria*, 47, 1907.)

Throughout the British Isles.

Most of Europe, Iceland and Faeroes. Western and central Asia. N. America: mostly scattered across the USA/Canada boundary. Greenland.

Fig. 64. Male-fern (*Dryopteris filix-mas*). Fertile upper pinna seen from beneath (about life-size). (*See also* Fig. 107*B*).

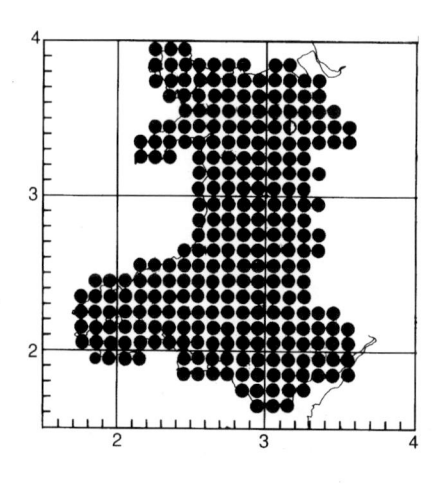

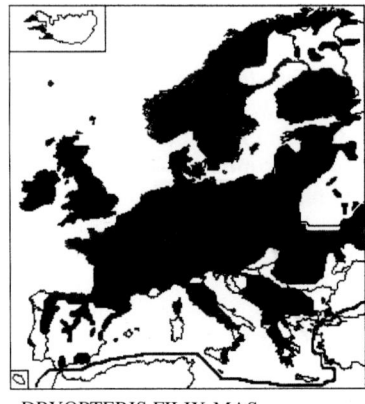

DRYOPTERIS FILIX-MAS

DRYOPTERIS OREADES Fomin

(*Dryopteris filix-mas* var. *abbreviata* (DC.) Newman; *D. abbreviata* auct., non DC.; *Lastrea propinqua* Woll., non J.Sm.)

Dwarf Male-fern; Marchredynen Fach y Mynydd

As *D. filix-mas* except: **Stock** when mature branched and tufted. **Fronds** usually 50cm or less, stiff; **stipe** $1/5$-$1/4$(-$1/3$) as long as rachis, clothed with pale-brown scales, the larger ovate, not hair pointed often curved at the base. **Blade**, when young, with minute glands beneath; **pinnae** with upcurled tips and edges resembling the hull of a rowing boat in transverse cross-section; **pinna segments** sinuate-crenate or crenately lobed, the teeth broad, obtuse, the apex often subentire, less deeply toothed, or with the teeth often appearing to be fanning outwards around the apex (rather than erect); lowest basiscopic pinna segment usually longer than its adjacent neighbour and bent outwards above the plane of the frond (resembling projecting steps up the rachis) and often partly overlapping the rachis; **rachis** with pale brown linear-to ovate-lanceolate scales. **Sori** 2-4(-8) on each side of the midrib of the longest pinna segments, though only 1 or 2 on each side of the midrib on the majority. **Indusium** usually less than 1mm diameter, the margin with minute glands on some sori and like similar species is tucked under at first, but in *D. oreades*, when the indusia uncurl and mature, some are seen to have a concentric outer rim when viewed directly from above (Fig. 60). **Spores** ripening July-September. Diploid. 2n=82.

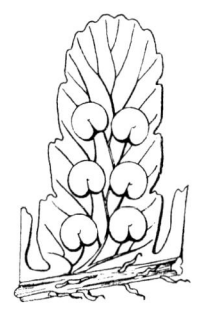

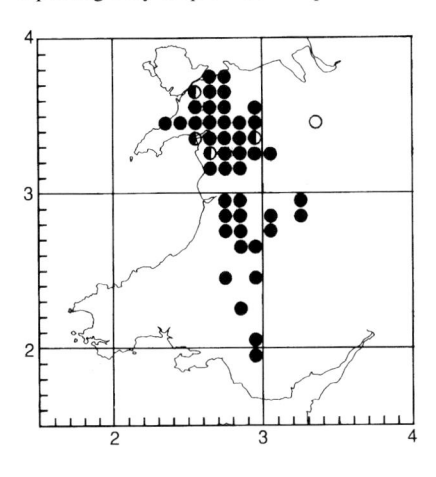

Fig. 65. Mountain Male-fern (*Dryopteris oreades*). Single fertile segment from a median pinna (×6). (*See also* Fig. 107C).

DRYOPTERIS OREADES

In rock crevices, block screes and on dry-stone retaining walls in mountain districts.

Northern England and Scotland. Very rare in Ireland.

W. Europe and west Central Europe southwards to Italy. Turkey and Caucasus. Full distribution inadequately known.

D. REMOTA (A.Braun ex Döll) Druce

(*D. woynarii* Rothm.)

Scaly Buckler-fern; Marchredynen Gennog

Intermediate between the putative parents (*D. affinis* and *D. expansa*) in frond-shape and dissection. Stipe and rachis clothed with golden to greyish brown scales with a dark central stripe. Presumed apogamous triploid hybrid derivative of *D. affinis* × *D. expansa*. 2n=123.

Extinct in the British Isles. Material from Galway and near Loch Lomond were examined cytologically. One of its parents, *D. expansa*, is not recorded from Ireland so its provenance there must remain uncertain. The material from Loch Lomond has both parents in the area but has not been refound. This controversial plant has been confused with *D. carthusiana* × *D. filix-mas* but German, Swiss and Hungarian records have been confirmed cytologically.

W. and central Europe extending eastwards to Rossia (W). Turkey.

DRYOPTERIS SUBMONTANA (Fraser-Jenk. & Jermy) Fraser-Jenk.

(*Dryopteris villarii* (Bellardi) Woyn. ex Schinz & Thell. subsp. *submontana* Fraser-Jenk. & Jermy; *Lastrea rigida* (Sw.) C.Presl)

Rigid Buckler-fern; Marchredynen Anhyblyg

Stock 10cm, horizontal to ascending, the older parts densely clothed with blackish-brown dead frond bases (the whole up to 5cm thick), and the younger with numerous bright brown narrow lanceolate scales. **Fronds** 20-60cm, numerous, tufted, erect or spreading, fragrant, in all their parts above and below more or less richly covered with short yellowish glandular hairs; **stipe** usually up to about ¹/₂ as long or rarely as long as the rachis, thickened and dark brown at the base, otherwise (like the rachis) straw-yellow to greenish-yellow, clothed (densely at the base, more sparingly so higher up) with lanceolate to ovate-lanceolate hair-pointed reddish-brown scales, the larger ones intermixed with numerous smaller and narrower, sometimes hair-like ones. **Blade** 15-40 × 5-13cm, lanceolate to narrow triangular in outline, dull green above, paler beneath, bipinnate; **pinnae** up to about 25 on

Fig. 66. Rigid Buckler-fern (*Dryopteris submontana*). Single pinnule from a median pinna (×6). (*See also* Fig. 103C).

either side, lower pairs opposite or sub-opposite, higher ones becoming alternate, the lowest pair the longest, the remainder gradually shortening to the apex, the lowest lanceolate to ovate-lanceolate, the remainder gradually narrowing to oblong-lanceolate, all shortly stalked, pinnate; **pinnules** opposite becoming alternate, the lowest very shortly stalked, the remainder sessile becoming decurrent, all in outline mostly oblong, obtuse or subacute, more or less deeply crenate-pinnatifid, the lobes typically rounded in outline with 1-6 acute but not spinulose teeth; **rachis** with numerous pale lanceolate or subulate scales. **Sori** large, round, arranged in two rows on each fertile pinnule, one on either side of the midrib, mostly situated on the lowest acroscopic tertiary venules, crowded and with the indusia touching or overlapping. **Indusium** convex, roundish kidney-shaped, its surface and margin glandular. **Spores** ripening July-August. Tetraploid. 2n=164. The British tetraploid is said to differ from the continental diploid (2n=82), *D. villarii* (Bellardi) Woyn. ex Schinz & Thell., in certain morphological characters. An artificial hybrid between the two proved to be sterile.

D. submontana is easily distinguished by its balsam-like fragrance, the numerous glands all over the frond and the neat crenately-lobed pinnules.

Crevices of limestone rocks in mountainous districts; very rare.

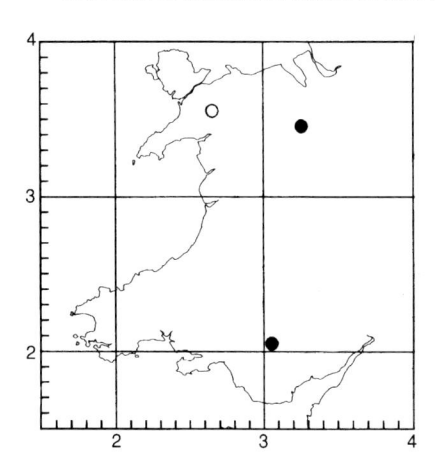

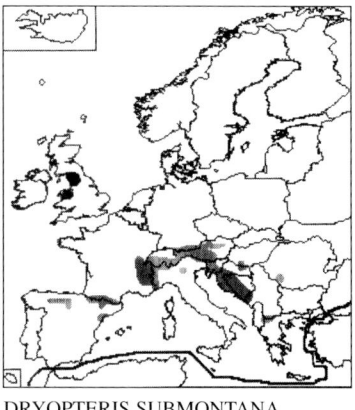

DRYOPTERIS SUBMONTANA

The mainland European distribution covers both *D. submontana* and *D. villarii*.

First recorded, 1868; G. R. Jebb, in *Science Gossip*, 139. 'North of Llangollen, Denbighshire.'

Local in NW England and one site in Derbys.

S. and south-central Europe.

HYBRIDS IN DRYOPTERIS

D. AEMULA × *D. DILATATA*

Resembling *D. dilatata* but with a more finely cut, glandular, hay-scented frond. The stipe is purplish at the base. Spores usually abortive.

British records have not been confirmed (Lake District, Kircudbrights., Argyll, Mid Ebudes and E. Mayo). Juvenile or immature plants from W. Scotland (Mull, v.c.103 and Soay, v.c.104) appear intermediate between the two species but such hybrids could only be distinguished from *D. dilatata* × *D. expansa* by the lack of chromosome pairing at meiosis.

D. AEMULA × *D. OREADES* = *D.* × *PSEUDOABBREVIATA* Jermy

Mull Fern; Marchredynen Mull

Rhizome branched, semi-erect, clad with pale rusty-brown scales. Frond 10-12cm, coumarin scented, oblong-lanceolate, mid- or yellowish-green, edges undulating. Stipe 2-3cm, purplish towards the base. Spores abortive. Diploid hybrid. 2n=82. Mull (v.c.103).

D. AFFINIS × *D. FILIX-MAS* = *D.* × *COMPLEXA* Fraser-Jenk.

(*D.* × *tavelii* auct., non Rothm.)

Hybrid Male-fern; Marchredynen Groesryw

Resembling *D. affinis*, but often more robust, with softer fronds and more strongly sickle-shaped pinnae. Pinna segments oblong-lanceolate, side-margins often toothed and apices rounded. Indusium flat, falling early. Spores mostly abortive and with a few large spores but the latter tend to fall early leaving what appear to be only abortive spores. Hybrids occur between *D. filix-mas* and two of the subspecies of *D. affinis* in the British Isles, but separating them morphologically is usually not possible: nothossp. *complexa* (with subsp. *affinis*), apogamous tetraploid, 2n=164; nothossp. *critica* Fraser-Jenk. (with subsp. *borreri*), apogamous pentaploid, 2n=205. What was thought to be the hybrid with subsp. *cambrensis* (nothossp.

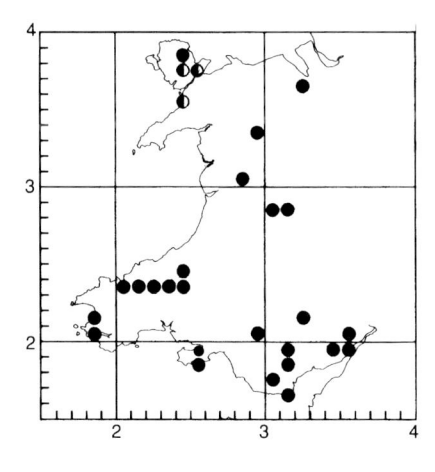

contorta Fraser-Jenk.), has since been shown to be a lobed form of nothossp. *complexa* and is an apogamous tetraploid. 2n=164.

Scattered records but probably throughout the range of the parents.

DRYOPTERIS × COMPLEXA

D. AFFINIS × D. OREADES

The specimens of this taxon at **NMW** have been redetermined as *Dryopteris affinis* subsp. *cambrensis*. The dried plants in the herbarium have the rich golden or brown scales, and a faint trace of the dark patch at the junction of the pinnae and the rachis of *D. affinis* subsp. *affinis*. Pinna segments have a rounded, not truncate, apex. They can also be more distinctly toothed. The lowest basiscopic pinna segment of each pinna is sometimes bent outwards above the plane of the frond (resembling projecting steps up the rachis) and often partly overlaps the rachis as in *D. oreades*. Apogamous.

D. CARTHUSIANA × D. CRISTATA = D. × ULIGINOSA (A.Braun ex Döll) Kuntze ex Druce

Hybrid Fen Buckler-fern; Marchredynen Gribog Groesryw

Resembles *D. cristata* in general appearance, differing in the bi- or tripinnate leaves and the greater acuteness of the lower pinnae. Tetraploid hybrid. 2n=164. Marshy places. Rare. East Anglia and formerly in other sites of *D. cristata*.

D. CARTHUSIANA × D. DILATATA = D. × DEWEVERI (J.T.Jansen) J.T.Jansen & Wacht.

Hybrid Narrow Buckler-fern; Marchredynen Gul Groesryw

Rhizome semi-erect, scales dark-striped, frond relatively narrow with the lowest pinnae shorter than the succeeding pair. Spores usually abortive. Tetraploid hybrid. 2n=164. Occurs frequently with the parents, sometimes apparently outnumbering both of them and probably much more frequent than the records suggest; especially in old and previously swampy woods, which are gradually drying out.

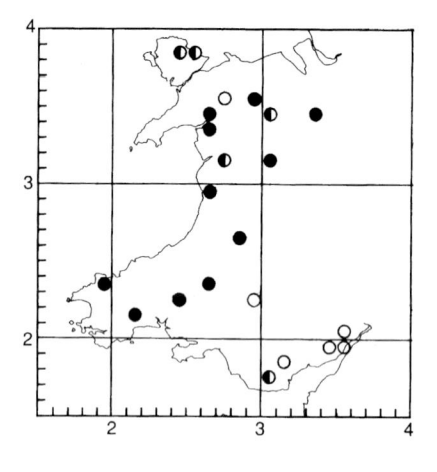

Scattered over Great Britain and Ireland.

DRYOPTERIS × DEWEVERI

D. CARTHUSIANA × *D. EXPANSA* = *D.* × *SARVELAE* Fraser-Jenk. & Jermy

Kintyre Buckler-fern; Marchredynen Kintyre

Triploid hybrid. 2n=123. Discovered in 1978. Occurs in Kintyre and Westerness in Scotland. Recorded from Finland.

D. CARTHUSIANA × *D. FILIX-MAS* = *D.* × *BRATHAICA* Fraser-Jenk. & Reichst.

(*Lastrea remota* sensu T.Moore excl. syn.)

Brathay Fern; Marchredynen Brathay

Resembling *D. filix-mas* in habit but intermediate in frond width and closer to *D. carthusiana* in the spinulose teeth on its pinnules. Sterile. Found once, by Lake Windermere. Long extinct there, but still in cultivation. Tetraploid hybrid. 2n=c.164. Recently found in France.

D. DILATATA × *D. EXPANSA* = *D.* × *AMBROSEAE* Fraser-Jenk. & Jermy

Gibby's Buckler-fern; Marchredynen Gibby

Resembling *D. dilatata* in its dark green frond and the dark central stripe on the scale but with the long basiscopic pinnule on the lowest pinnae usually found in *D. expansa*. Spores abortive. Difficult to identify because of the close affinity of the parents. Triploid hybrid. 2n=123. First recorded above Nant Ffrancon (SH66, Caerns.) by R. H. Roberts about 1976. In Scotland, scattered mostly on the western side; also in the Lake District.

D. DILATATA* × *D. FILIX-MAS* = *D.* × *SUBAUSTRIACA Rothm.

Resembling *D. remota* except in the much less scaly stipe and rachis and in the triangular-oblong blade, not narrowed at the base. Pinnae ovate-lanceolate as in *D. carthusiana* × *D. filix-mas* but shorter; pinna segments usually ovate-oblong, obtuse, pinnatifid; scales of stipe and rachis lanceolate, acuminate, with a dark centre. Spores abortive. Never confirmed for the British Isles. The record from S. Devon has not been confirmed and other records were shown to have arisen from plants of a different parentage (C. A. Stace ed., 1975).

D. FILIX-MAS* × *D. OREADES* = *D.* × *MANTONIAE Fraser-Jenk. & Corley

Manton's Male-fern; Marchredynen Manton

Resembling *D. filix-mas* but intermediate between the two parents in size, the margin of the indusium slightly inflexed, the spores all, or nearly all, abortive. Triploid hybrid. 2n=123. This triploid hybrid was also produced artificially by Prof. Manton in 1939.

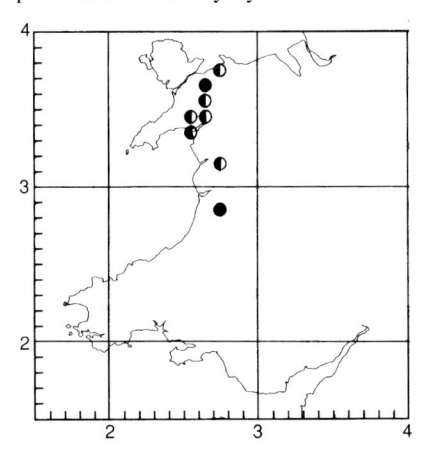

Scattered records in Lake District and Scotland.

DRYOPTERIS × MANTONIAE

Genus GYMNOCARPIUM Newman
Oak Ferns - Llawredyn y Derw

Rhizome widely creeping, slender. Fronds distant; stipe long, erect, sparsely scaly at the base only; blade deltoid or pentagonal, prominently jointed to the rachis and bent back, either simple deeply pinnatifid or compound, bi- or at the base even tri-pinnate, the lowest pair of pinnae much the largest and the most divided; veins forked, free. Sori circular to ovate-oblong, without indusia. 5 species. Europe, North America and south-eastern Asia.

KEY TO SPECIES

Each pinna of basal pair nearly as large as the apical pinna giving a symmetrically tripartite appearance; non-glandular **G. dryopteris**

Each pinna of basal pair noticeably smaller than the apical pinna giving an elongated tripartite appearance; glandular and smelling of apples, especially when crushed **G. robertianum**

GYMNOCARPIUM DRYOPTERIS (L.) Newman

(*Carpogymnia dryopteris* (L.) Á. Löve & D. Löve; *Phegopteris dryopteris* (L.) Fée; *Thelypteris dryopteris* (L.) Sloss.; *Dryopteris linnaeana* C.Chr.)

Oak Fern; Llawredynen y Derw

Rhizome creeping, slender, up to 20cm × 1-2mm, sparingly branched, and with few roots, black, glossy, the younger parts only clothed with pale brown ovate scales. **Fronds** 100cm, arising singly at irregular intervals from the rhizome, when very young resembling 3 little balls on wires; **stipe** 1.5-3 times as long as the rachis, slender, brittle, blackish-brown at the extreme base, otherwise straw coloured, with a few ovate-lanceolate pale brown scales at the base, otherwise glabrous. **Blade** 5-15 × 6-17cm, bent back into

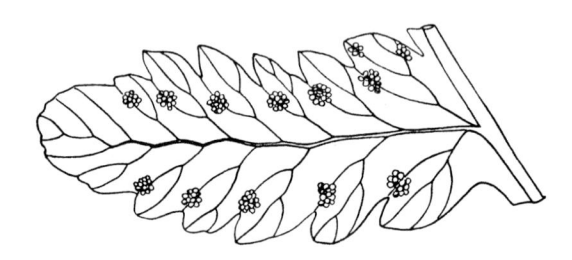

Fig. 67. Oak Fern (*Gymnocarpium dryopteris*). Third basiscopic pinnule of second pinna (×5.5). (*See also* Fig. 109*A*).

a horizontal position at right-angles to the vertical stipe, broadly triangular, bright yellowish-green, thin, glabrous, tripinnate, changing rapidly upward to simply pinnatifid; **pinnae** about 6 distinct pairs, opposite throughout, the lowest pair of pinnae distant from the others, long stalked (their stalks conspicuously jointed to the main rachis) and much the largest (often almost as large as the rest of the blade) unequally ovate in outline (wider on the side away from the apex of the frond, the lowest pinnule on this side being as large as the pinnae of the third pair), acute or obtuse, bipinnate, the second pair of pinnae stalked or sessile, attached by a knotty joint, oblong to oblong-lanceolate in outline, pinnate; the remaining distinct pinnae sessile and becoming decurrent, all acute or obtuse, pinnate becoming pinnatifid; **pinnules** of first pair of pinnae pinnate, their pinnulets ovate to oblong, entire or crenate-margined or pinnately toothed, rounded at the apex; pinnules of the upper pinnae resembling the lowest pinnulets of the first pair; **rachis** glabrous. **Sori** circular or the lowest ones somewhat elongated, situated fairly close to the margin, on the secondary veins, or, if these are forked, on the acroscopic branch or on both, the sporangia sometimes running together to form an apparently continuous band. **Indusium** absent. **Spores** ripening July-August. Tetraploid. 2n=160.

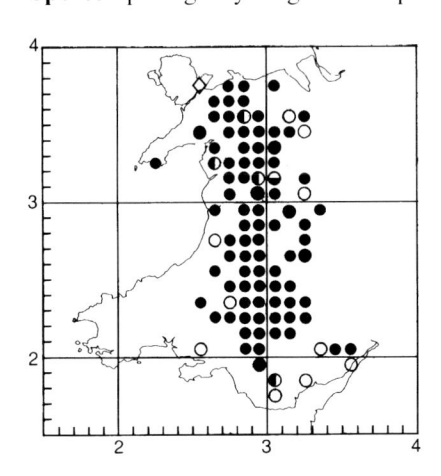

GYMNOCARPIUM DRYOPTERIS

Rocky woods, shaded rocky banks of streams, by waterfalls and on screes; locally frequent. First recorded, 1662 Ray, Itinerary III (*Memorials* ed. E. Lankester, 179, 1846). 'Near Tintern.'

Great Britain: north and west of line from Severn estuary to N Yorkshire Moors. Rare SE of this line and in Ireland.

Protected in Northern Ireland under the *Wildlife (NI) Order, 1985*.

Most of Europe from Scandinavia to N. Spain, including Russia but excluding much of the Mediterranean. Iceland and Faeroes. Asia: Turkey and round the Caucasus. Afghanistan and W. Himalaya. Across the middle of Siberia. China. Japan and across to Alaska and much of Canada to Newfoundland. NE USA and isolated sites elsewhere in USA. Greenland.

GYMNOCARPIUM ROBERTIANUM (Hoffm.) Newman

(*Carpogymnia robertiana* (Hoffm.) Á. Löve & D. Löve; *Phegopteris robertiana* (Hoffm.) A. Braun; *Thelypteris robertiana* (Hoffm.) Sloss.)

Limestone Fern; Limestone Polypody; Llawredynen y Calchfaen

Rhizome creeping, relatively stout, up to 20cm × 3.5mm, sparingly branched and with a few roots, the younger parts with lanceolate brown scales and golden-brown woolly hairs. **Fronds** 15-55cm, arising singly at irregular intervals from the rhizome; **stipe** 1.5-2 times as long as the rachis, soft and sappy when young becoming firm, dark brown and with a few brown lanceolate scales at the extreme base, straw coloured and rarely with scattered scales above, glandular when young. **Blade** 5-22 × 5-22cm, somewhat bent back at an angle to the vertical stipe, triangular in outline,

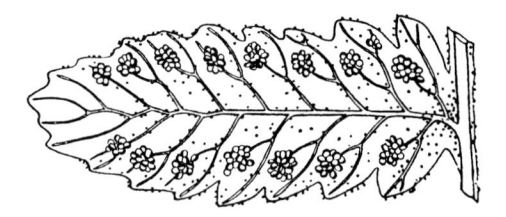

Fig. 68. Limestone Fern (*Gymnocarpium robertianum*). Second acroscopic pinnule of second pinna (×5.5). (*See also* Fig. 109*B*).

dull green, glandular-mealy, bi- or tri-pinnate changing rather rapidly upward to pinnatifid; **pinnae** opposite throughout, about 5-10 distinct pairs; the lowest pair of pinnae rather distant from the rest, stalked (their stalks strongly jointed to the main rachis) and much the largest, unequally ovate-lanceolate in outline (wider on the side away from the frond-apex, the lowest pinnule on this side being as large as the pinnae of the fifth pair) acute to almost obtuse, simply pinnate or at the base bipinnate; the second pair of pinnae usually stalked and well jointed to the rachis, broadly oblong-lanceolate, becoming linear, obtuse becoming rounded at the apex, pinnate or pinnatifid; **pinnules** of the first pair of pinnae pinnate or deeply pinnatifid; ultimate segments oblong, rounded or truncate, and entire or crenate at the apex; **rachis** glandular. **Sori** small, round, forming a row on either side fairly near the margin. **Indusium** absent. **Spores** ripening July-August. Tetraploid. 2n=160.

Rocky woods and ledges, stony slopes and screes and on walls; frequent in limestone districts.

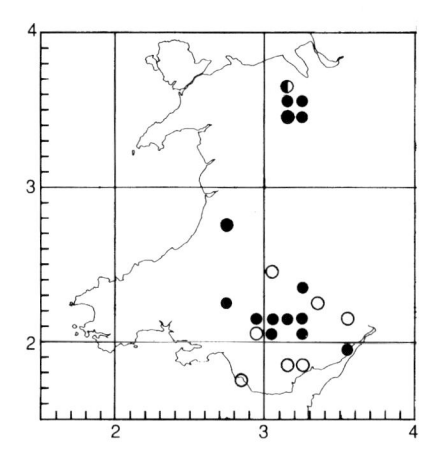

GYMNOCARPIUM ROBERTIANUM

First recorded, 1805; J. W. Griffith in Turner and Dillwyn, *Bot. Guide*, I, 96. Cwm Idwal, Caern. This is queried as a probable error. The first certain record is in 1833; W. Christy, in *Mag. Nat. Hist.*, VI, 52. 'A steep limestone declivity, on the left of the road just after leaving Llangollen, was completely covered with Polypodium calcareum.'

Protected in the Republic of Ireland under the *Flora Protection Order, 1987*.

Limestone districts. England: local; north of a line from N. Somerset to Thames estuary, and as far north as Lake District. Rare in Scotland and Ireland.

Most of Europe from Scandinavia to N. Spain. Corsica, Italy, the Balkan Peninsula and Rossia. Asia: Caucasus. Afghanistan and W. Himalaya. W. Siberia but needing confirmation across the rest of Siberia. China. Japan. N. America: from Alaska through to the Great Lakes and the Gulf of St Lawrence. Less frequent world-wide than *G. dryopteris*.

Hybrids between *G. dryopteris* and *G. robertianum* have been reported in Sweden, Finland, Rossia and North America.

Genus HYMENOPHYLLUM Sm.

Filmy-ferns - Rhedynach Teneuwe

Indusium deeply divided into two valves or lips (Fig. 69). Receptacle included inside the indusium (not projecting beyond it as in *Trichomanes*).

Small delicate ferns of very uniform appearance, closely resembling the leafy liverworts and mosses; rhizome usually thread-like, creeping; fronds in two pendulous, usually crowded to form cushion-like masses, dull

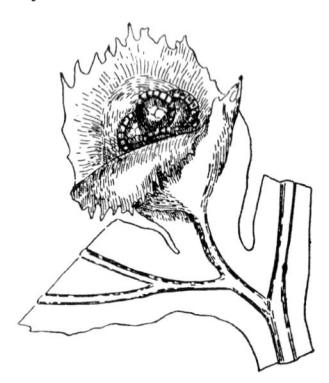

Fig. 69. Ripe sorus of Tunbridge Filmy-fern (*Hymenophyllum tunbrigense*) showing the two lips of the indusium, one of which has been slightly bent back in order to display the sporangia ($c.\times 15$).

brownish-green, translucent, simply or bi- to quadri-pinnatifid, the lobes narrow, elongated and ribbon-like. About 25 species. Chiefly in Tropical rain forests, extending far into the South Temperate zone, a few reaching as far north as Japan in the east, and the Atlantic coasts of Europe in the west.

For discussion of the gametophyte see under *Trichomanes*.

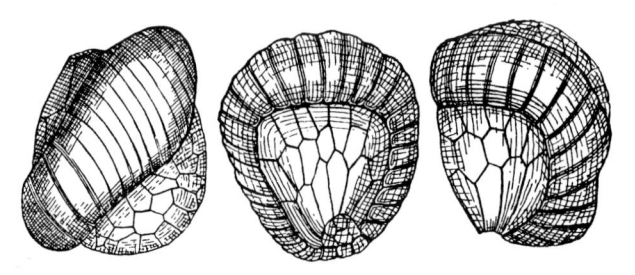

Fig. 70. Sporangium of a Filmy-fern (*Hymenophyllum tunbrigense*). Left: showing the stomium; middle: the same sporangium turned through about ninety degrees to the left; right: the same from the side away from the stomium. (After Luerssen).

KEY TO SPECIES

Leaves blue-green; blade usually elliptic-oblong hanging laxly over substrate; veins ending just short of leaf-margin (×10 lens); lips of indusium toothed; sorus held in plane of frond **H. tunbrigense**
Leaves olive-green; blade usually narrowly elliptic-lanceolate arching away from substrate; veins mostly reaching leaf-margin (×10 lens); lips of indusium entire; sorus held out from plane of frond **H. wilsonii**

Keys to sterile sporophyte material and gametophyte material of *Hymenophyllum* and *Trichomanes* in the British Isles have been published (Rumsey, Raine & Sheffield, 1993).

HYMENOPHYLLUM TUNBRIGENSE (L.) Sm.

Tunbridge Filmy-fern; Rhedynach Teneuwe

Fronds stalked, 2.5-11.5cm long, blue-green, more or less regularly divided on either side of the rachis into rhomboid or linear-rhomboid flat spreading segments (usually 5-11 on the largest pinnae), having spinulose teeth near the apex; **veins** ending slightly short of the margin, but near to the tip in young, vigorously expanding segments. **Sori** developed mostly in the upper half of the frond and in the angles between the main lobes and the axial portion of the lamina. **Indusium** a flattened pocket, slit ³/₄ of the way down into two orbicular valves which are irregularly toothed on their upper edges. **Spores** ripening June-July. 2n=26.

In mountainous districts on moist shady acidic rocks and rock ledges, more rarely on tree trunks. Rare or local.

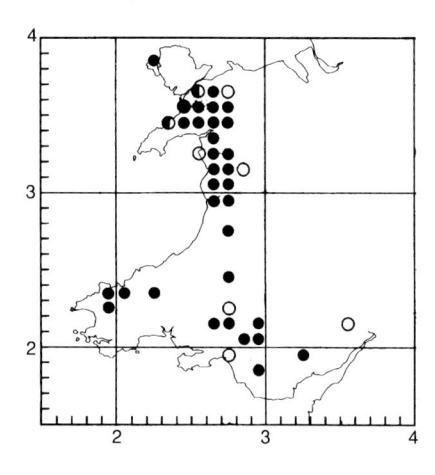

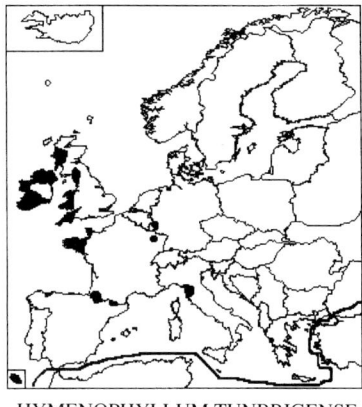

HYMENOPHYLLUM TUNBRIGENSE

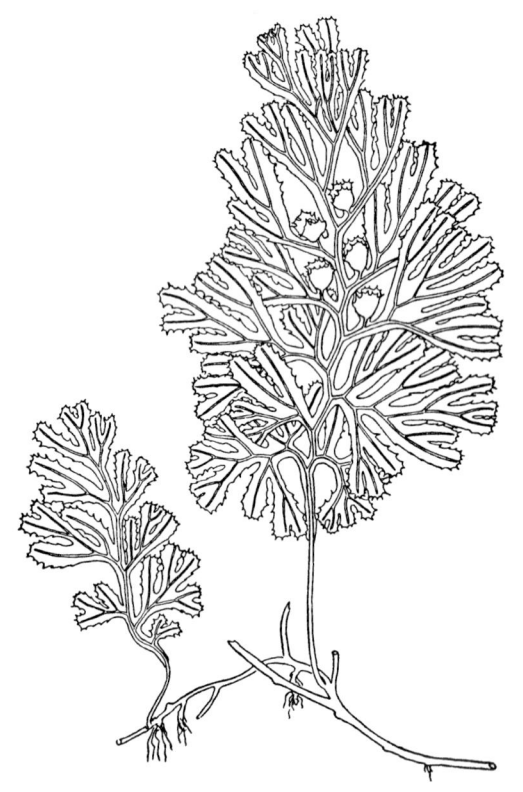

Fig. 71. Tunbridge Filmy-fern (*Hymenophyllum tunbrigense*), rather more than twice natural size.

First recorded, 1724; Dillenius in Ray, *Syn.*, edn 3, 123. '... on the left Hand as soon as you enter the Mountains to go to the old *castle*, near *Lhanperis* [Llanberis, Caerns.].' As the two British species were not distinguished until long after Dillenius' day, the record is ambiguous: it may refer either to *Hymenophyllum tunbrigense*, or to *H. wilsonii* which grows abundantly below Dolbadarn Castle (the 'old castle' referred to).

On the western side of Great Britain northwards to the inner Hebrides and in E. Sussex; in Ireland, chiefly in the west.

Cosmopolitan but markedly discontinuous. The Atlantic fringe of Europe with small outliers in Germany and the Mediterranean (Corsica and central Italy). Azores, Madeira and Canary Is. Africa: in Kenya, South Africa, Madagascar and Mauritius. Turkey. Caucasus. Only in S. Carolina, USA. Mexico, W. Indies and Central America. S. America: in Venezuela, Chile and NW Argentina. Temperate Australia and Tasmania. New Zealand.

HYMENOPHYLLUM WILSONII Hook.

(*H. peltatum* auct.)

Wilson's Filmy-fern; Rhedynach Teneuwe Wilson

Fronds very similar in most respects to those of *H. tunbrigense*, but more rigid, of a somewhat darker olive-green and recurved at the tip and margin, whereas those of *H. tunbrigense* are flat, main lobes of the lamina sometimes undivided or, if divided, the resulting segments usually 3-5 in number on the largest pinnae and situated on the sides of the lobes, towards the apex of the frond, giving the lobes a one-sided appearance (such one-

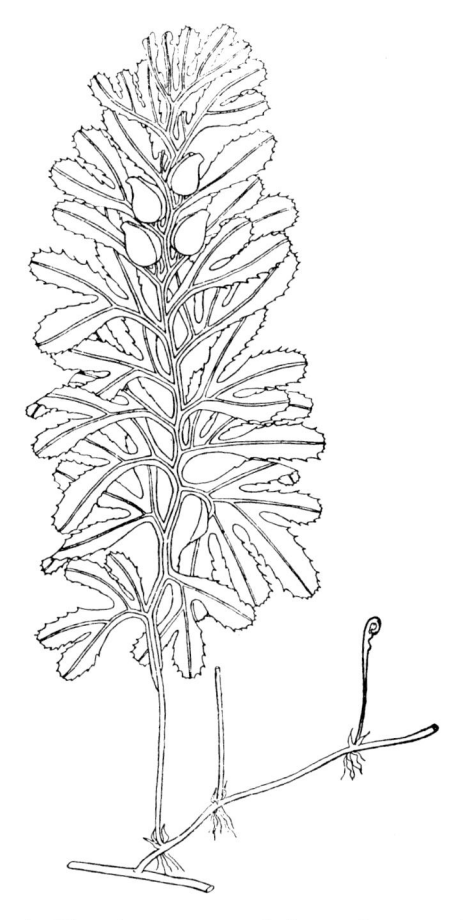

Fig. 72. Wilson's Filmy-fern (*Hymenophyllum wilsonii*), rather more than twice natural size.

sided lobes occur in *H. tunbrigense*, especially in the upper part of the frond, but in that species the lowest lobes are equally divided whilst in *H. wilsonii* all the lobes are unilateral); **veins** extending to the tips of the segments. **Indusium** more ovoid than in *H. tunbrigense* and projecting from the margin of the frond so as to appear stalked; margins of the valves entire. **Spores** ripening June-July. 2n=36.

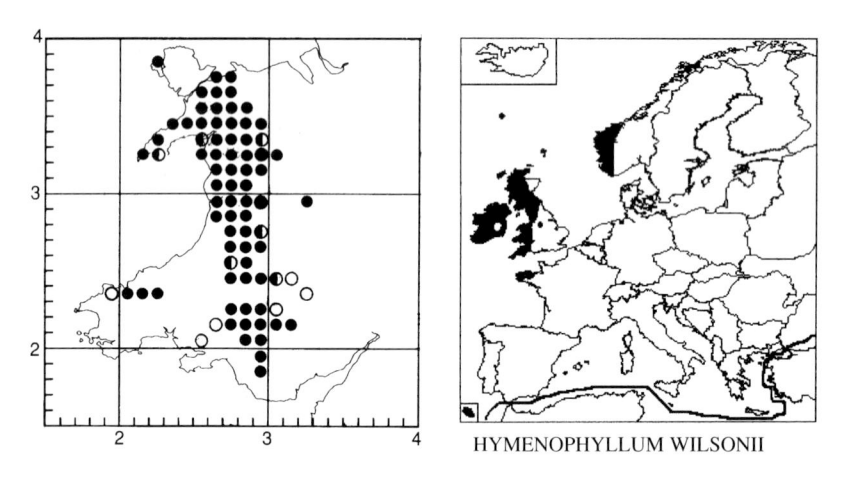

HYMENOPHYLLUM WILSONII

Grows in similar situations to the last species, but much more frequently, probably because it is less demanding in its requirements for shade and moisture. The two are sometimes found growing together.

First recorded in 1830; Hooker, *Brit. Flora*, edn 1, *1*, 450. 'Wales.'

In the south-west of England and throughout the mountainous districts of Great Britain to Shetland; in Ireland, mostly in the western half.

Cosmopolitan but markedly discontinuous. Western Norway, NW France, Iceland and Faeroes. Azores, Madeira and Canary Is. Tristan da Cunha. Central Africa, Réunion and Kerguelen. Chile including Cape Horn.

Genus MATTEUCCIA Tod.

Ostrich Ferns - Rhedyn Estrys

Stock stout, crowned with pinnate blades with lobed or pinnatifid pinnae; dimorphic, the fertile fronds much smaller, less dissected and longer stalked than the sterile. Sori globose, contiguous in longitudinal rows, covered by the inrolled, coriaceous margin of the blade. 3 species, North Temperate.

MATTEUCCIA STRUTHIOPTERIS (L.) Tod.

Ostrich Fern; Rhedynen Estrys

Stock short, erect, but with long underground rhizomes. **Sterile fronds** short-stalked, with large lanceolate scales at the base; **blade** bright green, broad-lanceolate, up to 170cm long and 35cm across, not persistent; **pinnae** 30-70, pinnatifid superficially resembling *Dryopteris filix-mas*. **Fertile fronds** up to 60cm long, with a long stipe densely scaly at the base; **pinnae** densely arranged, up to 6cm long, linear and subcylindrical due to the strongly inrolled margins, persistent, becoming dark brown. **Indusium** cup-shaped, thin, soon obsolete. 2n=80. *See* Fig. 113, C.

Naturalized in woods. Two sites in Wales: Llanelli (SN40, Carms.) and Lisvane, Cardiff (ST18, Glam.), the latter habitat destroyed by housing development in 1996.

Grown in gardens; naturalized in shady places in Scotland, NW England and N Ireland.

Native to Europe except in the west. Asia: Turkey through the Caucasus to Iran. Across the middle of Siberia. NE China and Japan. N. America: Alaska, Canada and NE USA.

Genus ONOCLEA L.

Sensitive Fern - Rhedyn Croendenau

A genus with a single species. Characters as for the species.

ONOCLEA SENSIBILIS L.

Sensitive Fern; Bead Fern; Rhedynen Groendenau

Rhizome creeping. **Sterile fronds** up to 100cm, with a long stipe and a broad, ovate-triangular blade and broadly winged rachis; **pinnae** 8-12 pairs, up to 8cm wide, deeply lobed to sinuate or entire. **Fertile fronds** with a long stipe terminated by a shorter, lanceolate fertile zone of short, erect pinnae with bead-like, globose segments formed by the inrolled margins enclosing small groups of sori, becoming dark brown. 2n=74. *See* Fig. 113, D.

Grown in gardens; occasionally naturalized in wet ground in western Scotland, NW England and the Channel Islands.

Native to eastern North America and eastern Asia.

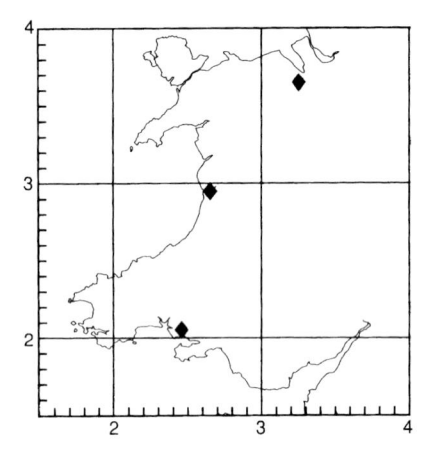

ONOCLEA SENSIBILIS

Genus OPHIOGLOSSUM L.

Adder's-tongues - Tafodau y Neidr

Sterile leaf-blade usually simple and linear to ovate in outline with an entire margin, but in some species forked or palmately lobed, and in one (purely saprophytic) species absent altogether; fertile spike usually simple but sometimes forked or pinnately branched; sporangia large, sunken, forming two closely set marginal rows one on either side of the midrib; dehiscence by transverse slits.

Small terrestrial herbs with short upright, unbranched stock usually growing underground and bearing one or several fronds. About 50 species, mostly local but together covering all regions (some larger epiphytes in the Tropics).

KEY TO SPECIES

1 Leaves appearing in autumn; dying down in spring; sterile leaf-blade linear to narrowly elliptic, 1.5-4(-6)mm wide, texture thick fleshy to almost waxy; spores shed in early spring **O. lusitanicum**
Leaves appearing in spring; dying down in summer or autumn; sterile leaf-blade broadly elliptic-lanceolate to ovate, wider than 4mm, texture thin to fleshy; spores shed in summer **2**

2 Sterile leaves convex, reflexed to almost horizontal; leaves usually arising in pairs; sporangia 6-14(-20) either side of spike **O. azoricum**

Sterile leaves flat, held erect; leaves arising singly; sporangia 10-45 either side of spike **O.vulgatum**

OPHIOGLOSSUM AZORICUM C.Presl

(*O. vulgatum* subsp. *ambiguum* (Coss. & Germ.) E.F.Warb.; *O. vulgatum* subsp. *polyphyllum* auct., non A.Braun)

Small Adder's-tongue; Tafod y Neidr Bach

Fronds (l)2 or 3 in number, 4-10cm tall, all or 1-2 fertile, seldom all sterile. **Sterile blade** 0.9-4(-5)×0.4-1.4(-3)cm, broadly elliptic-lanceolate to ovate, apex usually acute, narrowed at the base (cuneate); fleshy. **Fertile spike** up to twice as long as the blade, with about 6-14(-20) sporangia on each side. **Spores** ripening May-August. 2n=720.

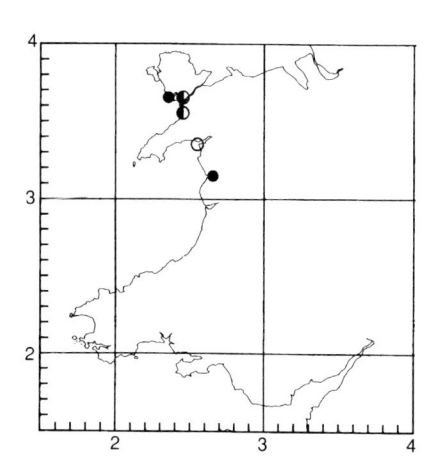

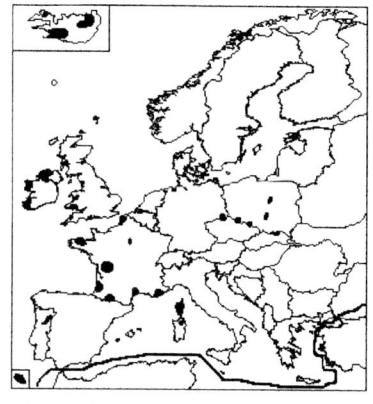

OPHIOGLOSSUM AZORICUM

First recorded, 1879; Charles Bailey, *Botl Exch. Club Br. Isl. Report* for 1879, 22 (1880). 'Damp sandy ground ... west of Dyffryn railway station, Merionethshire.'

Sandy ground and short turf near the sea. Rare on coastal islands and peninsulas from the Channel Islands to Scotland.

Western Europe and very local in central Europe. Iceland, Azores, Madeira, Canary Is. and Cape Verde Is.

OPHIOGLOSSUM LUSITANICUM L.

Least Adder's-tongue; Tafod y Neidr Lleiaf

Fronds usually 2-3 in number, 10cm tall at the most, usually much less; appearing in November (those of *O. vulgatum* in June) and dying down in April. **Sterile blade** 8-23(-38)×1.5-4(-6)mm linear to narrowly elliptic, narrowing to the base; thick, very fleshy (almost waxy). **Fertile spike** with

5-10 sporangia on each side. **Spores** ripening February-March. 2n=250-260. Scilly, Channel Islands and Mediterranean.

Protected in Great Britain under the *Wildlife and Countryside Act, 1981.*

Mainly coastal in Western Europe and the Mediterranean region. Caucasus.

OPHIOGLOSSUM VULGATUM L.

Adder's-tongue; Tafod y neidr

Stock underground, short, upright, covered above by the bases of old fronds and beset below with thick fleshy horizontal, usually unbranched, roots, some of which elongate like stolons and produce at the apex buds from which new plants arise. **Fronds** 1 or rarely 2 in number, 8-30cm overall with a hollow free-standing base enclosing a bud which will give rise to next year's frond; **common stalk** up to 15(-18)cm, more or less half underground. **Sterile blade** 3-15(-30) × 2-6cm, broadly ovate-lanceolate to ovate, widest below the middle, apex obtuse or subacute, base attenuate to more or less cordate, often clasping the fertile spike; glabrous, fleshy, easily wilting. **Fertile spike** 2-15(-27) × 0.2-0.3cm simple, usually much overtopping the sterile blade, the upper fertile portion 1-5(-7)cm bearing (12-)16-45 sunken **sporangia** on either side and ending in a drawn-out sterile point. **Spores** ripening May-August. 2n=480.

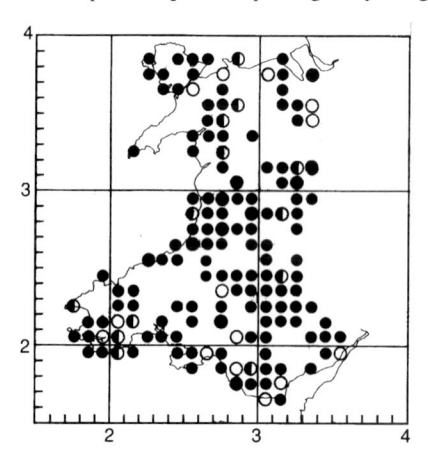

OPHIOGLOSSUM VULGATUM

First recorded in 1606-08 by Sir John Salusbury: MS. record in his copy of Gerard's Herbal now in the Library of Christ Church, Oxford. (*See* Gunther, R.W.T.: *Early British Botanists*, Oxford, 1922). 'The Herbe

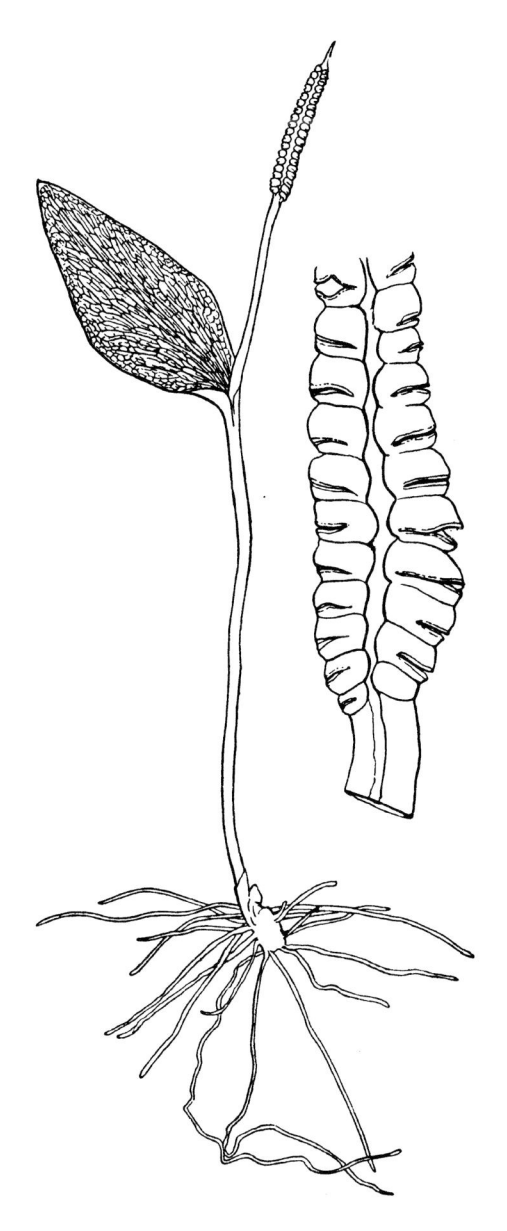

Fig. 73. Adder's-tongue (*Ophioglossum vulgatum*), two-thirds natural size. On the right: portion of fertile spike (×5).

Addertongue groweth likewise in the lower end of Ravenscroftes field [nr. Denbigh]'.

Throughout most of Great Britain and Ireland but less frequent in Scotland; Channel Islands.

The species has been split into other species, subspecies and varieties throughout its range and therefore only the distribution *sensu lato* can be given here. Circumpolar but with wide gaps. In the greater part of Europe except the extreme north. Africa: scattered on west side from Algeria to São Tomé, tropical east Africa. Asia: scattered, including Caucasus, western Siberia, India, Japan and Kamchatka peninsula. N. America: mostly eastern half of USA. Aleutian Is., Canada, Mexico.

Genus OREOPTERIS Holub

Lemon-scented Ferns - Marchredyn y Mynydd

Rhizome erect, with thin, scales. Fronds in tufts, pinnate, narrowing gradually towards the base, the lower pinnae very short; pinnae pinnatifid, veins free, simple or forked, the lower ultimate venules not quite reaching the margin; sessile spherical glands present on the lower surface. Indusia present. Number of species uncertain. North temperate regions.

OREOPTERIS LIMBOSPERMA (Bellardi ex All.) Holub

(*Thelypteris limbosperma* (All.) H.P.Fuchs; *T. oreopteris* (Ehrh.) Sloss.; *Dryopteris oreopteris* (Ehrh.) Maxon; *Lastrea oreopteris* (Ehrh.) Bory)

Lemon-scented Fern; Mountain Fern; Marchredynen y Mynydd

Stock about 10cm long, stout, branched, sometimes caespitose, ascending, the older parts clothed with the bases of dead fronds, the younger parts with brown, ovate to lanceolate, acuminate scales. **Fronds** tufted, 30-90cm or more, fragrant when bruised, with a lemon or balsam-like odour; **stipe** short, about ¹/₅ as long as rachis , rather sparsely clad with pale-brown ovate or ovate-lanceolate, acuminate scales. **Blade** 25-75×8-25cm lanceolate in outline, tapering at both ends, often bright yellowish-green, pinnate; **pinnae** about 20-30 on either side, opposite or sub-opposite at the base, becoming alternate towards the apex, linear-lanceolate (except the lower pairs which are very short and triangular) with numerous sessile golden-yellow glands beneath, sessile, pinnatifid; **pinna segments** oblong, equal-sided, obtuse or slightly sickle-shaped, acute, with the apex incurved towards that of the pinna, their margins slightly and irregularly crenate, the lowest acroscopic segments the longest and curved towards the main rachis; **rachis** and **midribs** with fine white glandular hairs intermixed (especially in

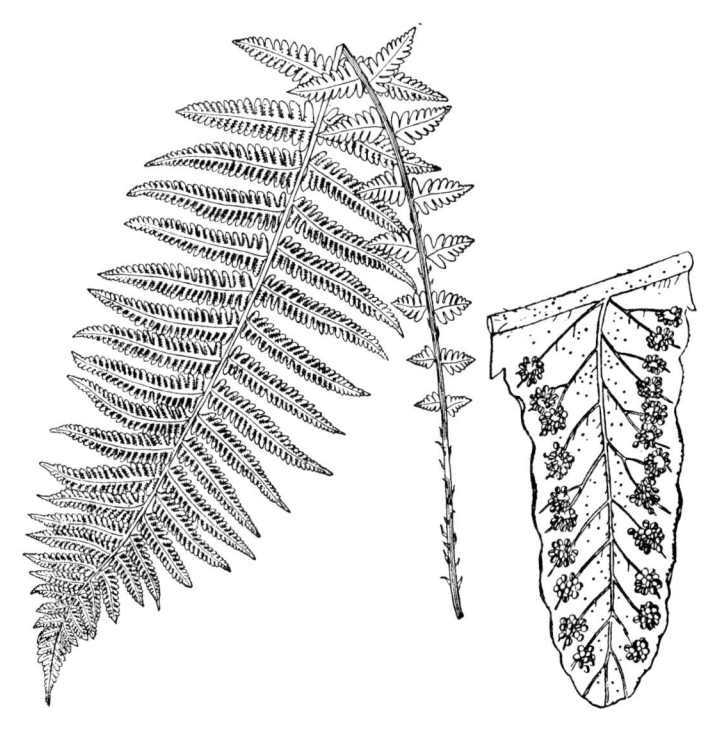

Fig. 74. Lemon-scented Fern (*Oreopteris limbosperma*). Left: single frond from an average-sized plant (×1/4). Right: single segment of median pinna (×5.5).

their lower parts) with a few scales and jointed hairs. **Sori** small, borne on the vein endings (on both branches when the secondaries are forked) forming a row near the margin of the segment. **Indusium** small, irregular in shape, dentate along the margin, often imperfect, always disappearing early and sometimes absent altogether. **Spores** ripening July-August. Diploid. 2n=68. For comparison with *Thelypteris palustris* see that species.

Stream sides and damp woods; common in mountainous districts.

First recorded, 1726; Dillenius, MS. diary of his journey into Wales. (Druce and Vines, The Dillenian Herbaria, xlix, 1907). Cader Idris.

Throughout the British Isles but local in Ireland and absent from much of southern central England and East Anglia.

Through most of Europe from western Scandinavia, just inside the Arctic Circle, to northern Spain, Rossia (W). Azores and Madeira. Asia: Turkey, Caucasus, northern Iran, outlier in central Siberia. Japan and Kamchatka

160　OSMUNDA

peninsula. Eastwards to western North America, from Alaska to Washington State, USA.

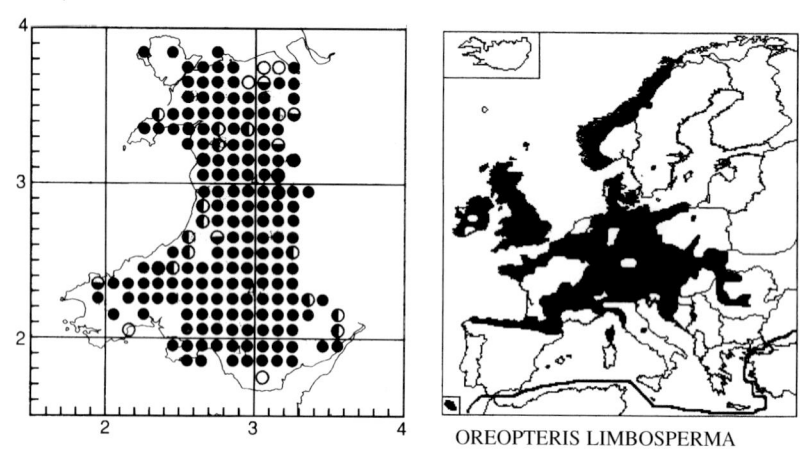

OREOPTERIS LIMBOSPERMA

Genus OSMUNDA L.

Royal Ferns - Rhedyn Cyfrdwy

Fertile leaves or pinnae distinct from the sterile ones and also more richly branched, the sporangia confined to the upper or middle part of the frond or occupying the whole of it and borne on ultimate segments which are reduced in width to skeletons having no normal leaf-tissue, and so appear like rows of tassels on the pinnae of the next higher order. 10 species. Handsome ferns found chiefly in swamps in the Temperate and Tropical regions, mainly of the Northern Hemisphere.

Unmistakably Osmundaceous stems occur as fossils from as far back as the Permian Period; these ancient plants possessed the upright stock surrounded by frond bases with broad flaps still seen in the Royal Fern of today.

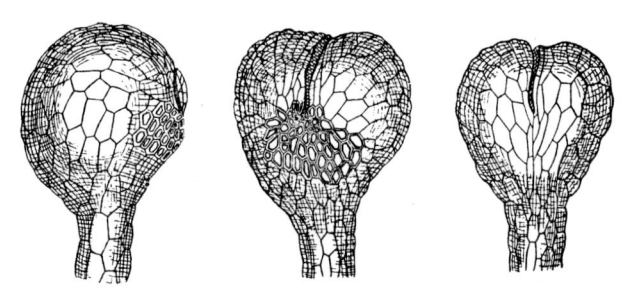

Fig. 75. Sporangia of Royal Fern (*Osmunda regalis*). Left: from one side. Middle: from behind showing the annulus. Right: from the front (×50). (After Luerssen).

OSMUNDA REGALIS L.
Royal Fern; Rhedynen Gyfrdwy

Stock short, upright, branched, clothed in a tangled covering formed by the densely set persistent bases of old leaves and the numerous much branched black roots, the whole forming a rounded ball-like mass up to 30cm in diameter. **Fronds** 60-160cm tufted, in a closely set spiral, the outer ones sterile, curved and nodding outwards, the inner ones fertile, upright; **stipe** in the purely sterile fronds $^1/_6$-$^1/_3$, in the fertile fronds up to $^1/_2$ their length, dark-coloured and winged below, the wings broadening out into broad membranous flaps, the stipe otherwise (like the channelled rachis) yellow-green to yellowish in the very young stage densely covered with yellow-brown cobweb-like hairs, later glabrous. **Blade** 40-120 × up to 40cm, lanceolate in outline, bipinnate (fertile portion when present tripinnate); **primary pinnae** usually 7-9 on either side, up to 30cm, opposite or nearly so, set rather wide apart at an acute angle to the rachis, shortly stalked, oblong, acute, pinnate; the **pinnules** on the side of the pinna directed towards the apex of the frond, somewhat shorter than the others; **sterile pinnules** 7-13 on either side, 2.5-6.5cm, wide apart, sessile or nearly so, oblong or oblong-lanceolate, obtuse unequal and often lobed at the base, the margin shallowly serrate. **Fertile fronds** usually with 2-3 pairs of sterile pinnae below followed by 7-14 pairs of fertile pinnae of gradually decreasing length, only the lowest 3 or 4 of which are again pinnate; ultimate **fertile segments** very narrow, more or less completely covered with the clustered sporangia. Frequently the lowest fertile pinnae are sterile and more or less leafy towards the apex or almost to the base. **Sporangia** pear-shaped. **Spores** ripening June-August. 2n=44.

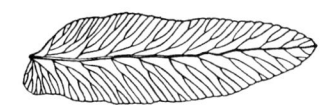

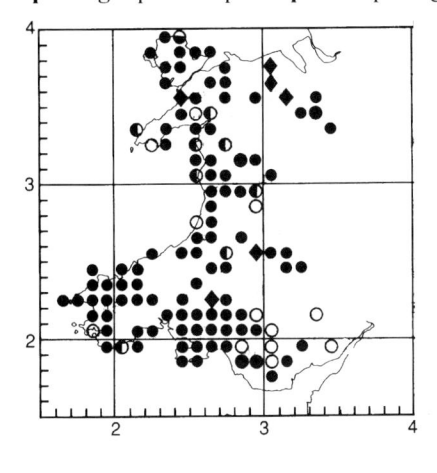

Fig. 76. Royal Fern (*Osmunda regalis*). Single pinnule about natural size. (*See also* Fig. 110).

OSMUNDA REGALIS

Bogs, marshy ground, marshy copses and borders of damp woods. Rare in Wales and now extinct in many of its former localities.

First recorded, 1670; Ray, Cat. Plant. Ang., 113. 'I have observed it in boggy places ... in Wales.'

Throughout the British Isles, especially western Ireland but unrecorded for several areas of eastern Great Britain. Has decreased in many areas due to land drainage. Introduced at a few sites.

A very wide but disjunct distribution. Europe: from S. Sweden to Portugal but scarce in eastern Europe. Azores. N. Africa and pockets in central and southern Africa to the Mascarene Islands. Asia: Turkey, Caucasus then Himalaya across to Japan. SW India. N. America: on eastern side of USA and Canada. Mexico. Central America and West Indies southwards to Uraguay and Argentina in S. America.

Genus PHEGOPTERIS (C.Presl) Fée

Beech Ferns - Rhedyn y Graig

Rhizome creeping, slender, in the only British species. Fronds pinnate with pinnatifid pinnae attached by a broad leafy base, veins free, the lower ones reaching the margin; scales on the midrib bearing scattered, inconspicuous, slender marginal hairs. Indusia absent. 3 species. North temperate regions and SE Asia.

PHEGOPTERIS CONNECTILIS (Michx.) Watt

(*Polypodium phegopteris* L.; *Phegopteris polypodioides* Fée; *Thelypteris phegopteris* (L.) Sloss.)

Beech Fern; Rhedynen y Graig

Rhizome creeping, slender, up to 20cm×1-2mm, sparingly branched and rooted, blackish-brown, clothed when young with whitish woolly hairs and

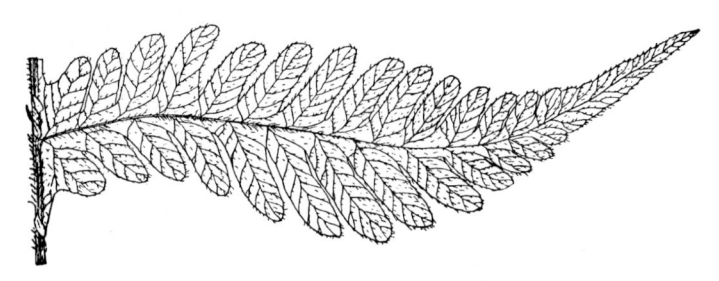

Fig. 77. Beech Fern (*Phegopteris connectilis*). Third pinna (rather larger than natural size).

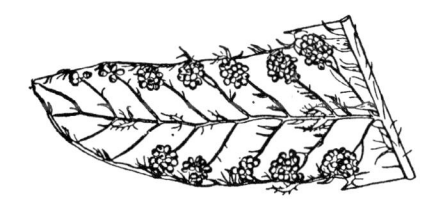

Fig. 78. Beech Fern (*Phegopteris connectilis*). Single segment of a median pinna (×5). (*See also* Fig. 109C).

scattered golden-brown ovate to oblong scales. **Fronds** 20-40cm, arising singly at irregular intervals from the rhizome; **stipe** as long to twice as long as the rachis, brittle, dark brown at the base otherwise greenish, usually clothed (at least in the upper half) with minute reflexed hairs, the base beset with lanceolate pale-brown scales, the rest of the stipe with or without scattered scales. **Blade** 10-20×8-15cm, triangular or triangular-ovate in outline, acute or acuminate, dull pale green, thin, pinnate; **pinnae** about 20 or more on either side, opposite or sub-opposite at the base, usually becoming alternate, all fringed with hairs and with scattered hairs on both surfaces especially on the midribs, deeply pinnatifid, the lowest pair rather distant and deflected downwards away from the rest, lanceolate, acute, attached by the midrib only, the next pairs horizontal, linear-lanceolate, acute or obtuse, attached by a broad leafy base, the rest directed more or less upwards, oblong, obtuse, similarly attached, running together towards the apex; **pinna segments** oblong, obtuse, entire or slightly (rarely strongly) crenate; **rachis** with acuminate scales and scattered hairs. **Sori** circular, situated close to the margins of the segments mostly on the secondary veins or, where these are forked, on the acroscopic or rarely on both branches. **Indusium** absent. **Spores** ripening June-August. Apogamous triploid. 2n=90.

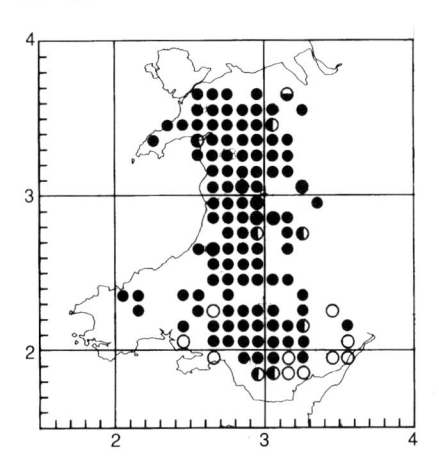

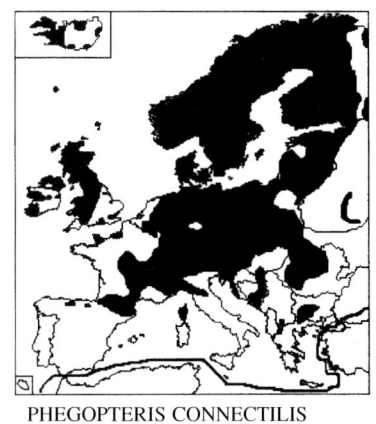

PHEGOPTERIS CONNECTILIS

Moist shady places: rocky woods, ravines, streamsides, and by waterfalls; locally frequent.

First recorded, 1726; Herb. Dillenius (Druce and Vines, The Dillenian Herbaria, 48, 1907), 'Hysvae in a cave [Caern.]' and 'Nascitur copiose in monte Snodon [Snowdon] Llanberys saxosis locis subudis.'

Western and northern Great Britain, including most of Scotland, but local in Ireland and scarce in S. England.

Most of Europe from northern Spain northwards. Iceland and Faeroes. Asia: Turkey, Caucasus, and across southern Siberia. Himalaya. Japan eastwards to Washington State, USA. North-eastern USA through to Greenland.

Genus PHYLLITIS Hill, see under ASPLENIUM L.

Phyllitis scolopendrium (L.) Newman, see ***Asplenium scolopendrium*** L.

Genus PHYMATODES C.Presl
Kangaroo Ferns - Rhedyn Cangarŵ

Fronds pinnately lobed to varying extent on one plant, but base of blade not lobed and gradually cuneate; sori in row on either side of rachis and on midribs of main lobes, sunken. A genus of about 10 species from the Old World tropics and Australasia.

PHYMATODES DIVERSIFOLIA (Willd.) Pic.Serm.
(*Microsorium diversifolium* (Willd.) Copel.)
Kangaroo Fern; Tongue Fern; Rhedynen Cangarŵ

Fronds (incl. stipe) to 60cm but usually much less. **Blade** varying from entire to pinnately lobed nearly to midrib, the lobing on one frond very uneven. 2n=74. *See* Fig. 114, C.

Grown in gardens, naturalized on shady walls and damp places in woods.

Guernsey (Channel Islands); Tresco (Scilly) and as a garden relic spreading by vegetative means at Rossdohan (S. Kerry).

Native to Australia and New Zealand.

Genus PILULARIA L.
Pillworts - Pelenllys

Leaves rush-like, without pinnae, each bearing one spherical pill-like sporocarp (hence the name) with 2-4 vertical compartments, each containing a sorus and splitting longitudinally when ripe, liberating the sporangia in a drop of mucilage. 6 species. Europe, America and Australia.

PILULARIA GLOBULIFERA L.

Pillwort; Pelenllys

Rhizome cylindrical, up to 50cm long and 1.5mm in diameter (usually much less), creeping or free floating, curved upwards at the tip. **Leaves** narrow and subulate, 3-10cm × *c*.0.5-1mm, circinately coiled in the bud, cylindrical, thread-like, glabrous, fresh green, crowded to distant (0.4cm to 2.0cm apart), arising alternately from the upper surface of the rhizome; close

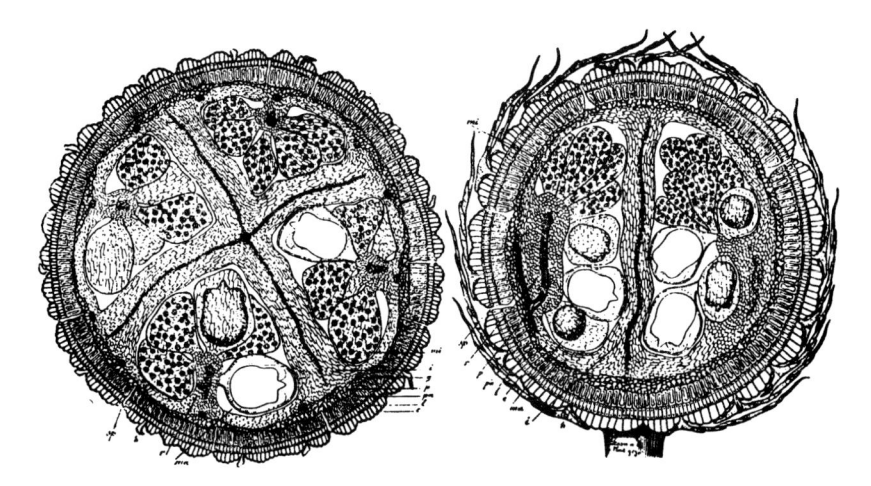

Fig. 79. Pillwort (*Pilularia globulifera*). Left: transverse section of the ripe but still closed sporocarp, about half way up. Right: vertical (and approximately median) section of the same. (Both ×20). The sporangia are produced in sori, microsporangia above and megasporangia below. (After Luerssen).

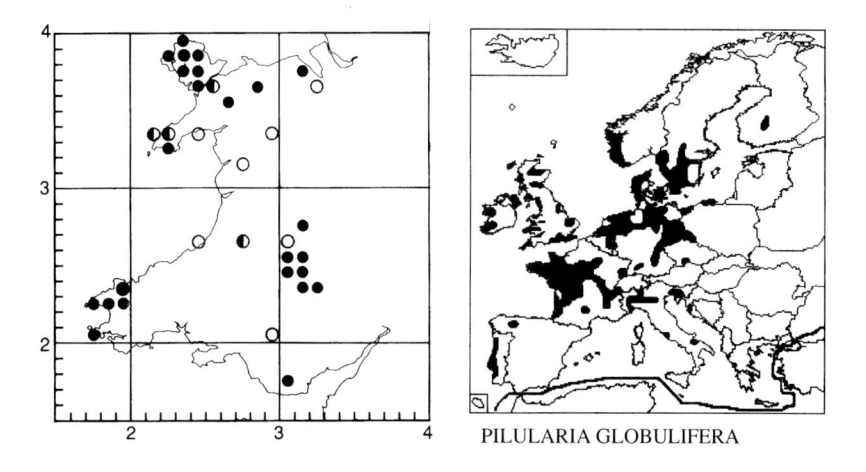

PILULARIA GLOBULIFERA

to the base of each are a root and a bud which grows into a branch shoot. **Sporocarps** (in 'fruiting' specimens) borne one at the base of each leaf on a stalk 0.5-1.0mm long, spherical (2.5-3.5mm in diameter) and covered with a dense mat of hair, yellowish-green at first changing to light brown and finally brownish-black; internally divided into four vertical compartments each containing a single sorus which comprises a swollen receptacle together with numerous **microsporangia** (in the upper half) and fewer (13-25) **megasporangia** below. Both kinds of sporangium sessile (some elongated cells near the tip of each may be vestiges of an annulus); megasporangia mostly ovoid or obovoid; microsporangia narrower, elongated obovoid to club-shaped, both kinds usually somewhat laterally compressed. The internal tissues of the sporocarp become mucilaginous when ripe and swell in the presence of water. The shell then breaks into quarters liberating the sporangia. **Spores** ripening June-September. 2n=26.

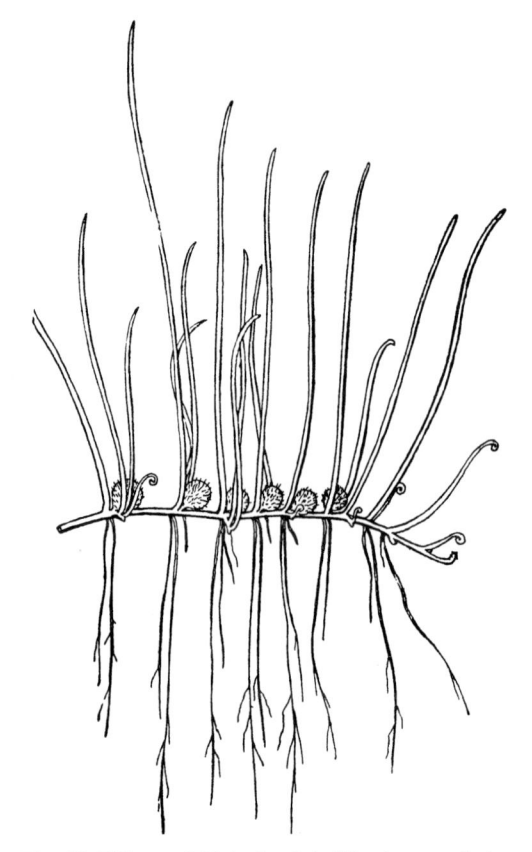

Fig. 80. Pillwort (*Pilularia globulifera*), natural size.

Mud or damp ground beside pools and lakes, or free floating in standing water. Rare.

First recorded, 1796, Griffith in Withering, Arr. Brit. Plants, edn 3, *3*, 760. 'About 2 miles from Mold, on the north side of the Chester road, near Clawdd Offa, or Offa's Dyke.'

Very scattered throughout Great Britain but has decreased significantly; frequent only in central southern England. A similar picture in Ireland, now frequent only in central western Ireland. Extinct in the Channel Islands.

Protected in Ireland under the *Flora Protection Order, 1987* and the *Wildlife (NI) Order, 1985*.

Western Europe from southern Scandinavia to Portugal; scattered in central and southern Europe, Rossia (E).

Genus POLYPODIUM L.

Polypodies - Llawredyn y Fagwyr

Rhizome creeping, fleshy, often glaucous and densely scaly. Fronds pinnatifid or pinnate, rarely simple, all similar in form, non-scaly. Veins free in some few species (e.g. *P. vulgare*) but usually reticulate. Sori terminal, usually in one row (less frequently 2-3 rows) on either side of the midrib. About 75 species, mostly epiphytic, in Tropical and Sub-Tropical America, Asia and Polynesia; only a few free-veined species in the North Temperate zone.

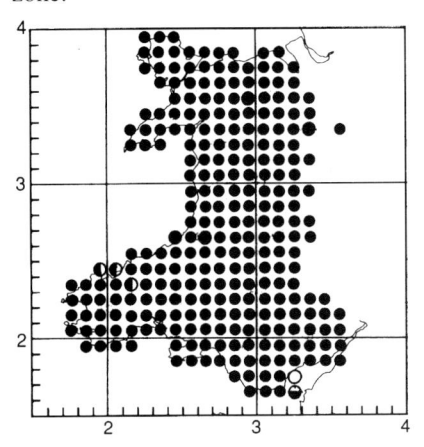

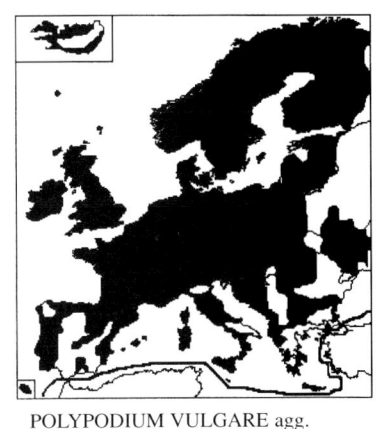

POLYPODIUM VULGARE agg.

Polypodium vulgare agg. is found throughout the British Isles.

The three polypodies in the British Isles (disregarding their hybrids) were first classified as distinct species by Shivas in 1961 (*J. Linn. Soc. (Bot.)*, **58**,

13-25, 27-38) following work published by Manton in 1947 and 1950. Rothmaler had ranked them as subspecies in 1929 and Willdenow had distinguished *P. vulgare* L. var. *serratum* (now *P.cambricum*) as long ago as 1810.

The records of *Polypodium vulgare* have to be treated with caution as many apply to *P. vulgare* sensu lato. Where accurate identification cannot be established records should be referred to *Polypodium vulgare* aggregate (*P. vulgare* agg.) not *P. vulgare*.

For table giving microscopic measurements for the species and hybrids, see after the following key.

KEY TO SPECIES

The key can only give an indication of what the plant is most likely to be. Microscopic examination is necessary to confirm the identification. See table for measurements.

1 Indurated annulus cells when mature with dark reddish-brown walls contrasting sharply with the rest of the annulus (×10 lens); rhizome scales not tapering to a long point; blades typically narrowly lanceolate to linear; sori usually orbicular. A weak calcifuge found in mosty acidic conditions (but considerable overlap with *P. interjectum*) **P. vulgare**
Indurated annulus cells when mature with pale buff or pale yellow to bright yellow, orange-yellow or golden brown walls (×10 lens); rhizome scales tapering to a long point; blades typically ovate to deltoid; sori often broadly elliptic. Preferring mildly basic to high-calcium conditions 2

2 Indurated annulus cells when mature usually with bright yellow walls not contrasting much with the rest of the annulus (×10 lens but microscope preferable); rhizome scales mostly linear-lanceolate. Preferring base-rich conditions with a high calcium content **P. cambricum**
Indurated annulus cells when mature with walls quite variable in colour in different sporangia: pale buff, pale yellow, bright yellow, orange-yellow or golden brown (× 10 lens but microscope preferable); rhizome scales mostly abruptly narrowed just above the dilated (rounded) relatively broader base. Preferring mild base-rich conditions thus overlapping the other two species requirements **P. interjectum**

MICROSCOPIC EXAMINATION OF *POLYPODIUM*

The table given refers to <u>mean</u> figures. Mean values are needed when checking individual plants of *Polypodium* as one or a few measurements of a character are meaningless and can result in incorrect identification.

	SPECIES		
<u>Mean</u> figures	**P. cambricum**	**P. interjectum**	**P. vulgare**
No. of indurated cells per annulus	(4-)5-10(-19)	(4-)7-9(-13)	(7-)10-14(-17)
Indurated cell length × breadth (µm)	21-26 × 81-100	28-35 × 76-86	22-28 × 60-80
No. of basal cells per sporangium	3 - 4	2 - 3	1
Width of basal cell(s) relative to annulus	much wider	much wider	as wide as to slightly wider than
Rhizome scale length	5 - 16 mm	3.5 - 11 mm	3 - 6 mm
Paraphysis length	0.5 - 1.4 mm	absent	absent

	HYBRIDS		
<u>Mean</u> figures	**P. × font-queri**	**P. × shivasiae**	**P. × mantoniae**
No. of indurated cells per annulus	(8-)11-14(-20)	(5-)7-8(-11)	(2-)9-11(-14)
Indurated cell length × breadth (µm)	18-24 × 60-80	24-26 × ?	28-35 × 74-80
No. of basal cells per sporangium	(1-)1.5-2.5(-3)	(1-)2.6-3.1(-4)	(1-)1.5-2.0(-3)
Paraphyses	absent from wild material	absent or to nearly as long as in *P. cambricum* but less freely branched and less abundant	absent

POLYPODIUM CAMBRICUM L.

(*P. australe* Fée; *P. vulgare* var. *serratum* Willd.; *P. vulgare* subsp. *serrulatum* Arcang.)

Southern Polypody; Llawredynen Gymreig

Rhizome horizontal, stout, up to 40cm × 6-7mm, creeping on or beneath the surface, branched, somewhat flattened in cross-section, yellowish to dark brown with age, the younger parts thickly clothed with reddish-brown, linear-lanceolate to lanceolate **rhizome-scales** 5-16mm long, generally longer and narrower than in the other two species. **Fronds** up to about 50cm long. New fronds produced in late summer and autumn. **Stipe** from ⅓ to as long as the rachis. **Blade** 10-37 × 8-15cm, broadly ovate to triangular, acute to almost acuminate, not coriaceous. **Pinnae** mostly alternate, usually acute and often prominently toothed or double toothed, the longest usually 2nd-4th pair from the base in unstunted plants; basal (and sometimes the succeeding) pair of pinnae usually inflexed; stomatal guard cells 48-65µm long. **Sori** oval; branched hairs (paraphyses*) intermixed with sporangia; **annulus** with a mean of (4-)5-10(-19) pale golden brown indurated cells which are broader than in the other two species, averaging 21-26µm in length and 81-100µm in width. Basal cells of the sporangium normally 3-4 separating it from *P. vulgare*. **Spores** less than 74µm long (average), ripening September-Spring. Diploid. 2n=74.

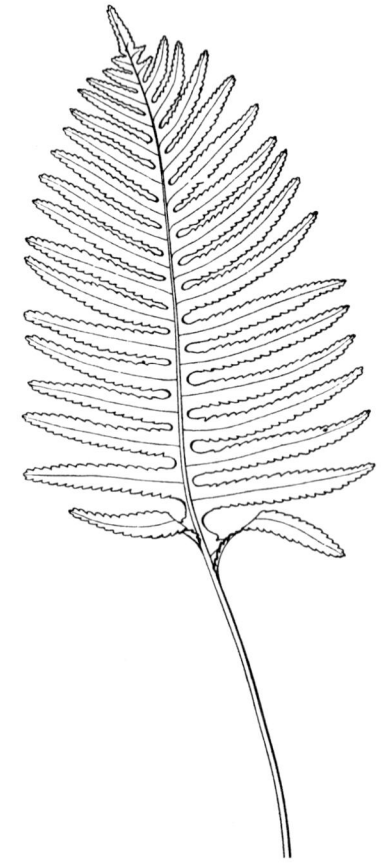

Fig. 81. Southern Polypody (*Polypodium cambricum*) (×1/3).

*The paraphyses of *P. cambricum* have a mean length just under 1 millimetre (970µm) with the terminal cells invariably much shorter than those of the basal ones which are mostly 100-200µm long. The paraphyses must not be confused with the minute glandular hairs which occur scattered over the lower surface of the frond in all three species and which sometimes can be found among the sporangia or close to the base of the sorus.

These hairs are mostly only 100-240µm in length (cf. mean of 970µm for paraphyses) and consist of only 3-4 cells with the glandular, inflated terminal cell often coloured a dark reddish-brown and as long as the preceding cells. It is these glandular hairs, not paraphyses, that have been reported for *P. interjectum* and *P. vulgare* by European botanists, and can sometimes be seen in the sori of all species from the British Isles.

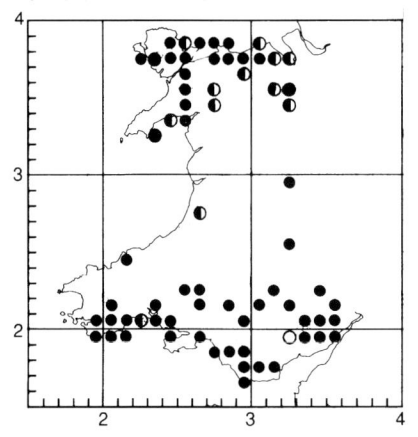

POLYPODIUM CAMBRICUM

Scattered in western Great Britain from the Channel Islands to Central Scotland; East Scotland; very local in SE England. Scattered in Ireland.

Southern and western Europe, Rossia (K). N. Africa; Turkey.

POLYPODIUM CAMBRICUM L. 'CAMBRICUM'

(*P. cambricum* L.; *P. australe* Fée 'Cambricum'; *P. vulgare* var. *cambricum* (L.) Lightf.)

Welsh Polypody

Blade ovate or broadly ovate-lanceolate, bipinnatifid; **segments of blade** ovate-lanceolate, pinnatifid, with oblong to linear lobes. Always sterile (complete absence of sporangia) and presumed diploid, according to Shivas (1961, *J. Linn. Soc.* (*Bot.*), **58**, 27-38).

Fig. 82. Welsh Polypody (*Polypodium cambricum* 'Cambricum') (×1/3).

First discovered by Richard Kayse of Bristol in 1668 and recorded in Ray's *Synopsis stirpium britannicarum*, edn 1, 22, 1690, as 'Polypodium Cambrobritannicum pinnulis ad margines laciniatis. Laciniated Polypody of Wales. On a rock in a wood near *Dennys Powis* Castle' [Dinas Powis Castle, Glam.].

There are other records claimed for Glam., Carms., Pembs., Cards., Caerns., Denbs. and Anglesey. Some of these plants were reported to bear spores, which nullifies the claim for them to be 'Cambricum'. Some records in the British Isles are of plants found on walls. It is likely that most of these were originally planted. There are other named barren dissected varieties of *P. cambricum* L. (Dyce, 1988).

POLYPODIUM INTERJECTUM Shivas

(*P. vulgare* subsp. *prionodes* (Asch.) Rothm.)

Intermediate Polypody; Llawredynen Rymus

As *P. cambricum* except: **New fronds** produced in summer. **Rhizome-scales** 3.5-6(-11)mm long, ovate-lanceolate. **Blade** ovate-lanceolate in outline. **Longest pinnae** 4th-6th pair from the base in unstunted plants; stomatal guard-cells 58-71µm long. **Sori** more or less oval; paraphyses absent, but see under *P. cambricum* for discussion about minute glandular hairs; **annulus** with a mean of (4-)7-9(-13) pale golden-brown indurated cells

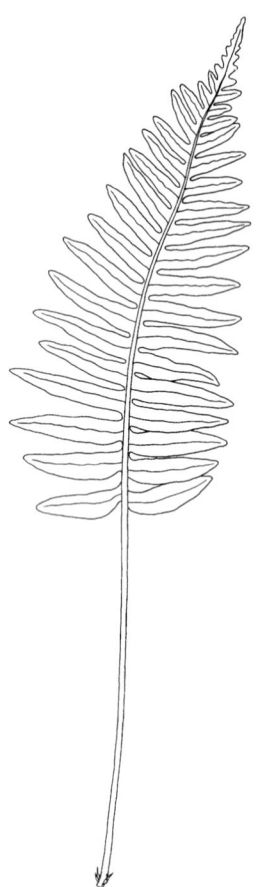

Fig. 83. Intermediate Polypody (*Polypodium interjectum*) (×1/3).

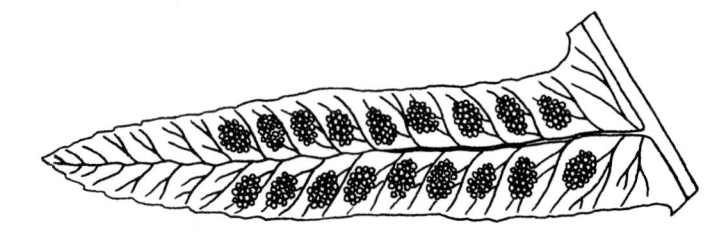

Fig. 84. Intermediate Polypody (*Polypodium interjectum*). Single segment from an average-sized frond (×2.5).

which are longer than in the other two species averaging 28-35µm in length and 76-86µm in width. **Basal cells** of the sporangium normally 2-3, separating it from *P. vulgare*. **Spores** more than 74µm long (average) ripening July-December. Hexaploid. 2n=222.

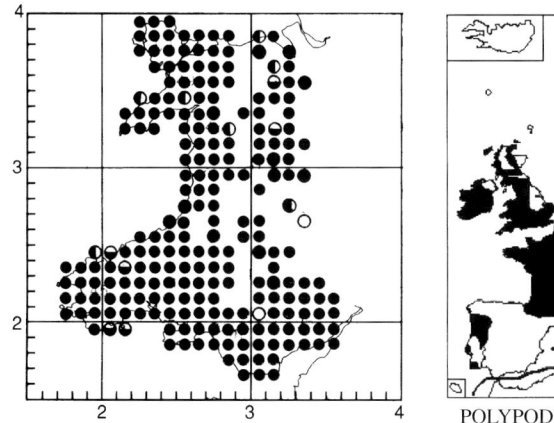

POLYPODIUM INTERJECTUM

Throughout much of the British Isles, less common in parts of central eastern England and Scotland; scattered in Ireland.

Western and central Europe northwards to Denmark and extending eastwards locally to Rossia (B,W). Turkey.

POLYPODIUM VULGARE L.

Polypody; Common Polypody; Llawredynen y Fagwyr

As *P. cambricum* except: **New fronds** produced in early summer. **Rhizome-scales** 3-6mm long, ovate, acuminate. **Blade** 5-38×3-9cm, usually lanceolate, somewhat coriaceous. **Pinnae** obtuse or acute, their margins obscurely or sometimes more or less deeply toothed, almost all more or less equal in length except near the apex, the basal pair not inflexed; stomatal guard-cells 43-58µm long. **Sori** round; paraphyses absent but see under *P. cambricum* for discussion about minute glandular hairs; **annulus** with a mean of (7-)10-14(-17) reddish-brown indurated cells which are narrower than in the other two species averaging 22-28µm in length and 60-80µm in width. Normally only one basal cell to the sporangium, separating it from the other two species. **Spores** less than 70µm long (average), ripening July-August. Tetraploid. 2n=148.

Throughout the British Isles; thinning out in central and eastern England and in Ireland.

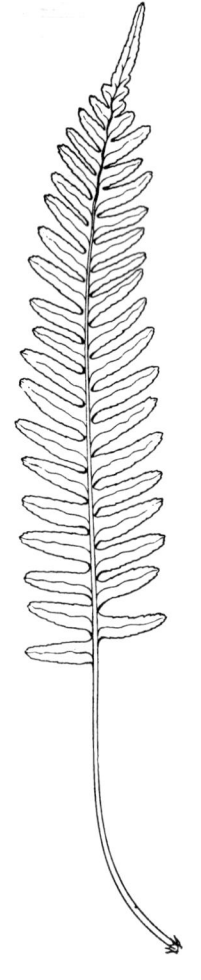

For World distribution, only *Polypodium vulgare* sensu lato (less *P. cambricum*) can be considered here, due to the lack of separation of other species in Floras etc. Throughout Europe except much of the Hungarian Plain. Iceland and Faeroes. Azores, Madeira and Canary Is., North Africa, southern Africa, Kerguelen. Asia: in Turkey, Caucasus, Iran, across the middle of Siberia to China and Japan. N. America: Aleutian Is., Alaska, mainland Canada, much of USA, Mexico, Greenland. Many of the North American *Polypodium* were once considered as just varieties of *P. vulgare* L. but are all now treated as separate species as is the case in many other areas of the world outside Europe.

Fig. 85. Polypody (*Polypodium vulgare*) (×1/3).

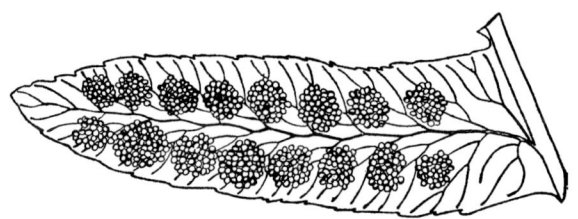

Fig. 86. Polypody (*Polypodium vulgare*). Single segment from an average-sized frond of an acutely-segmented form (×2.5).

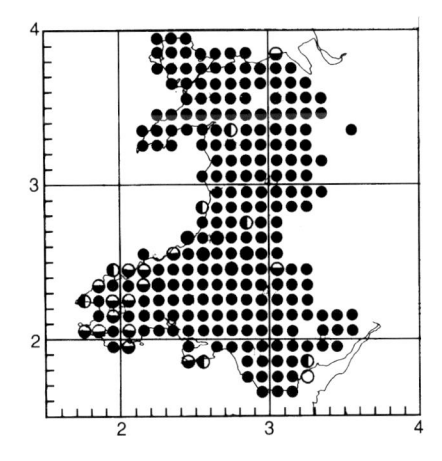

POLYPODIUM VULGARE

HYBRIDS IN POLYPODIUM

For table giving microscopic measurements for the species and hybrids, see after the key to the species (p. 169).

P. CAMBRICUM × *P. INTERJECTUM* = *P.* × *SHIVASIAE* Rothm.

(*P.* × *rothmaleri* Shivas)

Shivas' Polypody; Llawredynen Shivas

Very robust, with fronds up to *c*.60cm long. Intermediate between the parents in frond shape, pinnae obscurely toothed, or pinnatifid like *P. cambricum* 'Cambricum'. Sorus shape and number of indurated cells in the annulus also intermediate. Sporangia and spores usually shrivelled, or some sporangia may be abnormally large, highly sterile. Paraphyses variously reported present or absent. Tetraploid hybrid. 2n=148.

Very scattered in W. Great Britain and Ireland; also Guernsey.

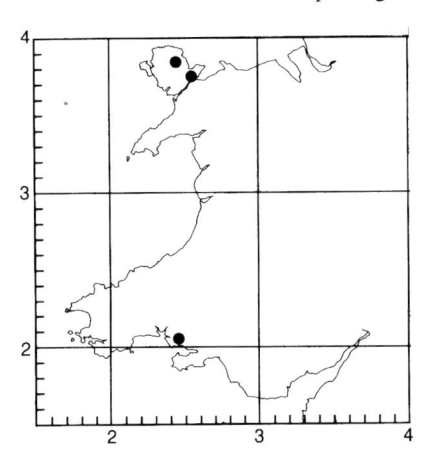

POLYPODIUM × SHIVASIAE

P. CAMBRICUM × *P. VULGARE* = *P.* × *FONT-QUERI* Rothm.

Font-Quer's Polypody; Llawredynen Font-Quer

Robust. Intermediate between the parents in frond shape, the pinnae generally long in proportion to their width, acute to acuminate. Sorus shape and number of indurated cells also intermediate. Sporangia and spores mostly shrivelled, with some abnormally large. Paraphyses not found. Triploid hybrid. 2n=111.

Rare, northwards to SW Scotland.

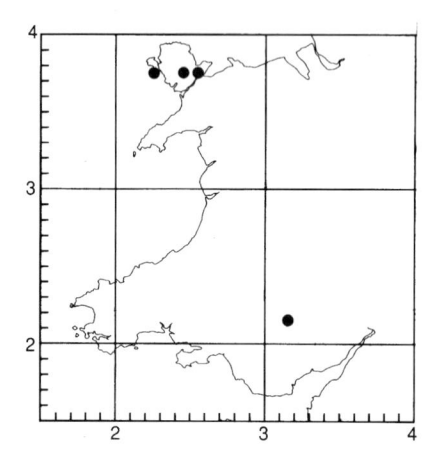

POLYPODIUM × FONT-QUERI

P. INTERJECTUM × *P. VULGARE* = *P.* × *MANTONIAE* Rothm. & U.Schneid.

Manton's Polypody; Llawredynen Manton

Robust, with fronds up to *c*.45cm long. Intermediate between the parents in frond shape and number of indurated cells. Sori more or less circular. Sporangia and spores usually shrivelled or abnormally large and pale, perhaps not completely sterile. Paraphyses absent. Pentaploid hybrid. 2n=185. By far the commonest hybrid of the three.

Well recorded in north-west and south-west Wales.

Scattered throughout Great Britain; rare in Ireland. Under-recorded.

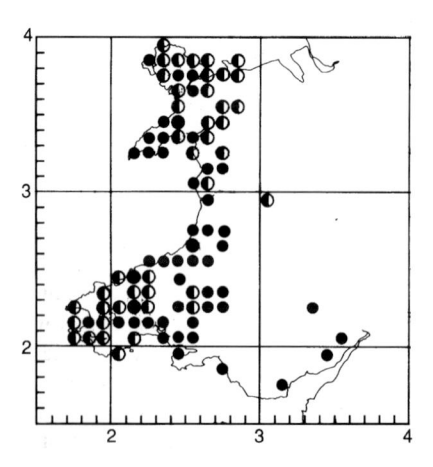

POLYPODIUM × MANTONIAE

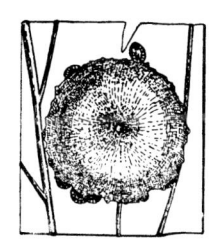

Fig. 87. Soft Shield-fern (*Polystichum setiferum*). Part of a fertile pinnule (×13), showing a single sorus.

Genus POLYSTICHUM Roth
Shield-ferns - Gwrychredyn

Sori globose, situated on the backs of the veins, rarely at the ends. Indusium circular, peltate, attached by its centre to the apex of the receptacle. Handsome ferns with (in our species) short erect stock and abundant broad soft scales; fronds tufted with short stipes; blade mostly stiff and leathery, simply pinnate to tripinnate with distinctly unequal-sided pinnae having their largest basal pinnules (or lobes) on the side nearer to the frond apex, the ultimate segments with acuminate bristle-pointed teeth. About 135 species.

KEY TO SPECIES

1 Fronds with shallowly toothed (not lobed) pinnae **P. lonchitis**
  Fronds with deeply lobed pinnae to 2(-3)-pinnate 2
2 Basal pinna lengths strikingly smaller than those of the middle pinnae. Hard texture to the green lamina of pinnae **P. aculeatum**
  Basal pinna lengths nearly half that of the middle pinnae giving the blade outline a truncate base. Soft texture to the green lamina of pinnae **P. setiferum**

POLYSTICHUM ACULEATUM (L.) Roth
(*P. lobatum* (Huds.) Chevall.)
Hard Shield-fern; Gwrychredynen Galed

Stock similar to that of *P. setiferum*. **Fronds** 30-90cm arranged in a shuttlecock-like manner, rigid, erectish or more or less spreading, generally winter-green; **stipe** usually about ¹/₅ as long as rachis, densely clothed especially at the base with large brown ovate-lanceolate acuminate scales, sometimes intermixed with some hair-like and/or some scurfy scales. **Blade** 25-75 × 5.5-22cm, lanceolate in outline, narrowed above and below, firm in texture, stiff, its upper surface dark to yellowish-green, somewhat glossy, the under surface paler with scattered hair-like scales, pinnate or bipinnate; **pinnae** up to about 50 on either side, opposite or sub-opposite below, becoming alternate, or alternate almost throughout, very shortly stalked, unequal at the base, linear, lanceolate-acuminate to ovate-lanceolate in outline, pinnate in the lower half of the frond and pinnatifid above or pinnatifid throughout, becoming less divided towards the apex of the frond

Fig. 88. Hard Shield-fern (*Polystichum aculeatum*). Second pinnule of a media pinna (×5.5). (*See also* Fig. 111A).

and ultimately sickle-shaped, simple and toothed; **pinnules** or segments sessile and all but the lowest decurrent mostly incurved towards the apex of the pinna, narrowly and obliquely ovate-lanceolate or oblong, very unequal at the base (much broader and almost truncate on the side towards the pinna apex, cuneate on the other, but as a whole more acutely angled than in *P. setiferum*) rather suddenly acute with a long bristle at the apex, serrate with shorter bristle-pointed teeth along the margin (except towards the base); the first acroscopic segment of each pinna with its edge flattened to the main rachis, much larger than the rest, and with a strong triangular long-bristle-pointed acroscopic lobe at the base, this lobe being less often developed (though more weakly) also in the remaining pinnules or only represented by the long bristle or not at all; **rachis** clothed with broadly lanceolate scales intermixed with narrowly subulate to hair-like ones. **Sori** small, usually confined to the upper half of the frond, situated much below the vein endings and forming a line on either side of the midrib of each pinnule and also that of its basal lobe. **Indusium** circular, peltate. **Spores** dark brown, papillate, 40-48μm, ripening July-August. Tetraploid. 2n=164.

P. aculeatum may be distinguished from *P. setiferum* by means of (a) the outline of the frond particularly at the basal pinnae (see key) (b) the stiff leathery rigid fronds, (c) the pinnules which more often run together or if distinct are sessile and usually acute not obtuse at their base, (d) the fertile veins which run beyond the sori to the margin of the segment.

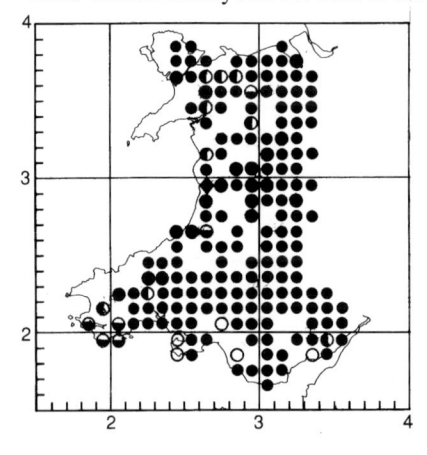

POLYSTICHUM ACULEATUM

Woods and shady hedge banks and on shady rock ledges in mountainous districts; frequent to common in the south, rarer in the north.

First recorded, 1696; Llwyd in Ray, *Syn.*, edn 2, 48. '... in montosis Cambrobritannicis.' This record is of forma *cambricum*.

Throughout the British Isles but less widespread than *P. setiferum* in the south-west.

Throughout Europe except northern countries, and only in Rossia (K). Canary Is. Distribution world-wide difficult to assess due to confusion with other species especially *P. setiferum* under which it is considered.

POLYSTICHUM ACULEATUM (L.) Roth forma *CAMBRICUM*

(*Polystichum aculeatum* (L.) Roth var. *cambricum* (Gray) Hyde & A.E. Wade; *Polystichum aculeatum* (L.) Roth var. *lonchitidoides* (Hook.) Deakin

Almost all pinnae simple obliquely ovate or sickle-shaped, and coarsely toothed with a marked basal acroscopic lobe. This former variety is merely a growth form of *P. aculeatum* and is liable to be mistaken for *P. lonchitis*, but may usually be distinguished therefrom by (a) the two distinct lobes, basiscopic as well as acroscopic, at the base of the lowest pinnae, (b) the relatively coarse main serrations which are almost lobes, (c) the sharp teeth in between or on the sides of the main serrations, (d) the thinner and less coriaceous texture of the frond. Sori are usually absent but they have been found even on very small plants.

POLYSTICHUM LONCHITIS (L.) Roth

Holly-fern; Celynredynen

Stock short, thick, ascending, densely covered with the dead bases of old fronds, the younger part clothed with scales similar to those of the stipes. **Fronds** 15-60cm, tufted, winter-green; **stipe** about 1/6 as long as the rachis, clothed especially towards the base with ovate or ovate-lanceolate brown scales, large and small intermixed. **Blade** 12-50 × 2-6cm, linear-lanceolate in outline, tapering gradually to a very narrow base, rigid and somewhat leathery, deep green on the upper surface, paler beneath with scattered hair-like scales, pinnate; **pinnae** up to about 40 on either side, sometimes opposite or sub-opposite at the base otherwise alternate throughout, rather closely set (except towards the base) and often overlapping one another from below upwards, very shortly stalked, undivided, the margin serrate (except towards the base), the most prominent teeth having sharp spinous or bristle-like points and smaller blunt teeth on their sides or in between, the lowest and smallest pinnae deltoid to ovate-deltoid. not or only indistinctly lobed at

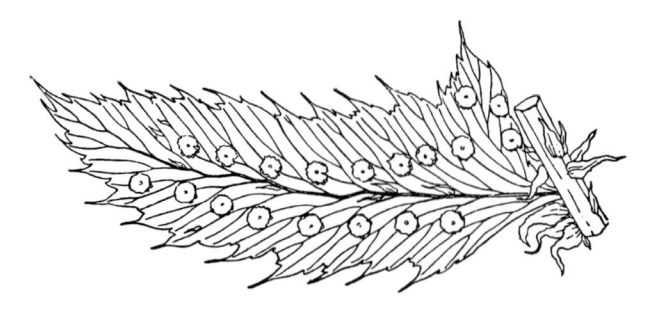

Fig. 89. Holly-fern (*Polystichum lonchitis*). Median pinna from an average-sized frond (×3). (*See also* Fig. 112).

the base, the middle pinnae broadly lanceolate (convexly rounded on the side turned to the base of the frond and almost straight or concave on the other), very unequal at the base (broadly truncate with a well-marked lobe on the side turned towards the apex of the frond, cuneate on the other), the uppermost pinnae becoming relatively narrower and less markedly lobed at the base; **rachis** sharply channelled and with numerous narrow-lanceolate or subulate scales. **Sori** large, usually confined to the upper half of the frond, occupying the middle of the acroscopic fork of each secondary vein and forming a line on either side of the midrib of each pinna halfway between it and the margin, also on either side of the midrib of the basal lobe, and occasionally of the next one or two secondary veins. **Indusium** circular, peltate, its margin irregularly crenate or toothed. **Spores** ripening June-August. Diploid. 2n=82. *P. aculeatum* forma *cambricum* is sometimes mistaken for this species.

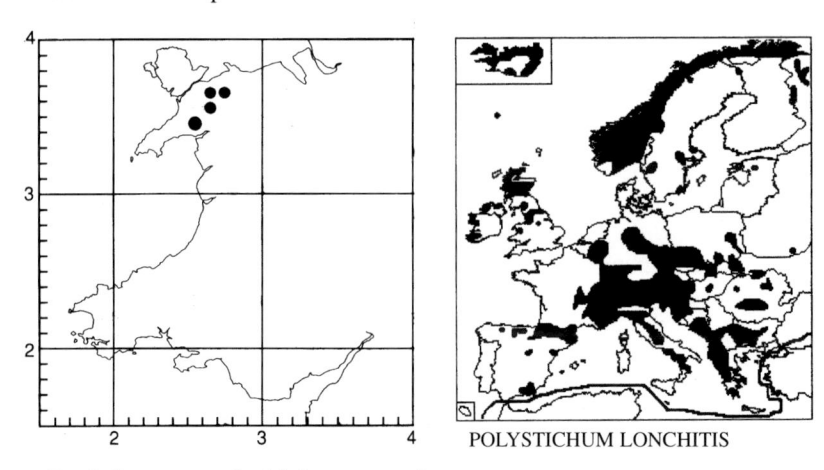

POLYSTICHUM LONCHITIS

Rock fissures on the higher mountains; very rare.

First recorded, 1690, Llwyd in Ray, *Syn.*, edn 1, 27. 'Clogwyn y Garnedh y Crîb Gôch Trygvylchau [Caerns.].'

Recorded in error from Glam., Monts. and Merioneth., *P. aculeatum* forma *cambricum* having been mistaken for it.

Local in northern England, central and northern Scotland and west of Ireland.

Protected in Northern Ireland under the *Wildlife* (NI) *Order, 1985*.

Scandinavia especially the western side, and other mountainous areas of Europe as far south as Crete. Very local in Rossia. Iceland and Faeroes. As n m0oia Minor. Caucasus. W. Himalaya and north-eastwards through the middle of Siberia. Kamchatka peninsula. N. America: on west side south to Arizona. The Great Lakes and round the Gulf of St Lawrence. Greenland.

POLYSTICHUM MUNITUM (Kaulf.) C.Presl

Western Sword-fern; Gwrychredynen Gledd Orllewinol

1-pinnate like *P. lonchitis* but the pinnae are linear and the basal ones are nearly as long as the longest. Pinnae have spinulose teeth and an eccentric lobe ('auricle') at the base of the acroscopic side of each pinna. Diploid. 2n=82.

Formerly naturalized on a hedge bank in Surrey.

Native to western N. America.

POLYSTICHUM SETIFERUM (Forssk.) Woyn.

(*P. angulare* (Willd.) C.Presl)

Soft Shield-fern; Gwrychredynen Feddal

Stock ascending or erect, up to 10cm, thick, woody, surrounded by the dead bases of old fronds, the younger parts clothed with scales similar to those on the stipes. **Fronds** 30-120 cm, arranged in a shuttlecock-like manner, spreading and more or less drooping, winter-green in sheltered places; **stipe** $1/4$-$1/2$ as long as the rachis, markedly channelled on its inner (upper) face, dark

Fig. 90. Soft Shield-fern (*Polystichum setiferum*). Median pinnule from a median pinna (×5.5). (*See also* Fig. 111*B*).

brown at the base, green above, densely clothed, especially at the base, with large pale brown ovate-lanceolate acuminate scales intermixed with smaller hair-like scales and adpressed ciliate scurf-like ones. **Blade** 20-100 × 9-25cm, lanceolate in outline, rather thin in texture, soft and flaccid when young, its upper surface rich green, the under surface paler and often bluish-green with scattered hair-like scales, bipinnate; **pinnae** up to 40 on either side, opposite or sub-opposite becoming alternate, or alternate almost throughout, shortly stalked, linear-lanceolate in outline, unequally broad-based, pinnate, the upper pinnae pinnatifid or pinnately toothed and varying from lanceolate to sickle-shaped; **pinnules** shortly stalked (except where they run together towards the apex of the pinna) obliquely ovate or somewhat crescent-shaped, broadly-acutely to obtusely angled and very unequal at the base (broadly truncate and usually with a well marked lobe on the side towards the apex of the pinna, cuneate on the other) acute or rounded at the apex, deeply serrate with long hair-pointed teeth along the margin; basal acroscopic pinnule usually longer than the rest and, together with other adjacent pinnules, sometimes deeply pinnatifid; **rachis** shaggy with scales and hairs like those of the stipe; **midribs of the pinnae** with hair-like scales only, all with a narrow well-marked channel down the middle of the upper surface. **Sori** small, round, usually confined to the upper half of the frond, situated on or near the endings of (mostly) tertiary venules and forming a line on either side of the midrib of each pinnule and also of that of its basal lobe. **Indusium** circular, peltate. For comparison with *P. aculeatum* see under that species. **Spores** ripening July-August. Diploid. 2n=82.

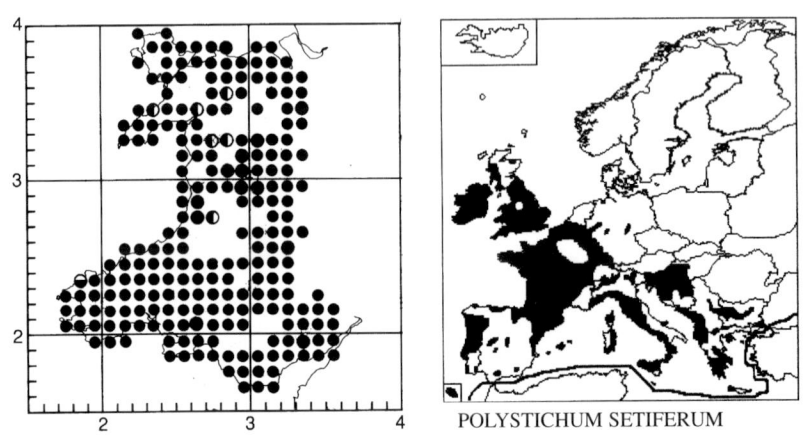

POLYSTICHUM SETIFERUM

Woods and hedge banks: frequent in the south, less so in the north.

First recorded, 1842, Gutch. In: Flower and Lees. Additions to the List of Plants met with in the neighbourhood of Swansea. *Phytologist*, **1**, 380. 'Hedge going from Swansea to Cromlyn [Crymlyn] Bog.'

Most frequent in the south-west of the British Isles, thinning out north-eastwards; rare in Scotland.

Europe: chiefly western and southern from the British Isles to the Mediterranean region. Azores, Madeira and Canary Is. Turkey, Caucasus. Distribution world-wide difficult to assess due to confusion with other species especially *P. aculeatum*. Therefore only *P. setiferum* sensu lato is considered here which includes *P. aculeatum*. N. Africa, Cameroon, tropical east Africa. Asia Minor, Caucasus through Himalaya and India to Japan, E. of Caspian Sea. SE Australasia, Pacific Islands including Samoa and Hawaii.

HYBRIDS IN POLYSTICHUM

P. ACULEATUM × *P. LONCHITIS* = *P.* × *ILLYRICUM* (Borbás) Hahne

Alpine Hybrid Shield-fern; Gwrychredynen Groesryw Alpaidd

Frond shape and dissection intermediate between the parents; rather rigid. Spores abortive. Triploid hybrid. 2n=123. Co. Leitrim, W. Ross and W. Sutherland. Common on the Continent and may be expected to be found elsewhere in the British Isles.

P. ACULEATUM × *P. SETIFERUM* = *P.* × *BICKNELLII* (H.Christ) Hahne

Lowland Hybrid Shield-fern; Gwrychredynen Groesryw yr Iseldir

Hybrid occurring in mixed populations of the parent species. Differences from *P. aculeatum*: stipe and rachis with scales of two kinds (as in *P. setiferum*), pinnules broader and more oval; from *P. setiferum*: fronds more rigid and nearly winter green, pinnae less numerous (12-15 free pinnae on either side) and longer in proportion to their width. Sterile. Triploid hybrid. 2n=123. Scattered throughout the British Isles to central Scotland. W. and central Europe.

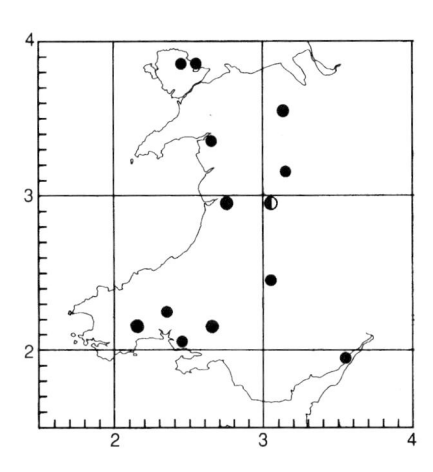

POLYSTICHUM × BICKNELLII

P. LONCHITIS × *P. SETIFERUM* = *P.* × *LONCHITIFORME* (Halácsy) Bech.

Atlantic Hybrid Shield-fern; Gwrychredynen Groesryw y Gorllewin

Intermediate between the parents and strongly resembling *P.* × *illyricum*, from which it is difficult to distinguish. Diploid hybrid. 2n=82. Co. Leitrim. Greece, Hungary.

Genus PTERIDIUM Gled. ex Scop.

Brackens - Rhedyn Cyffredin

Sori linear on a marginal receptacle. Indusia following the course of the sorus, the inner very thin, the outer thicker and somewhat wider, appearing as if it were a reflexed membranous fringe of the blade. 1 cosmopolitan species with several geographical subspecies.

PTERIDIUM AQUILINUM (L.) Kuhn

(*Pteris aquilina* L.)

Bracken; Rhedynen Gyffredin

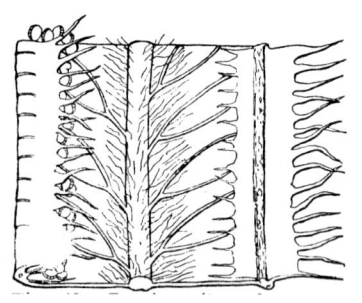

Fig. 91. Bracken (*Pteridium aquilinum*): portion of an ultimate segment of the frond seen from below and magnified about 10 diameters. On the left the outer indusium is shown in its natural position, fringing the margin and reflexed so as to cover over and protect the young sporangia. On the right the outer indusium has been flattened out and the sporangia have been removed so as to allow the inner indusium to be seen.

The description that follows is of what is now referred to as *Pteridium aquilinum* subsp. *aquilinum* (Common Bracken). Details of the current view of Bracken in the British Isles follow the main entry.

Rhizome widely creeping and branching underground, somewhat flattened-oval in section up to 2-3cm diameter, brownish-black to black, covered with short jointed, bluntly pointed, rust-coloured to dark brown glossy hairs. Strictly the underground parts of bracken consist of shoots of two kinds - thick long shoots which grow deep in the soil, and thinner short shoots which arise therefrom and grow up towards the surface, then turning parallel with it. Only these short shoots bear fronds. They are beset with frond bases and with a dense covering of fine roots (Smith, W. G., 1928; Watt, A. S., 1940).

Fronds produced as a rule singly in two rows along the rhizome; when mature 0.5-1.5(-3)m overall; **stipe** erect, up to about as long as the rachis, swollen, spindle-shaped, black and hairy at the base, otherwise nearly semicircular in section, about 1cm thick, channelled (as is the rachis) down the upper side, yellow-green to straw-coloured and glabrous. **Blade** often very large, 30-110×28-90cm, always more or less bent back, triangular-ovate (less often oblong), bi- to tri-pinnate; **pinnae** sub-opposite (except sometimes the upper ones), the lowest pair the longest, the lower pairs stalked, ovate or oblong-ovate to oblong, usually bipinnate, the middle pairs sessile, oblong to lanceolate, pinnate (with pinnatifid pinnules), the upper pairs linear-lanceolate to linear, simply pinnate; **pinnules** of the lower pinnae alternate, narrow-lanceolate, acuminate; **ultimate segments** sessile, set close together like the teeth of a comb, broad based and running together at the base, oblong to linear, often slightly sickle-shaped, blunt, those of the lower and middle pinnae often lobed at the base, all segments otherwise entire margined, leathery in texture, shining green and glabrous above, paler and more or less covered (especially on the midribs and other veins) with yellowish hairs below. **Sori** as described above for the genus, running all round the margins of the pinnules, outer indusium always present, inner absent when leaves are barren, both fringed with hairs. **Spores** ripening July-August. Diploid. 2n=104. However, a triploid plant (2n=156) has been reported from Denbs. (Sheffield *et al.*, 1993).

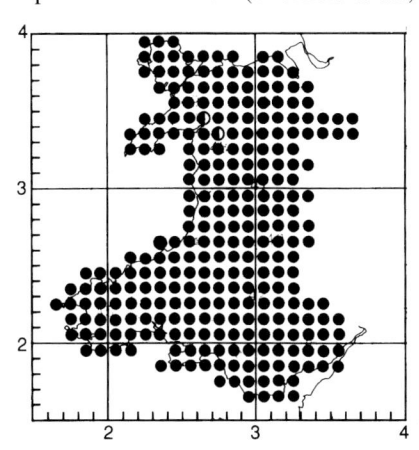

PTERIDIUM AQUILINUM

Common and abundant on deep well-drained siliceous soils in woods, grass heaths, sand dunes, neglected pastures, hedgerows and even as stunted plants on walls.

First recorded, 1726; Dillenius, *Diary of a Journey into Wales*. Druce and Vines, *The Dillenian Herbaria*, p. 1, 1907. 'Near Trawsfynedd [Merioneth].'

Throughout the British Isles.

Cosmopolitan. Virtually the whole of Europe except the extreme north. Azores, Madeira, Canary Is. and Cape Verde Is. Africa: North and West, Central to South Africa and the Mascarene Islands. Asia Minor, Caucasus and in a broad band across southern Siberia. A triangle from Afghanistan to the Kamchatka peninsula to Indonesia including India and Ceylon. N. America: on west side from western Canada to Central America and eastern halves of southern Canada and USA. Down through the West Indies to S. America as far as Uruguay. SW Australia and eastern side including Tasmania. New Zealand, Pacific islands including Hawaii.

Current view of Bracken in the British Isles

The views described below are those of C. N. Page and R. H. Mill. Other pteridologists disagree with this interpretation of Bracken. It is anticipated that future work will clarify the situation.

Bracken has traditionally been regarded as a single species world-wide although various regional subspecies and varieties have been recognised. In recent years the monotypic view of the species has been increasingly challenged. The current view is that two species exist in the British Isles: *Pteridium aquilinum*, divided into three subspecies, and *P. pinetorum* divided into two subspecies (Page, 1995; Page & Mill, 1995a,b). These two species belong to two different northern hemisphere bracken complexes.

P. aquilinum complex is distributed in a broad band across most mid-latitudes of the northern hemisphere and spreads southwards into Africa. Its main areas are the north temperate deciduous rain forests.

P. latiusculum complex is the most northern bracken group with an extensive range across all the northern continents. The complex largely follows the northern conifer forests although it occurs further south in similar habitats in the more mountainous, less oceanic districts. *P. latiusculum* (Desv.) Hieron. is the North American species and molecular evidence shows it to be distinct from the high latitude bracken of northern Europe (Rumsey, Sheffield, & Haufler, 1991; Wolf *et al.*, 1995), the latter having been renamed *P. pinetorum* (Page & Mill, 1995).

The keys to species and subspecies essentially follow Page & Mill (1995b) and the distributions given follow that paper and Page (1995). However it must be noted that fertile intermediates occur.

KEY TO SPECIES

Blades more or less triangular to triangular-ovate, never ternate; pinna midribs horizontally to almost downwardly curved when flushing, usually long remaining so; croziers and young stipes with a tomentum of white hairs, often overlain (in subsp. *aquilinum*) with red hairs, whole

indumentum lost slowly and progressively or occasionally long-persistent
P. aquilinum

Blades triangular, ternate or semi-ternate; pinna midribs more or less rising upwards and straight when flushing and throughout life; croziers and young stipes without or with few white hairs but a relatively large number of ephemeral red hairs, rapidly becoming totally glabrous **P. pinetorum**

KEY TO THE SUBSPECIES OF PTERIDIUM AQUILINUM

1 Stipes bright orange-brown to pale straw yellow, very short, semi-slender and wiry (to date, only in Perthshire) subsp. **fulvum**
 Stipes pale green, more or less long and stout 2

2 Crozier surface with a mixed tomentum of white hairs overlain with ephemeral, cinnamon-coloured hairs giving crozier a brindled appearance close-up and a reddish-brown hue at a distance; frond expansion moderately rapid, complete by end of season
 subsp. **aquilinum**
 Crozier surfaces with a tomentum mainly or entirely of white hairs, with sparse white tomentum long-persistent along stipe, rachis and midrib; frond expansion slow, strongly sequenced and almost indefinite
 subsp. **atlanticum**

PTERIDIUM AQUILINUM (L.) Kuhn subsp. *AQUILINUM*
Common Bracken; Rhedynen Gyffredin

Only this subspecies is an aggressive weed and a widespread problem to man, the other subspecies are rare in the British Isles.

Throughout the European range of the species with its northern boundary generally the interface with *P. pinetorum* and its southern boundary on the south side of the Mediterranean but extending further south along the Atlantic side of Africa, to W. Africa and Macaronesia.

PTERIDIUM AQUILINUM (L.) Kuhn subsp. *ATLANTICUM* C.N.Page
Atlantic Bracken; Rhedynen Gyffredin y Gorllewin

Wales: so far it is known from Brecs. (SO 00 or 01, 1912, **NMW**), Caerns. (SH 46, 1991) and Carms. (SN 30 or 31, 1930, **NMW**). All plants have been determined by Dr C. N. Page. In the British Isles it is known from SW Scotland especially, and also NW and SW England. It is a low-altitude subspecies of oceanic mild winter and moist summer climates and is confined to markedly base-rich soil. It occurs in a number of thinly scattered stations in mainly Atlantic-seaboard sites from western Scotland southwards to the Ivory Coast in W. Africa (specimen **NMW**, determined C. N. Page). Intermediates between this subspecies and subsp. *aquilinum* are currently treated as taxonomically part of the variable subsp. *aquilinum*.

PTERIDIUM AQUILINUM (L.) Kuhn subsp. *FULVUM* C.N.Page
Perthshire Bracken; Rhedynen Gyffredin Perth

So far, it is only known from several localities in Perthshire, Scotland but further work is needed to assertain its true distribution.

PTERIDIUM PINETORUM C.N.Page & R.R.Mill
(*P. aquilinum* (L.) Kuhn subsp. *latiusculum* (Desv.) C.N.Page)
Pinewood Bracken; Rhedynen Gyffredin Pîn

It ranges from the high northerly latitudes of Scandinavia across Siberia to at least central China. Its southern border extends to the mountains of Macedonia, in Greece, and the Caucasus. It is mainly associated with cold winter conditions in mainly pinewood vegetation.

KEY TO THE SUBSPECIES OF PTERIDIUM PINETORUM

Blades about as broad as long or broader than long, held at *c*.30 degrees to horizontal from point of insertion of first pinna-pair   subsp. **pinetorum**

Blades as long as broad or longer than broad, held in a strongly ascending, sometimes nearly vertical plane ('*Osmunda*-like')   subsp. **osmundaceum**

PTERIDIUM PINETORUM C.N.Page & R.R.Mill
subsp. *OSMUNDACEUM* (H.Christ) C.N.Page

Known from central Perthshire; also Switzerland and northern Italy.

PTERIDIUM PINETORUM C.N.Page & R.R.Mill
subsp. *PINETORUM*

So far, known from Inverness-shire, Scotland, in the vicinity of the native Caledonian pinewood (*Pinus sylvestris*). Intermediates between it and subsp. *aquilinum* occur in the area.

Genus PTERIS L.
Brake Ferns - Rhedyn Ruban

Small, medium or large epiphytes. Rhizome short-creeping, thin, bearing fronds in clusters. Fronds of one kind, simple, linear, elongate, tapering, ribbon-like, firm, leathery, yellow-green. Sori linear in a continuous almost marginal line along the edge of the frond, immersed into it and lacking an indusium. A genus of about 280 species mostly from tropical and subtropical regions.

KEY TO SPECIES

1 Pinnae entire, except lowest one or two pairs sometimes appearing to be forked into two — **2**
  Pinnae deeply pinnatifid to 2-pinnate — **3**

2 Pinnae with cuneaete bases, the lowest one or two pairs of pinnae often appearing to be forked into two — **P. cretica**
  Pinnae with cordate to truncate bases, frond 1-pinnate — **P. vittata**

3 Pinnules mostly 3-5mm broad — **P. incompleta**
  Pinnules mostly less than 2mm broad — **P. tremula**

PTERIS CRETICA L.
Brake Fern; Cretan Brake; Ribbon Fern; Rhedynen Creta

Fronds to 75cm, ascending to erect; **stipe** up to *c.*30cm; **blade** triangular to broadly ovate, 1-pinnate; **pinnae** finely toothed, oblong-linear, *c.*7-16 × 0.5-2cm and cuneate at the base, not decurrent to the rachis, deep green. Less than 10 pairs of pinnae. 2n=58. *See* Fig. 115, A.

Grown in gardens and naturalized on walls and rocks in very sheltered places. Scattered in southern and central England; Channel Islands.

Native to S. Europe, Africa, Asia, and scattered world-wide in tropical and subtropical regions. Its native range world-wide is difficult to assess as it escapes from cultivation easily in warmer regions.

PTERIS INCOMPLETA Cav.
(*P. palustris* Poir.; *P. serrulata* Forssk., non L.f.)

Spider Brake; Rhedynen Ruban y Corryn

Fronds to 1.5m arching-ascending; **blade** triangular-ovate, 2- to 3-pinnate or pinnatisect, soft, hairless, pale green; **pinnules** oblong-tapering, mostly 2-3.5cm × 3-5mm, the margins very finely toothed. 2n=58. *See* Fig. 115, B.

A greenhouse escape formerly naturalized in W. Gloucs.

Native to Azores, Spain, Canary Is.

PTERIS TREMULA R.Br.
Tender Brake; Rhedynen Ruban Croendenau

Fronds to 2m ascending; **blade** more or less triangular, soft, hairless yellow-green, 2- to 3-pinnate or pinnatisect; **pinnules** narrow, oblong, mostly 1.5-3cm × 2mm, the margins very finely toothed. **Fertile fronds** with slightly narrower segments. 2n=232. *See* Fig. 115, C.

A casual greenhouse escape.

Native to Australia, New Zealand and Fiji.

PTERIS VITTATA L.
(*P. longifolia* auct., non L.)
Ladder Brake; Chinese Brake; Rhedynen Ruban Ysgolddail

Differs from *P. cretica* in having pinnae with cordate to truncate bases. **Fronds** to 60cm or more but with short stipes (up to *c.*10cm), ascending or erect, arching at the tips, ovate-lanceolate, 1-pinnate; **pinnae** regular, closely spaced, narrowly tapering, with auricled bases, becoming gradually shorter towards the base of the frond, more or less leathery, dark green, glossy. Pinnae on some of the longer fronds with more than 9 pairs of pinnae. 2n=116. *See* Fig. 115, D.

Brickwork near greenhouses at Oxford Botanic Garden and Chelsea Physic Garden. Formerly naturalized on hot colliery tip in W. Gloucs.

Native to Mediterranean, South Africa, Madagascar, Malaysia, Japan, New Guinea and Australia and other tropical and subtropical areas.

Genus THELYPTERIS Schmidel
Marsh Ferns - Marchredyn y Gors

Rhizomes long and slender, creeping, with small, soft, hair-like scales. Fronds pinnate; pinnae pinnatifid to almost pinnate; veins free, the secondaries mostly forked, the ultimate venules all reaching the margin; stipe at first bearing a few small, thin scales and minute hairs, becoming glabrous with age; blade not conspicuously glandular. Indusia present. Probably 4 species, in north temperate regions, southern Africa, SE Asia and New Zealand.

THELYPTERIS PALUSTRIS Schott

(*T. thelypteroides* Michx. subsp. *glabra* Holub; *Dryopteris thelypteris* (L.) A.Gray; *Lastrea thelypteris* (L.) Bory)

Marsh Fern; Marchredynen y Gors

Rhizome long (up to 1m) and slender (0.25cm or less), branched at intervals and creeping extensively just below the surface, bearing (at the apex and on the leaf buds only) a few small yellow to grey-brown lanceolate hair-pointed scales which are very delicate and soon lost. **Fronds** borne singly or sometimes in small tufts at longer or shorter intervals on the rhizome, of two kinds, shorter sterile ones 15-60cm (developed in May or

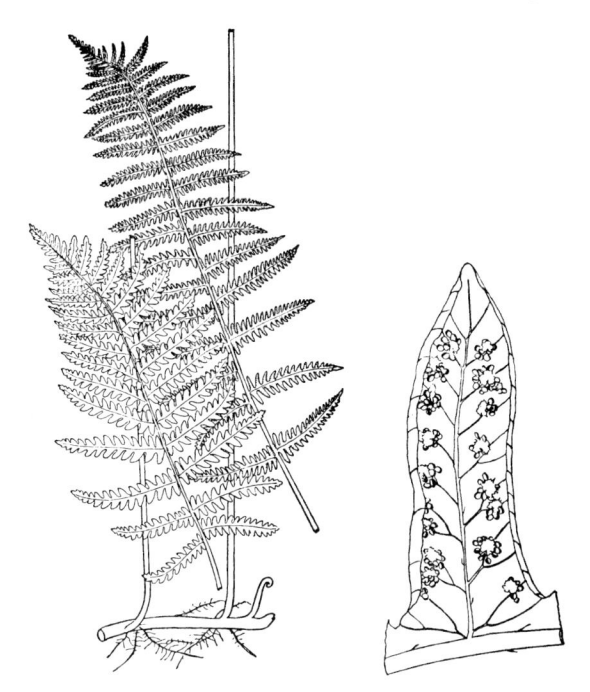

Fig. 92. Marsh Fern (*Thelypteris palustris*). Left: sterile frond and fertile frond both attached to the rhizome (1/4). Right: single segment of a median pinna from a fertile frond (×5.5).

early June), and taller, stouter and stiffer fertile ones 30-100cm (developed a month later); **stipe** of sterile frond as long as or somewhat shorter than the rachis, that of fertile frond usually somewhat longer than its rachis, both kinds brittle, black at the base, otherwise straw-yellow to greenish, usually without scales or with a few ovate-lanceolate scales below. **Blade** of sterile frond 7.5-37.5 × 5-15cm, of fertile frond 15-50 × 7.5-10cm, both lanceolate in outline tapering at both ends (but the apex sometimes narrowing suddenly), soft and membranous in texture, yellow- to bluish-green, without conspicuous glands, pinnate; **pinnae** up to about 25 on either side, opposite or sub-opposite below and sometimes nearly to the apex, linear, lanceolate, shortly stalked to practically sessile, pinnatifid to almost pinnate, **pinna segments** of the sterile fronds oblong-ovate, obtuse or acute, those of the fertile fronds narrower and more acute (resembling the teeth of a comb) with the margins slightly wavy and reflexed over the sori, the lowest pair sometimes much longer than the rest, **rachis** with scattered minute white hairs, those of the pinnae at first with a few very small scales, all becoming glabrous with age. **Sori** borne on fertile fronds to the extreme tip and often also to the base; small, rounded, situated just above the forks of the veins (on both branches when the secondary veins are forked), forming two rows one on either side of the midrib of the segment and midway between it and the margin (apparently marginal owing to the reflexing of the actual margin), the under surface of the segment ultimately covered with an uninterrupted mass of sporangia. **Indusium** small, delicate, roundish kidney-shaped, deciduous, its margin irregularly toothed or torn and glandular. To some extent it resembles *Oreopteris limbosperma* but may be distinguished by the creeping rhizome and scattered fronds, the long stipes, the relatively longer (not short and triangular) lower pinnae, the invariably inrolled edges of the fertile segments (in *O. limbosperma* the edges are sometimes recurved) and the absence of glands. **Spores** ripening July-August. Diploid. 2n=70.

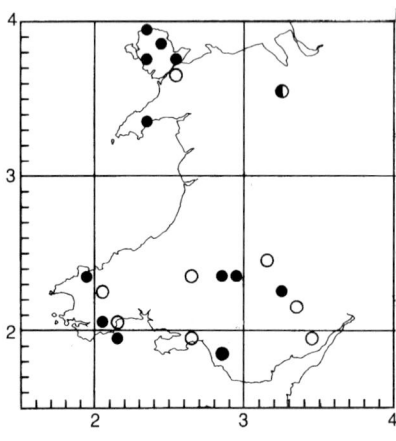

THELYPTERIS PALUSTRIS

Fens and marshy thickets; rare.

First recorded, 1796; Aikin in, *Arr. Brit. Plants*, edn 3, *3*, 776. 'In a moist dell at the foot of Snowdon near Llanberris.'

Scattered throughout the British Isles north to central Scotland, frequent only in East Anglia.

Cosmopolitan. Throughout almost the whole of Europe but becoming less frequent in the Mediterranean region. Azores. N. Africa, tropical and South Africa. Asia: Turkey and Caucasus and in temperate Asia eastwards to China, Korea, Japan and the Kamchatka peninsula. Himalaya and Southern India. South-eastern Canada and eastern half of USA. New Zealand.

Genus TRICHOMANES L.
Killarney Ferns - Rhedyn Gwrychog

Indusium tubular, flask-shaped or bell-shaped, its margin either indistinctly bivalved or entire. Receptacle thin, elongated and bristle-like, often projecting beyond the indusium (Fig. 93). Ferns similar in general appearance to *Hymenophyllum* but of firmer texture and varying greatly in size and development.

Gametophyte: most other ferns produce gametophytes as a single unit but in *Hymenophyllum* and *Trichomanes* they branch repeatedly forming mats of prothallial tissue. In *Hymenophyllum* they resemble tiny mats but in *Trichomanes* they consist of a network of branched filaments (×10 lens). In both genera the mat can continue branching to cover several square centimetres. Growing this way the gametophytes can persist for many years independently of the sporophyte. In most other ferns the gametophyte would die after producing the new sporophyte. Further details about the gametophytes of *Trichomanes* and *Hymenophyllum* in the British Isles have been published (Rumsey & Sheffield, 1990; Rumsey, Raine & Sheffield, 1993). About 25 or more species, nearly all Tropical.

TRICHOMANES SPECIOSUM Willd.

(*T. radicans* auct.; *Vandenboschia radicans* (Sw.) Copel.)

Killarney Fern; Bristle-fern; Rhedynen Wrychog

Rhizome stout, greater than 1.5mm diameter, creeping extensively over the substratum, relatively strong, black, covered with dark-coloured, jointed hairs. **Fronds**

Fig. 93. Killarney Fern (*Trichomanes speciosum*). Portion of primary pinna, showing a sorus with elongated bristle-like re-ceptacle projecting from within the cup-like indusium (*c.*×10). (*See also* Fig. 99*C*).

arising at intervals of 0.5-4cm; **stipe** as long as or rather shorter than the blade, erect, wiry and winged above (as is the rachis). **Blade** 5-20 × 2.5-10cm, ovate-lanceolate in outline, dark green, somewhat firm in texture, persisting for three years, pinnate; **pinnae** alternate to almost opposite lanceolate, twice or thrice pinnatifid. **Sori** situated on the margins of the upper pinnae, usually on the sides of the primary pinna segments which face towards the apex of the pinna, and replacing the lowest secondary pinna segments. **Indusium** 1.5-2mm, narrowly bell-shaped. **Receptacle** (when perfect) projecting 5mm or more. **Spores** ripening July-September. 2n=144.

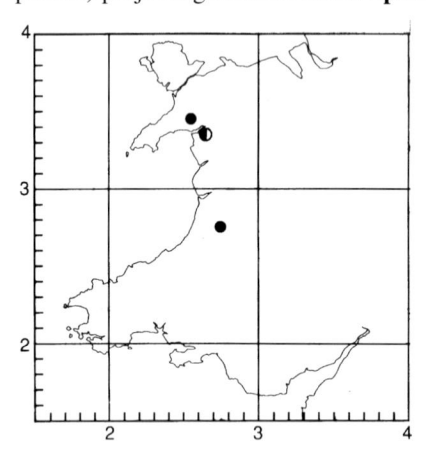

TRICHOMANES SPECIOSUM
sporophyte

Gametophyte: for description see under genus. In recent years this stage of the fern, which exists independently of the sporophyte, has been found at several sites in Wales.

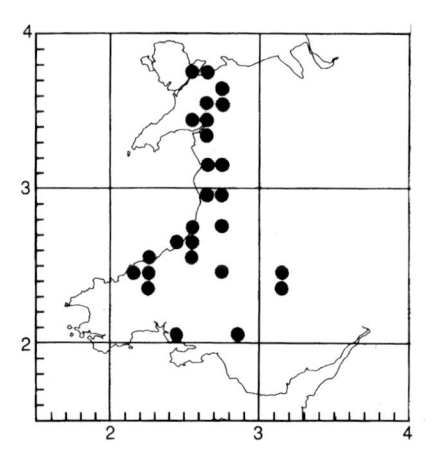

On rocks by shady streams and waterfalls; rare. First recorded, 1863; J. F. Rowbotham, ex A. M. Gibson, in *Phyt.*, 2nd ser. 6, 608. North Wales.

Rare in the west of Great Britain from the SW of England to Arran (v.c.100); mainly in SW Ireland.

Protected in Great Britain under the *Wildlife and Countryside*

TRICHOMANES SPECIOSUM
gametophyte

Act, 1981; **and in Ireland** under the *Flora Protection Order, 1987* and the *Wildlife* (NI) *Order, 1985.*

Extreme west of Europe eastwards to the Pyrennes. Azores, Madeira and Canary Is. W. Africa. Asia: in Himalaya, Burma, China and Japan. West Indies to Ecuador and Brazil.

TRICHOMANES VENOSUM R.Br.
(*Polyphlebium venosum* (R.Br.) Copel.)

Veiled Bristle-fern; Rhedynen Wrychog Orchuddiedig

Rhizome filiform, less than 1.5mm in diameter. **Blade** simply pinnate or pinnate-pinnatifid.

Established in a garden in W. Cornwall for over a century, perhaps brought in with *Dicksonia antarctica* its characteristic host in New South Wales, Australia.

Australasia.

Genus WOODSIA R.Br.
Woodsias - Coredyn

Sori on the backs of the veins and near their endings. Indusium surrounding the receptacle at the base, cup-shaped and deeply laciniate. Sporangia few, developing basipetally.

Small ferns of high mountains and rocky places, with scaly rhizomes and the leaves pinnate or bipinnate with free-ending veins, the stipe (in the subgenus *Woodsia* to which the Welsh species belong) jointed some distance above its attachment to the rhizome. 40 species.

Immature fronds of *Athyrium* species and *Cystopteris fragilis* can be confused with sterile *Woodsia*, but the former two have fine scales or hairs on the unfurling fronds (× 10 lens).

KEY TO SPECIES

Longest pinnae *c.*1.5-2 times as long as wide, oblong or ovate-oblong. Jointed hairs on both surfaces of pinnae. Scales (*c.*2-3mm) on blade underside **W. ilvensis**

Longest pinnae *c.*1-1.5 times as long as wide, triangular-ovate. Pinnae almost glabrous apart from sometimes having sparse jointed hairs on underside. Scales (*c.*1-2mm) only on stipe, rachis and costas **W. alpina**

WOODSIA ALPINA (Bolton) Gray

Alpine Woodsia; Coredynen Alpaidd

Stock short, 2-3 × 0.2cm, upright or ascending, more or less branched, thickly clothed with persistent frond-bases and numerous fine roots and having a few scales above. **Fronds** 3-15cm; **stipe** 1/4-2/3 as long as rachis, jointed near or below the middle, pale reddish-brown, with a few broadly

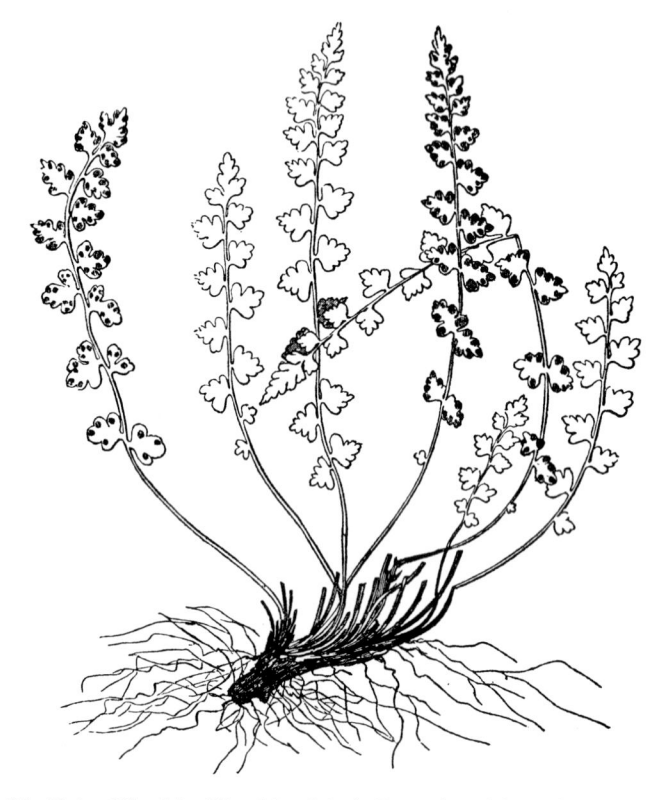

Fig. 94. Alpine Woodsia (*Woodsia alpina*). Reproduced from the woodcut in Edward Newman's *History of British Ferns*, 1854 (*c.*×2/3).

subulate scales near the base. **Blade** 2-10 × 0.8-2.3cm, linear-lanceolate to lanceolate, bluntly pointed to rounded at the apex, not always narrowed at the base, pinnate; **pinnae** about 8-12 on either side, usually opposite or sub-opposite and (in strong plants) wide apart at the base, becoming alternate and closer together towards the apex, ovate or triangular, pinnatifid, with 3-7 rounded, entire or faintly wavy-margined segments; **rachis** and under surface of veins and edges of pinnae with a few narrowly subulate to hair-like scales intermixed with zigzag jointed hairs. **Sori** and indusia as in *W. ilvensis*. Resembles *W. ilvensis* very closely but the fronds are smaller, narrower and more delicate and have fewer scales; the pinnae also are ovate in outline, whereas in *W. ilvensis* they are oblong. **Spores** ripening July-August. Tetraploid. 2n=164.

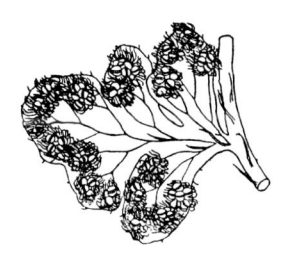

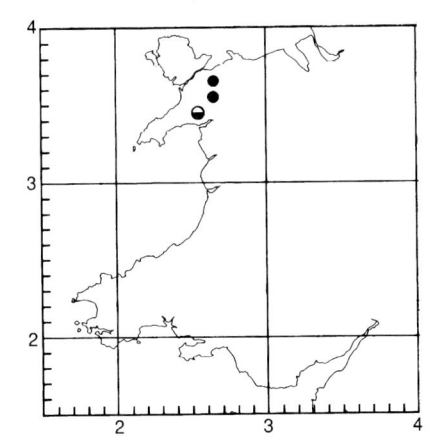

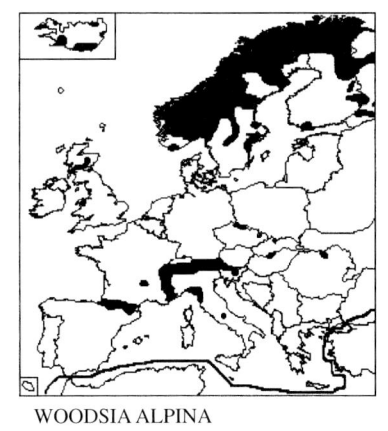

Fig. 95. Alpine Woodsia (*Woodsia alpina*). Single pinna (c.×5).

WOODSIA ALPINA

Damp rock crevices on high mountains; very rare.

First recorded, 1790; Knowlton in Bolton, *Filices Brit.*, Pt 2, 76. 'From the mountains of Wales.'

Very rare in the rest of Great Britain - only in central Scotland.

Protected in Great Britain under the *Wildlife and Countryside Act, 1981*.

Arctic and mountainous regions of the North Temperate zone. Europe: mostly Scandinavia, the Carpathians, Alps and Pyrennes, Rossia (N,C,W), Iceland. Asia: Turkey, Caucasus, Himalaya and across southern Siberia to the Kamchatka peninsula. N. America: from Alaska across northern Canada to NE USA and round Lake Superior. Greenland.

WOODSIA ILVENSIS (L.) R.Br.

Oblong Woodsia; Coredynen Hirgul

Stock short, up to 4mm thick, erect or ascending, branched thickly clothed with the persistent frond bases and having a few scales above. **Fronds** 5-10(-15)cm, **stipe** half as long to as long as the blade, jointed near the middle, pale reddish-brown, clothed (as is the rachis) with similarly coloured broadly subulate scales intermixed with jointed zigzag hairs. **Blade** 2.5-10 × 0.8-3.5cm, lanceolate, oblong-lanceolate or linear in outline, obtuse

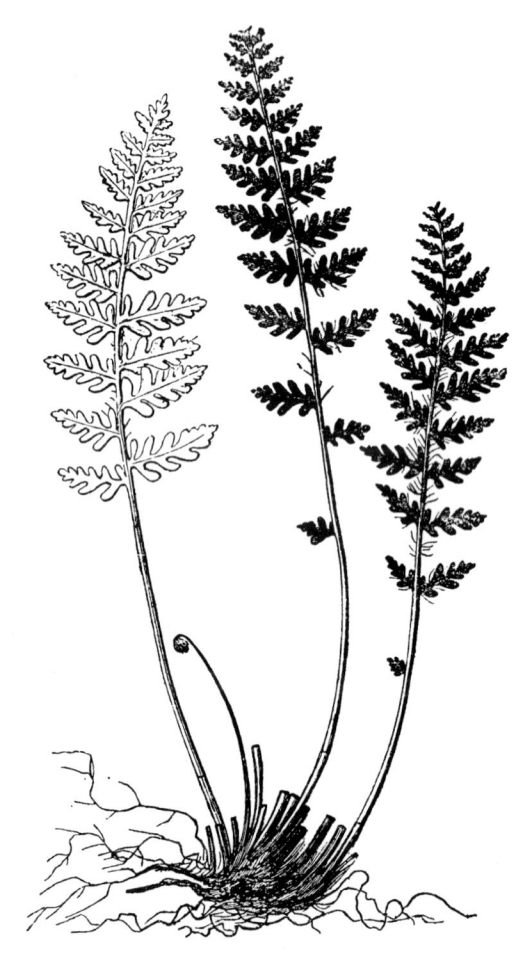

Fig. 96. Oblong Woodsia (*Woodsia ilvensis*). Re-produced from the wood-cut in Edward Newman's *History of British Ferns*, 1854 (*c.*×2/3).

to almost acute at the apex, as a rule somewhat narrowed at the base, pinnate; **pinnae** about 7-15 on either side, opposite or sub-opposite at the base, becoming alternate towards the apex, widely spaced except at the apex, oblong to ovate-oblong, rounded at the apex, deeply pinnatifid, with 7-13 oblong, obtuse segments, the margins of which are ciliate and (more especially on the acroscopic side) more or less crenate; **rachis** and under surface of pinnae densely covered with lanceolate to subulate or hair-like scales intermixed with jointed zigzag hairs. **Sori** circular, situated on the backs of the venules, a little short of their endings. **Indusium** membranous, cup-shaped, basal, completely but narrowly encircling the receptacle below its margin, irregularly lobed and beset with long, jointed hairs which at first completely enclose the sporangia. For comparison with *W. alpina* see under that species. **Spore**s ripening July-August. Diploid. 2n=82.

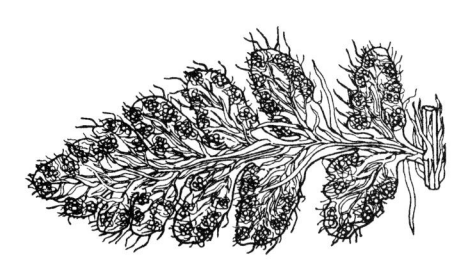

Fig. 97. Oblong Woodsia (*Woodsia ilvensis*). Single pinna (*c.*×3).

Damp rock crevices on high mountains; very rare.

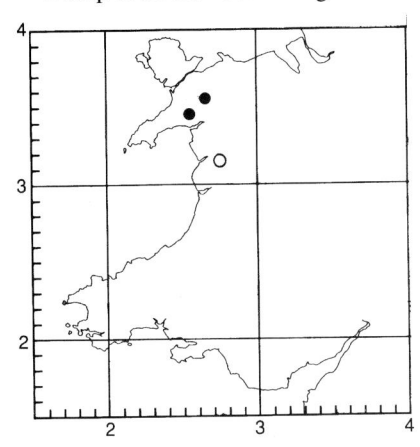

WOODSIA ILVENSIS

First recorded, 1690; Llwyd in Ray, *Syn.*, edn 1, 27 .'CIogwyn y Garnedh' [Caern.].

Very rare in the rest of Great Britain: Cumberland, Dumfriess. and Angus; formerly more widespread.

Protected in Britain under the *Wildlife and Countryside Act, 1981.*

Arctic and mountainous regions of the North Temperate zone. Europe: Scandinavia through central and eastern Europe as far south as the Alps. Iceland. Asia: Caucasus but mostly across northern and mid Siberia. N. America: from Alaska across Canada and NE USA and south through the Appalachians. Greenland.

Genus WOODWARDIA Sm.

Chain-ferns - Rhedyn Cadwyn

Fronds coriaceous, borne in tufts at ends of rhizome branches, spirally coiled when young, sparsely scaly, of one kind (two kinds in *Blechnum*) with deeply acutely lobed pinnae (1-pinnate with entire pinnae in *Blechnum*). Sori stout, discrete, in row either side of pinna-lobe (linear, continuous either side of pinna midrib in *Blechnum*). A genus of 10 species from the northern hemisphere.

WOODWARDIA RADICANS (L.) Sm.

European Chain-fern; Rhedynen Gadwyn Ewropeaidd

Fronds to 2m that reproduce by spores, and vegetatively by roots at the tips of the frond, produced from scaly buds. 2n=68. For differences between this genus and *Blechnum* see under description of the genus *Woodwardia*. A garden relic. *See* Fig. 114, D.

Regenerates in gardens in W. Cornwall and S. Kerry; formerly in the Channel Islands.

Native to the Mediterranean as far east as Crete. Azores, Madeira and Canary Islands.

OTHER ALIEN SPECIES

This list consists of other species which survive or did survive as greenhouse escapes or spread from planted origins in defunct estates or old gardens. Some could have been introduced in their lower reproductive state as spores or gametophytes unintentionally with other pteridophytes such as imported tree-ferns, and have managed to survive in the milder parts of the British Isles.

Adiantum venustum D.Don (Evergreen Maidenhair). Native to Himalaya.

Culcita macrocarpa C.Presl. Native to SW Europe and Azores.

Dennstaedtia punctiloba (Michx.) T.Moore (Hay-scented Fern). Native to eastern half of N. America.

Hymenophyllum flabellatum Labill. (Shiny Filmy-fern): pinnule margins are entire (toothed in the two native species); on *Dicksonia antarctica* in a garden in S. Kerry, Ireland. Native to Australasia.

Hypolepis tenuifolia (G.Forst.) Bernh. (Soft Ground Fern). Native to Australasia and Polynesia.

Leptolepia novae-zelandiae (Colenso) Mett. ex Diels (*Davallia zetlandia* Colenso) (Lace Fern): an 1874 record from Yorkshire. Native to New Zealand.

Osmunda claytoniana L. (Interrupted Fern). Native to India and eastern N. America.

Polystichum acrostichoides (Michx.) T.Moore (Christmas Fern). Native to eastern N. America.

Pyrrosia rupestris (R.Br.) Ching (Rock Felt Fern). Native to Australia.

Rumohra adiantiformis (G.Forst.) Ching (Leathery Shield Fern): on *Dicksonia antarctica* in a garden in S. Kerry, Ireland. Native to Australasia, Africa and America.

Tmesipteris Bernh. species (A Fork-fern): established on tree-ferns in S. Kerry and Dublin. The genus belongs to the Psilotaceae, a primative family allied to ferns. Genus native to Australasia.

ALIEN SPECIES RECORDED IN ERROR FOR THE BRITISH ISLES

Adiantum reniforme L. for *Asplenium ruta-muraria* L.

Botrychium lanceolatum (S.G.Gmelin) Ångstr. for *B. lunaria* (L.) Sw.

Botrychium matricariifolium A.Braun ex W.D.J.Koch for *B. lunaria* (L.) Sw.

Botrychium multifidum (S.G.Gmel.) Rupr. for *B. lunaria* (L.) Sw.

Polystichum braunii (Spenn.) Fée for *P. setiferum* (Forssk.) Woyn.

Pteris longifolia L. for *P. vittata* L.

Selaginella denticulata (L.) Link for *S. kraussiana* (Kunze) A. Braun.

Selaginella helvetica Link doubtfully from the Mendip Hills in 1726 and Co. Down in 1896.

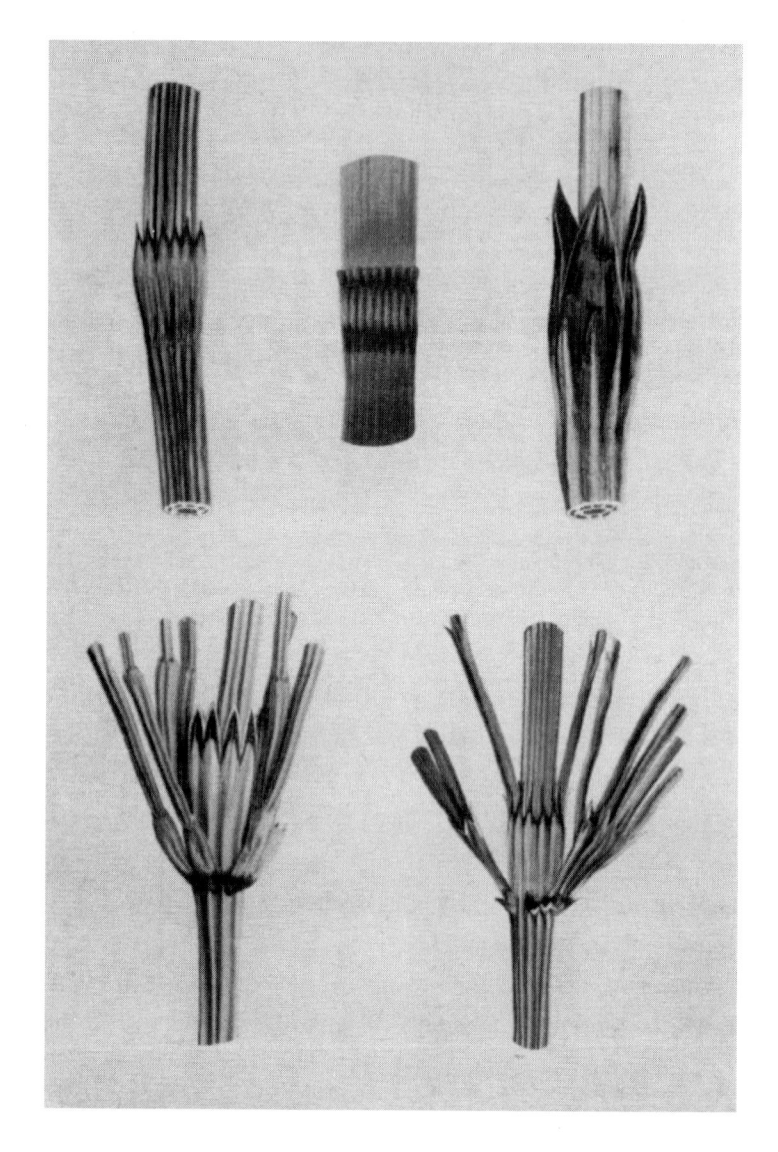

Fig. 98. Horsetails (*Equisetum*). Single nodes, showing characteristic sheaths and teeth.
Above: *Equisetum fluviatile*, *E. hyemale* and *E. sylvaticum*.
Below: *E. palustre* and *E. arvense*.

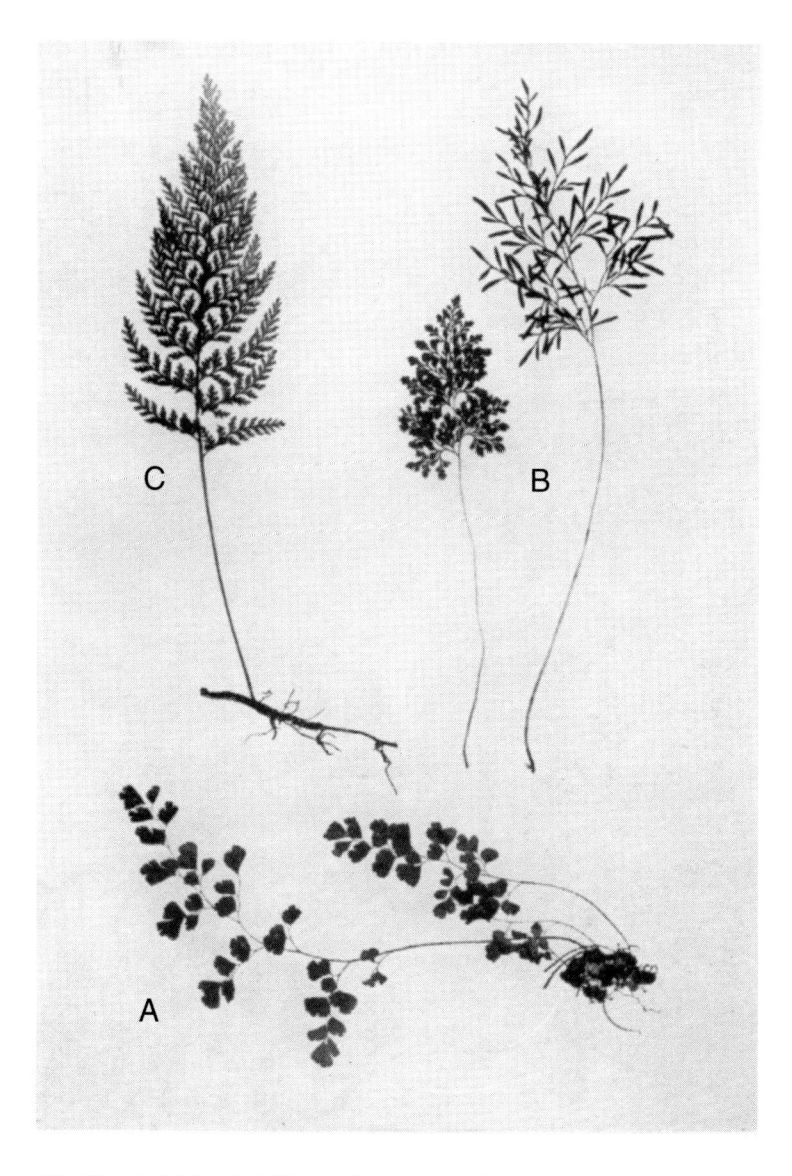

Fig. 99. A. Maidenhair Fern (*Adiantum capillus-veneris*).
B. Parsley Fern (*Cryptogramma crispa*) (sterile and fertile fronds).
C. Killarney Fern (*Trichomanes speciosum*).
All ×1/3.

Fig.100. *A.* Black Spleenwort (*Asplenium adiantum-nigrum*).
B. Sea Spleenwort (*A. marinum*).
C. Lanceolate Spleenwort (*A. obovatum* subsp. *lanceolatum*).
All ×1/2.

Fig. 101. Above: Rustyback (*Asplenium ceterach*) growing in a dry and sunny situation on a mortared limestone wall.
Below: Hart's-tongue (*Asplenium scolopendrium*) growing in a damp and shady situation on old limestone walling in a wood.

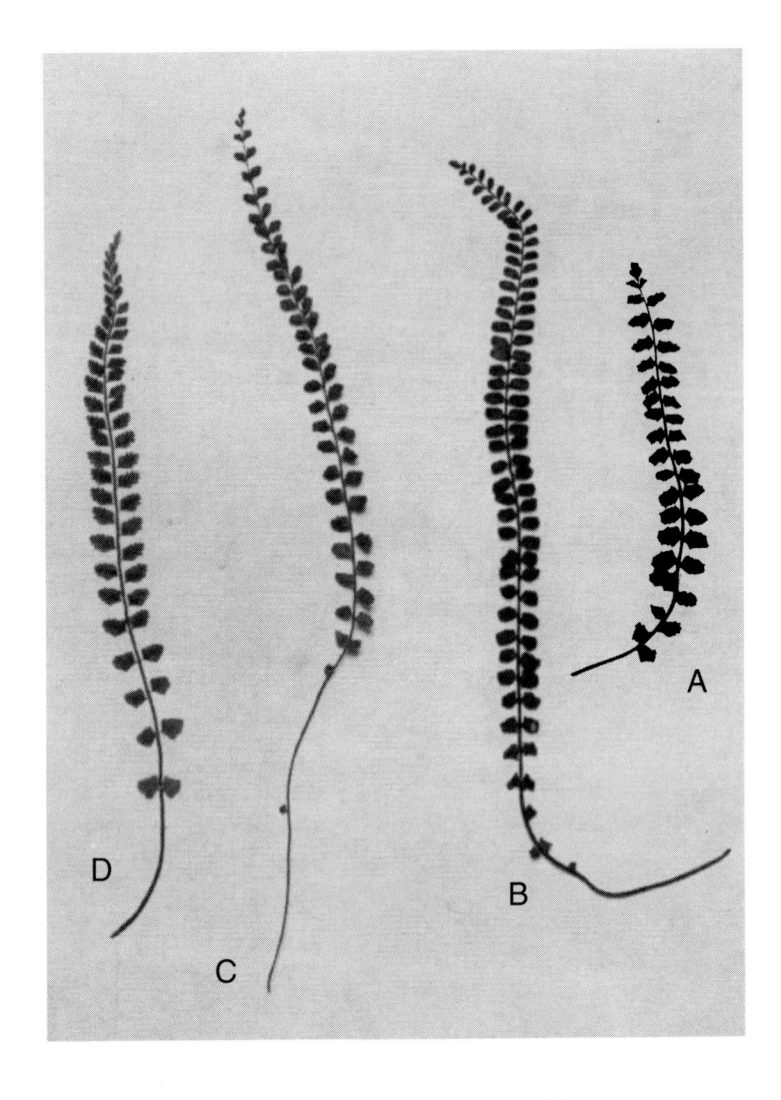

Fig. 102. A. Lobed Maidenhair Spleenwort (*Asplenium trichomanes* subsp. *pachyrachis*) (×3/4).
B. Common Maidenhair Spleenwort (*Asplenium trichomanes* subsp. *quadrivalens*) (B-D, ×1/2).
C. Delicate Maidenhair Spleenwort (*A. trichomanes* subsp. *trichomanes*).
D. Green Spleenwort (*A. trichomanes-ramosum*).

Fig. 103. A. Lady-fern (*Athyrium filix-femina*).
B. Brittle Bladder-fern (*Cystopteris fragilis*).
C. Rigid Buckler-fern (*Dryopteris submontana*).
All ×1/3.

Fig. 104. Hard-fern (*Blechnum spicant*). Rhizome with two sterile fronds and one fertile frond; the sterile fronds have been raised into an upright position for the purpose of making the illustration, and the remaining fronds have been removed. Reproduced from Moore's *Ferns of Great Britain and Ireland* (Nature-printed by Henry Bradbury), Plate XLIII, C 1. (*c.*×1/4).

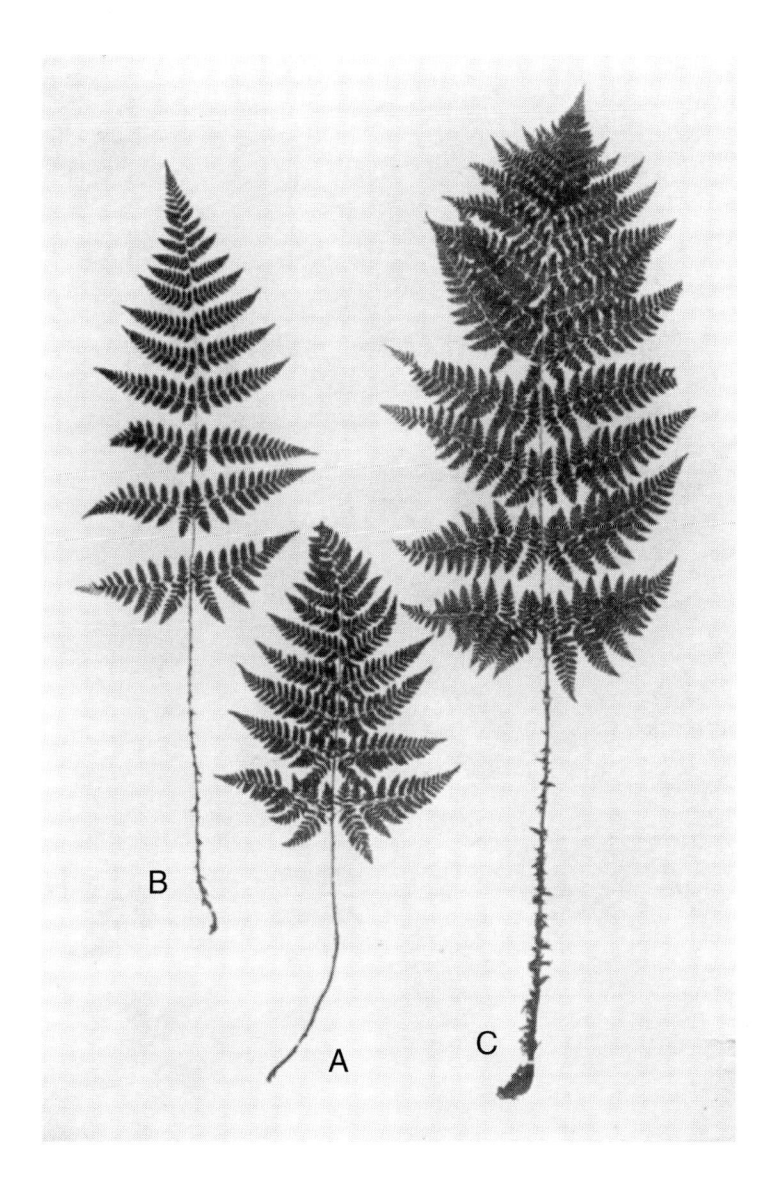

Fig. 105. *A*. Hay-scented Buckler-fern (*Dryopteris aemula*).
B. Narrow Buckler-fern (*D. carthusiana*).
C. Broad Buckler-fern (*D. dilatata*).
All ×1/4.

Fig. 106. Northern Buckler-fern (*Dryopteris expansa*) (×1/2).

Fig. 107. *A*. Scaly Male-fern (*Dryopteris affinis*).
B. Male-fern (*D. filix-mas*).
C. Dwarf Male-fern (*D. oreades*).
All ×1/5.

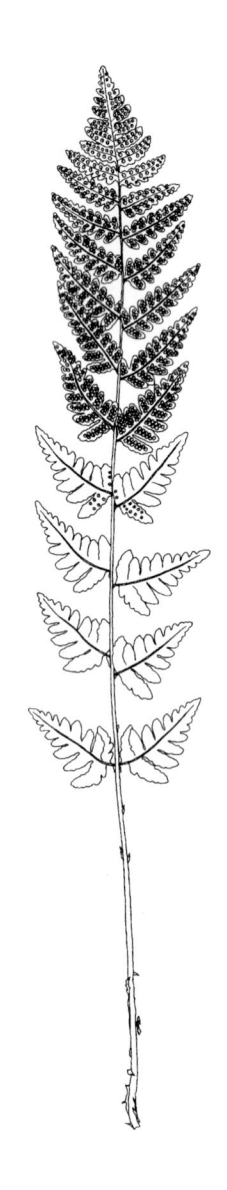

Fig. 108. Crested Buckler-fern (*Dryopteris cristata*). Underside of fertile frond (×1/4).

Fig. 109. A. Oak Fern (*Gymnocarpium dryopteris*).
B. Limestone Fern (*G. robertianum*).
C. Beech Fern (*Phegopteris connectilis*).
All ×1/4.

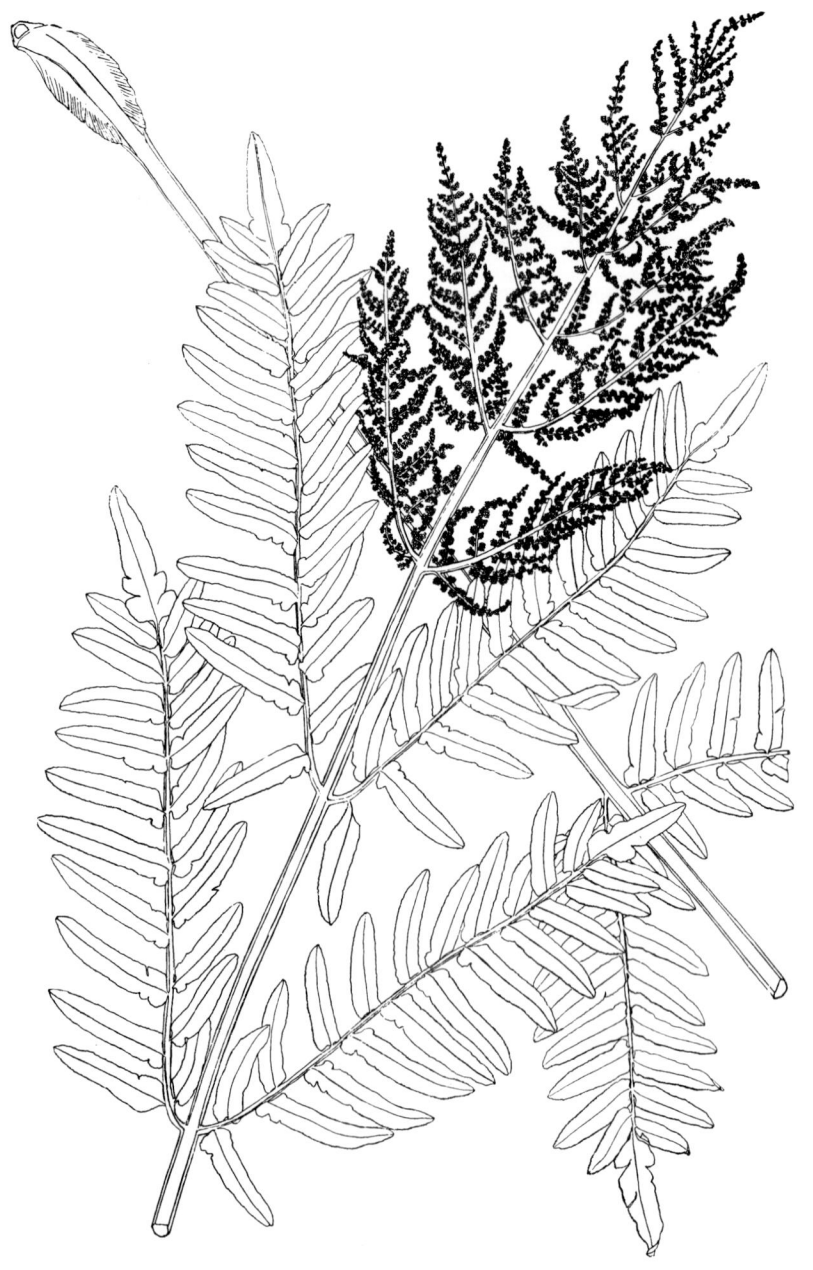

Fig. 110. Royal Fern (*Osmunda regalis*) complete frond (×1/3).

Fig. 111. *A*. Hard Shield-fern (*Polystichum aculeatum*).
B. Soft Shield-fern (*P. setiferum*).
Both ×1/4.

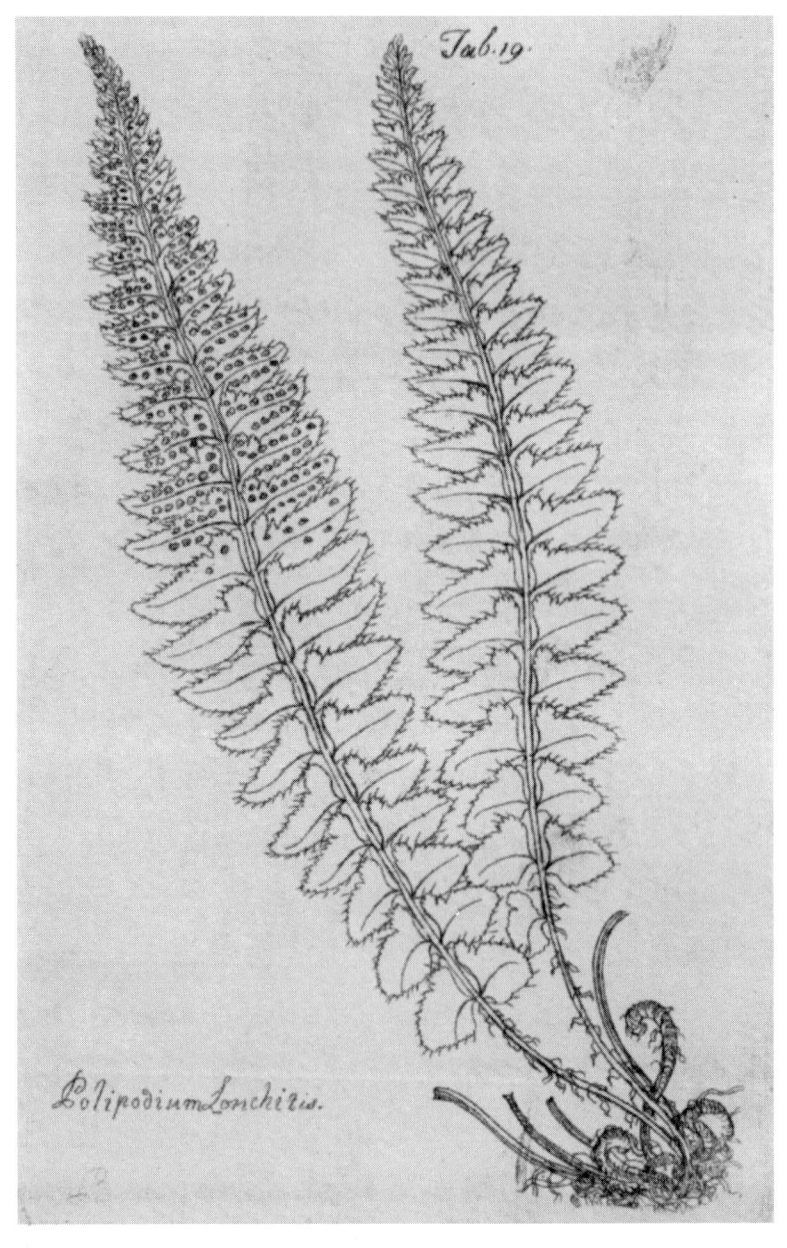

Fig. 112. Holly-fern (*Polystichum lonchitis*).Reproduced from the engraving in Bolton's *Filices Britannicae* (1785). 'Polipodium lonchitis grows ... on the rocks of *Glydar*, near Llanberris, where I gathered the specimen here figured.' Natural size.

217

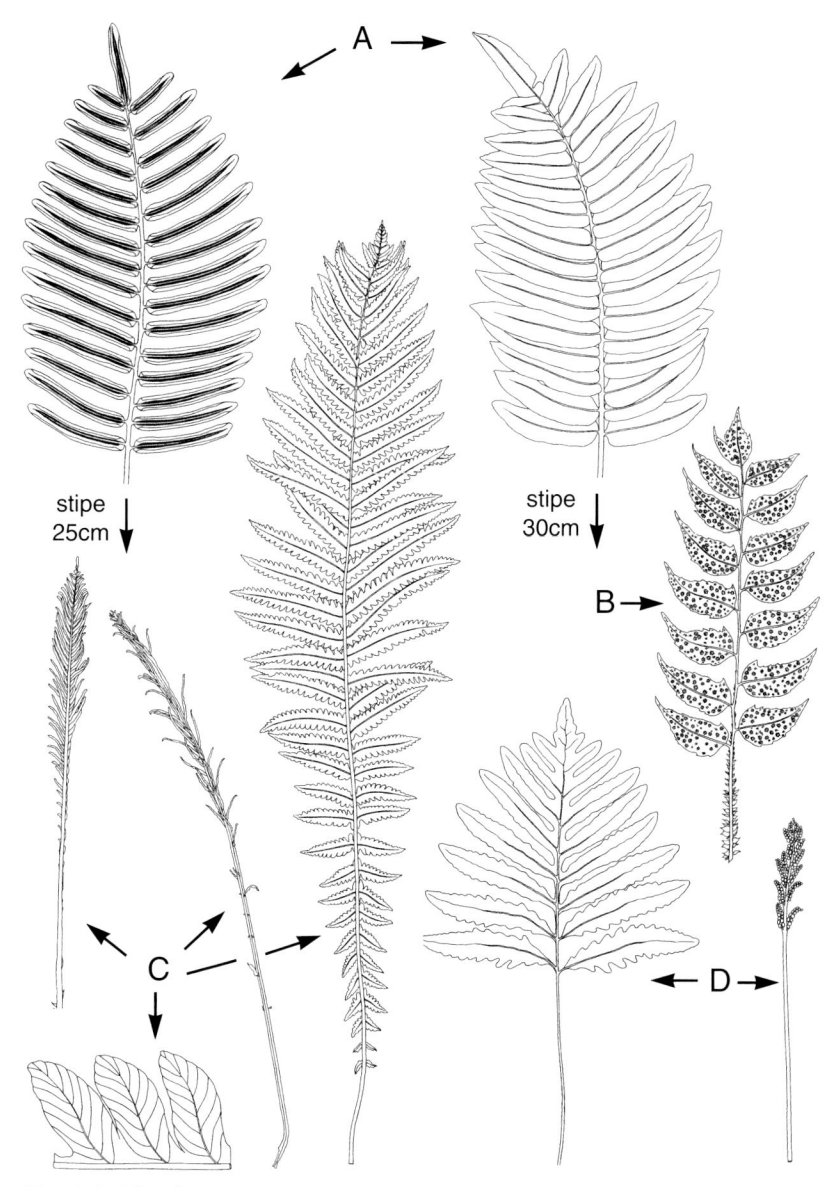

Fig. 113. Alien ferns recorded from Wales (Fronds all ×1/5). For a long stipe, the life-size length is given. *A*. Chilean Hard-fern (*Blechnum cordatum*). Fertile and sterile fronds. *B*. House Holly-fern (*Cyrtomium falcatum*) (pinnae fertile). *C*. Ostrich Fern (*Matteuccia struthiopteris*). 2 fertile & 1 sterile frond; 3 pinna segments (×2). *D*. Sensitive fern (*Onoclea sensibilis*). Sterile and fertile fronds.

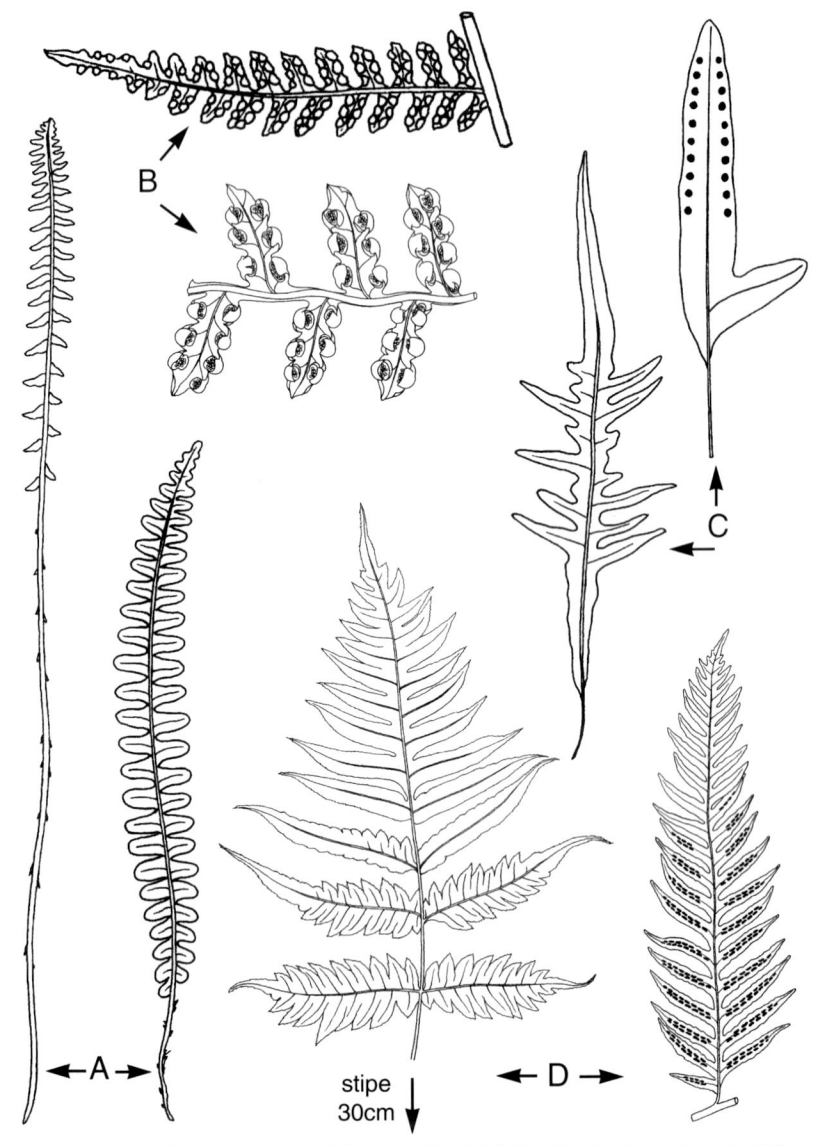

Fig. 114. Other alien ferns recorded from the British Isles. For the long stipe, the life-size length is given. *A*. Antarctic Hard Fern (*Blechnum penna-marina*). Fertile & sterile fronds (×3/4). *B*. Australian Tree-fern (*Dicksonia antarctica*). Fertile pinnule (×1.5) & section of fertile pinnule (×3). *C*. Kangaroo Fern (*Phymatodes diversifolia*). 2 fronds, one fertile (Both ×1/4). *D*. European Chain-fern (*Woodwardia radicans*). Young frond & fertile pinna (Both ×1/4).

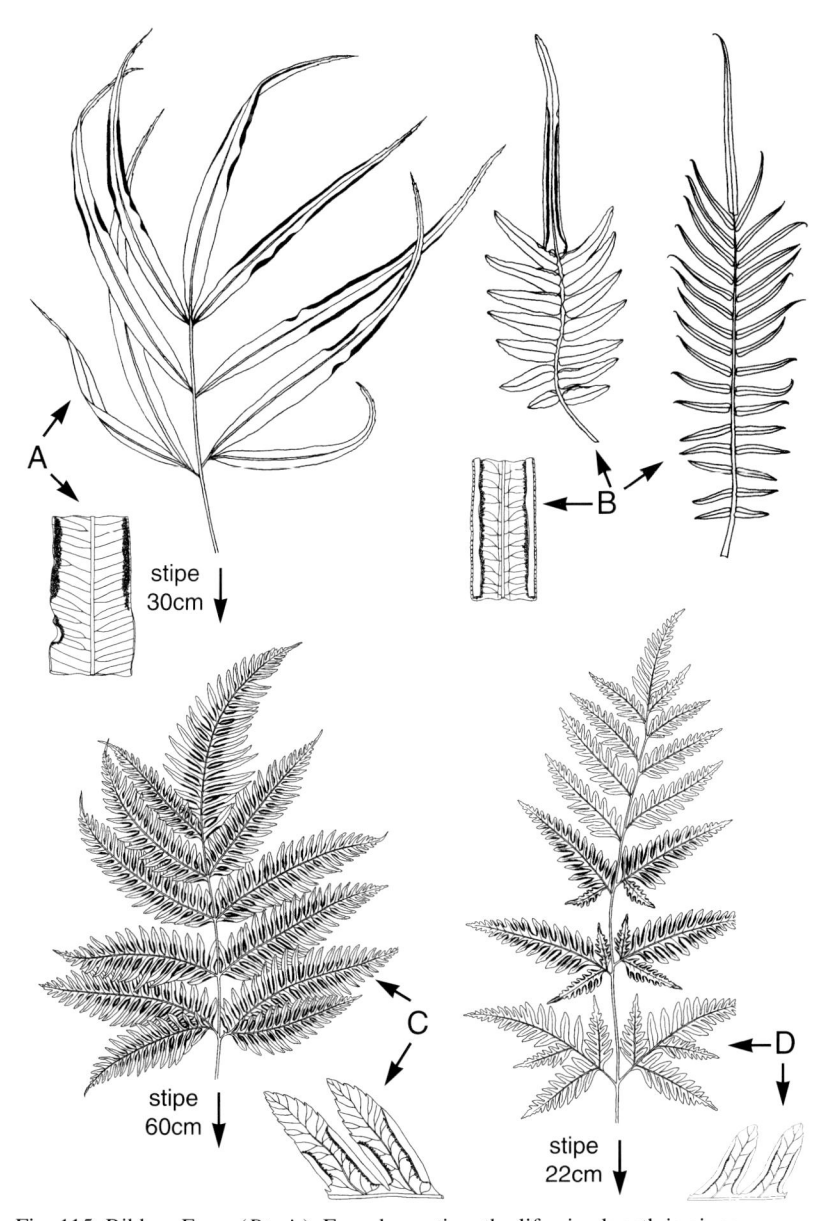

Fig. 115. Ribbon Ferns (*Pteris*). For a long stipe, the life-size length is given. A. Ribbon Fern (*Pteris cretica*) & B. Ladder Brake (*P. vittata*) (Both ×1/3; section of fertile pinna, both ×1.3). C. Spider Brake (*P. incompleta*) & D. Tender Brake (*P. tremula*) (Both ×1/5; fertile pinnules from upper part of pinna, both ×1).

BIBLIOGRAPHY

Anon. (1968). *Asplenium adiantum-nigrum* × *septentrionale*: a hybrid new to Britain. *British Fern Gazette*, **10**, 37.

Barker, T. W. (1905). *Handbook to the Natural History of Carmarthenshire*. W. Spurrell & Son, Carmarthen.

Bennett, A. (1906). *Supplement to 'Topographical Botany, edn 2'*. West, Newman & Co., London.

Bennett, A., Salmon, C. E., and Matthews, J. R. (1930). *Second Supplement to Watson's Topographical Botany*. London.

Benoit, P. M. & Richards, M. (1963). *A Contribution to a Flora of Merioneth*. Ed. 2. West Wales Naturalists' Trust, Haverfordwest.

Bingley, W. (1800). *A Tour round North Wales performed during the summer of 1798*.... 2 vols. Smeeton, London.

Bolton, J. (1785-1790). *Filices Britannicae*. Pt. 1 (1785). J. Binns, Leeds; Pt. 2 (1790). J. Brook, Huddersfield.

Boudrie, M., Garraud, L. & Rasbach, H. (1994). Discovery of *Dryopteris brathaica* in France (Dryopteridaceae: Pteridophyta). *Fern Gazette*, **14**, 237-244.

Bower, F. O. (1923-28). *The Ferns (Filicales)*. 3 vols. Cambridge University Press. Cambridge.

British Pteridological Society (1991). *The British Pteridological Society Abstracts of Reports and Papers Read at Meetings 1894-1905*. (The British Pteridological Society: Special Publication No. 5). British Pteridological Society, London.

Brummitt, R. K. & Powell, C. E. eds (1992). Authors of Plant Names. Royal Botanic Gardens, Kew.

Butcher, R. W. & Strudwick, F. E. (1930). *Further Illustrations of British Plants*. Reeve & Co., Kent.

Callé, J., Lovis, J. D. & Reichstein, T. (1975). *Asplenium* × *contrei* (*A. adiantum-nigrum* × *A. septentrionale*) hybr. nova et la Vraie ascendance de l' *Asplenium souchei* Litard. *Candollea*, **30**, 189-201.

Camus, J. M. ed. (1991). *The History of British Pteridology 1891-1991*. (British Pteridological Society Publication No. 4). British Pteridological Society, London.

Camus, J. M., Jermy, A. C. & Thomas, B. A. (1991). *A World of Ferns*. Natural History Museum Publications, London.

Christ, H. (1897). *Die Farnkräuter der Erde*. Fischer, Jena.

Christ, H. (1910). *Die Geographie der Farne*. Jena.

Christensen, C. (1906-34). *Index Filicum* (1906, for 1753-1905). *Supplement 1* (1913, for 1906-12); *2* (1917, for 1913-16); *3* (1934, for 1917-33). Hagerup, Copenhagen. [*see also* Pichi-Sermolli *and* Jarrett].

Christensen, C. (1938). Filicinae. In: *Manual of Pteridology* (ed. Fr. Verdoorn), pp. 522-550. M. Nijhoff, The Hague.

Clarke, J. H. (1868). *The Flora of Monmouthshire*. Usk.

Copeland, E. B. (1947). *Genera Filicum*. Waltham, Massachusetts.

Dandy, J. E. (1958). *List of British Vascular Plants*. British Museum, London.
Dandy, J. E. (1969). *Watsonian Vice-counties of Great Britain*. Ray Society, London. [Includes a North and South map of Great Britain showing vice-county boundaries].
Davies, D. & Jones, A. (1995). *Enwau Cymraeg ar Blanhigion - Welsh Names of Plants*. National Museum of Wales, Cardiff.
Davies, H. (1813). *Welsh Botanology*. W. Marchant, London.
Davis, T. A. W. (1970). *Plants of Pembrokeshire*. West Wales Naturalists Trust, [Haverfordwest].
Derrick, L. N. , Jermy, A. C. & Paul, A. M. (1987). Checklist of European Pteridophytes. *Sommerfeltia*, **6**, 1-94.
Dillwyn, L. W. (1848). *Materials for a Fauna and Flora of Swansea and the Neighbourhood*. D. Rees, Swansea.
Druce, G. C. (1907). *The Dillenian Herbaria*. (ed. S. H. Vines). Clarendon Press, Oxford.
Druce, G. C. (1928). *British Plant List*. T. Buncle & Co., Arbroath.
Druce, G. C. (1932). *The Comital Flora of the British Isles*. T. Buncle & Co., Arbroath.
Druery, C. T. (1910). *British Ferns and their Varieties*. G. Routledge & Sons, London.
Duval-Jouve, J. (1864). *Histoire Naturelle des Equisetum de France*. Paris.
Dyce, J. W. (1988). *Polypodium cambricum* 'Cambricum'. *Pteridologist*, **1**, 217-220.
Dyce, J. W. (1991). *The Cultivation and Propagation of British Ferns* (British Pteridological Society Special Publication No. 3). British Pteridological Society, London.
Engler, A. & Gilg, E. (1919). *Syllabus der Pflanzenfamilien*. Ed. 8, Berlin.
Engler, A. & Prantl, K. (1902). *Die Natürlichen Pflanzenfamilien*. Vol. **1(4)**. Engelmann, Leipzig.
Evans, Rev. J. (1804). *Letters written during a tour through North Wales in the year 1798 and at other times*. Ed. 2. London.
F[alconer], R. W. (1848). *Contributions towards a Catalogue of Plants Indigenous to the Neighbourhood of Tenby*. London.
Fraser-Jenkins, C. R. (1988). Some comments on the nomenclature of *Dryopteris*. *Indian Fern Journal*, **5**, 69-77.
Fraser-Jenkins, C. R. (1996). A reaffirmation of the taxonomic treatment of *Dryopteris affinis* (Lowe) Fraser-Jenkins. *Fern Gazette*, **15**, 77-81.
Gibson, E. (1695). *Camden's Britannia*. London.
Griffiths, J. E. [1895]. *The Flora of Anglesey and Carnarvonshire*. Nixon & Jarvis, Bangor.
Gunther, R. W. T. (1922). *Early British Botanists*. Oxford University Press, Oxford.
Hamilton, S. (1909). *The Flora of Monmouthshire*. J. E . Southall, Newport.
[Hanbury, F. J. ed.] (1925). *The London Catalogue of British Plants*. Ed. 2. London.

Hooker, W. J. (1830). *The British Flora.* Ed. 1. London.
Hooker, W. J. (1846-64). *Species Filicum.* 5 vols. W. Pamplin, Dulau & Co., London.
Hooker, W. J. (1861). *The British Ferns.* L. Reeve, London.
Howe, W. (1650). *Phytologia Britannica.* London.
Hultén, E. & Fries, M. (1986). *Atlas of North European Vascular Plants.* Vol. 1. Koeltz Scientific Books, Königstein, Federal Republic of Germany.
Hutchinson, G. & Thomas, B. A. (1992). Distribution of Pteridophyta in Wales. *Watsonia,* **19**, 1-19.
Jackson, B. D. (1928). *A Glossary of Botanic Terms.* Ed. 4. Duckworth & Co., London.
Jalas, J. & Suominen, J. eds (1972). *Atlas Florae Europaeae.* Vol. 1. The Committee for Mapping the Flora of Europe & Societas Biologica Fennica Vanamo, Helsinki.
Jarrett, F. M. (1985). *Index Filicum. Supplement 5* (1961-1975). Clarendon Press, Oxford. [*see also* Christensen (1906-1934) *and* Pichi-Sermolli].
Jermy, A. C., Crabbe, J. A. & Thomas, B. A. eds (1973). *The Phylogeny and Classification of the Ferns.* Academic Press, London.
Jermy, A. C., *et al.* eds (1978). *Atlas of Ferns of the British Isles.* Botanical Society of the British Isles and British Pteridological Society, London.
Jermy, C. & Camus, J. (1991). *The Illustrated Field Guide to Ferns and Allied Plants of the British Isles.* Natural History Museum Publications, London.
Johnson, C. & Sowerby, J. E. (1859). *The Ferns of Great Britain.* J. E. Sowerby, London.
Johnson, T. (1641). *Mercurii Botanici Pars Altera.* London.
Jones, D. L. (1987). *Encyclopaedia of Ferns.* British Museum (Natural History), London.
Komarov, V. L. *et al.* eds (1934-1964). [*Flora of the USSR*]. Academii Nauk SSSR, Leningrad & Moscow.
Lankester, E. ed. (1846). *Memorials of John Ray.* Ray Society, London.
Lawalrée, A. (1950). Pteridophytes. (*Flore Générale de Belgique*). Bruxelles.
Löve, Á., Löve D. & Pichi Sermolli, R. E. G. (1977). *Cytotaxonomical Atlas of the Pteridophyta.* Cramer, Vaduz.
Lowe, E. J. (1866-67). *Ferns: British and Exotic.* 8 vols. Groombridge & Sons, London.
Lowe, E. J. (1867-69). *Our Native Ferns.* 2 vols. Groombridge & Sons, London.
Luerssen, C. (1889). *Die Farnpflanzen oder Gefässbündelkryptogamen (Pteridophyta).* In: L. Rabenhorst, *Kryptogamen-Flora von Deutschland, Oesterreich und der Schweiz.* Part 2, vol. 3. ed. E. Kummer, Leipzig.
Manton, I. (1950). *Problems of Cytology and Evolution in the Pteridophyta.* Cambridge University Press, Cambridge.
May, R. F. (1967). *A List of the Flowering Plants and Ferns of Carmarthenshire.* West Wales Naturalists' Trust Ltd, [Haverfordwest].

Meusel, H., Jäger, E. & Weinert, E. (1965). *Vergleichende Chorologie der Zentraleurop ischen Flora.* 2 vols. VEB Gustav Fischer Verlag, Jena, German Democratic Republic.
Milde, J. (1865). *Monographia Equisetorum. Dresden.*
Moore, T. (1848). *A Handbook of British Ferns.* Ed. 1. Groombridge & Sons, London.
Moore, T. (1855). *The Ferns of Great Britain and Ireland* (ed. J. Lindley). Nature-printed by Henry Bradbury. Bradbury & Evans, London.
Moore, T. (1859-60). *The Nature Printed British Ferns.* 2 vols. Bradbury & Evans, London.
Nelson, E. C. (1988). Some Australasian ferns in Irish gardens. *Kew Magazine,* **5**, 129-136.
Newman, E. (1840-44). *A History of British Ferns and Allied Plants.* Ed. 1 (1840); Ed. 2 (1844). John van Voorst, London.
Newman, E. (1854). *A History of British Ferns.* John van Voorst, London.
Page, C. N. (1982). *The Ferns of Great Britain and Ireland.* Cambridge Univerity Press. Cambridge.
Page, C. N. (1988). *Ferns - Their Habitats in the British and Irish Landscape.* Collins, London.
Page, C. N. (1989). Three subspecies of Bracken, *Pteridium aquilinum* (L.) Kuhn, in Great Britain. *Watsonia,* **17**, 429-434.
Page, C. N. (1995). Structural variation in Western European Bracken: an updated taxonomic perspective. In: *Bracken: an environmental issue.* Contribution to an International Conference (Aberystwyth, 1994) (Ed. by R. T. Smith & J. A. Taylor), pp. 13-15. The International Bracken Group, Aberystwyth & Leeds.
Page, C. N. & Mill, R. H. (1995a). Scottish Bracken (*Pteridium*): new taxa and a new combination. *Botanical Journal of Scotland,* **47**, 139-140.
Page, C. N. & Mill, R. H. (1995b). The Taxa of European Bracken in a European Perspective. *Botanical Journal of Scotland,* **47**, 229-247.
Palmer, M. A. & Bratton, J. H. eds (1995). *UK Nature Conservation No. 8: A Sample Survey of the Flora of Great Britain and Ireland.* (The B.S.B.I. Monitoring Scheme 1987-1988, based on a report for the Nature Conservancy Council by T. C. G. Rich & E. R. Woodruff). JNCC, Peterborough.
Paul, A. M. (1987). The status of *Ophioglossum azoricum* (Ophioglossaceae : Pteridophyta) in the British Isles. *Fern Gazette,* **13**, 173-187.
Perring, F. H. ed. (1968). *Critical Supplement to the Atlas of the British Flora.* Botanical Society of the British Isles, London.
Perring, F. H. & Walters, S. M. eds (1962). *Atlas of the British Flora.* Botanical Society of the British Isles, London.
Petiver, J. (1716). *Graminorum ... et Britannicorum Concordia.* London.
Pichi-Sermolli, R. E. G. (1965). *Index Filicum. Supplement 4* (1934-1960). International Bureau for Plant Taxonomy and Nomenclature, Utrecht, Netherlands. [*see also* Christensen (1906-1934) *and* Jarrett].

Ray, J. (1670). *Catalogus Plantarum Angliae.* London.
Ray, J. (1690-1724). *Synopsis Methodica Stirpium Britannicarum.* Ed. 1 (1690); Ed. 2 (1696); Ed. 3 (1724). London.
Rickard, M. H. (1989). Two Spleenworts new to Britain - *Asplenium trichomanes* subsp. *pachyrachis* and *Asplenium trichomanes* nothosubsp. *staufferi. Pteridologist,* **1**, 244-248.
Riddelsdell, H. J. (1907). A Flora of Glamorganshire. *Journal of Botany* [Supplement], **45**, 1-88.
Roberts, R. H. (1970). A revision of some of the taxonomic characters of *Polypodium australe* Fée. *Watsonia,* **8**, 121-134.
Roberts, R. H. (1982). *The Flowering Plants and Ferns of Anglesey.* National Museum of Wales, Cardiff.
Rumsey, F. J., Raine, C. A. & Sheffield, E. (1993). *Trichomanes venosum* R. Br. (Hymenophyllaceae: Pteridophyta) in a Cornish garden - with a key to the Filmy-ferns established in Great Britain and Ireland. *Fern Gazette,* **14**, 155-160.
Rumsey, F. J., & Sheffield, E. (1990). British Filmy-fern Gametophytes. *Pteridologist,* **2**, 40-42.
Rumsey, F. J., Sheffield, E. & Haufler, C. H. (1991). A re-assessment of *Pteridium aquilinum* (L.) Kuhn in Britain. *Watsonia,* **18**, 297-301.
Rumsey, F. J., Thompson, P. & Sheffield, E. (1993). Triploid *Isoetes echinospora* (Isoetaceae: Pteridophyta) in Northern England. *Fern Gazette,* **14**, 215-221.
Salesbury, William (1916). *Llysieulyfr Meddyginiaethol.* (ed. E. Stanton Roberts). Liverpool.
Salter, J. H. (1935). *The Flowering Plants and Ferns of Cardiganshire.* University Press Board, Cardiff.
Sheffield, E. et al. (1993). Spacial Distribution and Reproductive Behaviour of a Triploid Bracken *(Pteridium aquilinum)* Clone in Great Britain. *Annals of Botany,* 72, 231-237.
Shivas, M. G. (1961). Contributions to the cytology and taxonomy of *Polypodium* in Europe and America: 2. Taxonomy. *Journal of the Linnean Society (Botany),* **58**, 27-38.
Shoolbred, W. A. (1920). *The Flora of Chepstow.* London.
Smith, J. E. (1800-4). *Flora Britannica* (vols. I-III), III, 1102.
Smith, R. T. & Taylor, J. A. eds (1986). *Bracken: ecology, land use and control technology.* Proceedings of the International Conference (Leeds, 1985). Parthenon Publishing, Carnforth, Lancs.
Smith, R. T. & Taylor, J. A. eds (1995). *Bracken: an environmental issue.* Contribution to an International Conference (Aberystwyth, 1994). The International Bracken Group, Aberystwyth & Leeds.
Smith, W. G. (1928). *Transactions of the Botanical Society of Edinburgh,* **30**, 3.
Sowerby, J. & J. E. (1902). *English Botany.* Ed. 3, by J. T. Boswell Syme. London.

Sporne, K. R. (1962). *The Morphology of Pteridophytes.* Hutchinson & Co., London.

Stace, C. A. ed. (1975). *Hybridization and the Flora of the British Isles.* Published in collaboration with the Botanical Society of the British Isles by Academic Press, London, New York and San Francisco.

Stearn, W. T. (1992). *Botanical Latin.* 4th edn. David & Charles, Newton Abbot.

Storrie, J. (1886). *The Flora of Cardiff.* Cardiff.

Taylor, P. (1960). *British Ferns and Mosses.* Eyre & Spottiswoode, London.

Tennant, D. J. (1995). *Cystopteris fragilis (L.) Bernh.* var. *alpina* Hook. in Britain. *The Naturalist,* **120**, 45-50.

Tennant, D. J. (1996). *Cystopteris dickieana* R. Sim. in the central and eastern Scottish Highlands. *Watsonia,* **21**, 135-139.

Thompson, J. A. & Smith, R. T. eds (1990). *Bracken Biology and Management.* Papers from International Conference: Bracken 1989 (Sydney). (The Australian Institute of Agricultural Science Occasional Publication No. 40). Sydney.

Trow, A. H. (1911). *The Flora of Glamorgan.* Vol. 1. Cardiff Naturalists' Society, Cardiff.

Trueman, I., Morton, A. & Wainwright, M. eds (1995). *The Flora of Montgomeryshire.* The Montgomeryshire Field Society & The Montgomeryshire Wildlife Trust, Welshpool.

Tryon, A. F. & Lugardon, B. (1991). *Spores of the Pteridophyta.* Springer-Verlag, New York.

Turner, D. (1835). *Extracts from the Literary and Scientific Correspondence of Richard Richardson, M.D., F.R.S.* Yarmouth.

Turner, D. & Dillwyn, L. W. (1805). The Botanist's Guide. 2 vols. Phillips & Fardon, London.

Tutin, T. G., *et al.* (1993). *Flora Europaea.* Vol. 1. Ed. 2. Cambridge University Press, Cambridge.

Usher, G. (1966). *A Dictionary of Botany.* Constable, London.

Vachell, E. (1936). Flowering Plants and Ferns. In: *Glamorgan County History.* Vol. 1: Natural History. (ed. W. M. Tattersall), pp. 123-178. Cardiff.

Verdoom, Fr. ed. (1938). *Manual of Pteridology.* M. Nijhoff, The Hague.

Wade, A. E. ed. (1952). *A Supplement to Dr J. H. Salter's The Flowering Plants and Ferns of Cardiganshire.* University of Wales Press, Cardiff.

Wade, A. E. (1970). *The Flora of Monmouthshire.* National Museum of Wales, Cardiff.

Wade, A. E., Kay, Q. O. N. & Ellis, R. G. (1994). *Flora of Glamorgan.* HMSO, London.

Watson, H. C. (1873-1883). *Topographical Botany.* Ed. 1. 2 parts (1873-4). Thames Ditton; Ed. 2 (1883) B. Quaritch, London. For supplements *see under* Bennett.

Watt, A. S. (1940). Contributions to the ecology of bracken *(Pteridium aquilinum)* 1. The rhizome. *New Phytologist,* **39**, 401-422.

Williams, H. J. (1988). *Bracken in Wales.* Nature Conservancy Council, Bangor.

Winkler, H. (1938). Geographie. In: *Manual of Pteridology* (ed. Fr. Verdoorn), pp. 451-473. M. Nijhoff, The Hague.

Withering, W. (1792-1801). *Arrangement of British Plants.* Ed. 2 (1792); Ed. 3 (1796); Ed. 4 (1801). London.

Wolf, P. G. et al. (1995). Bracken Taxa in Britain: a molecular analysis. In: *Bracken: an environmental issue.* Contribution to an International Conference (Aberystwyth, 1994) (Ed. by R. T. Smith & J. A. Taylor), pp. 16-20. The International Bracken Group, Aberystwyth & Leeds.

Woods, R. G. (1993). *Flora of Radnorshire.* National Museum of Wales in association with the Bentham - Moxon Trust, Cardiff.

Wynne, G. (1993). *Flora of Flintshire.* Gee & Son, Denbigh, Clwyd.

Young, Edward (1856). *The Ferns of Wales.* Thomas Thomas, Neath.

SOME FERN SOCIETIES

The British Pteridological Society, c/o Miss A. M. Paul, Department of Botany, The Natural History Museum, Cromwell Road, LONDON SW7 5BD

American Fern Society, Botany Department, Milwaukee Public Museum, 800 W Wells Street, Milwaukee Wisconsin 53233, USA, or in Great Britain details from Miss A. M. Paul, Department of Botany, The Natural History Museum, Cromwell Road, London SW7 5BD

Fern Society of the Philippines, c/o National Museum, P. Burgess Street, Manila, PHILIPPINES

Fern Society of Victoria, P.O. Box 711, Heidelberg, Victoria 3081, AUSTRALIA

Indian Fern Society, c/o Prof. S. S. Bir, Department of Botany, Punjabi University, Patiala 147 002, INDIA

Japanese Pteridological Society, Botanical Gardens, Univerity of Tokyo, Hakusan 3-7-1, Bunkyo-Ku, Tokyo 112, JAPAN

Los Angeles International Fern Society, P.O. Box 90943, Pasadena, California 91109-0943, USA

Nederlandse Varenvereniging, The Secretary, J. G. Greep, van Remagenlaan 17, Arnhem 6824 LX, NETHERLANDS

Nelson Fern Society Inc., 9 Bay View Road, Atawhai, Nelson, NEW ZEALAND

Schweizeris Vereinigung der Farnfreunde, c/o Dr J. J. Schneller, Institüt für Systematische Botanik, Zollikerstrasse 107, Zürich CH 8008, SWITZERLAND

S.G.A.P. (Fern Study Group), c/o Moreen Woollett, Hon. Sec., 3 Currawang Place, Como West, New South Wales 2226, AUSTRALIA

GLOSSARY

The definitions given apply to the words as used in this book; for more complete treatments of botanical terminology, reference should be made to *Botanical Latin* by W. T. Stearn, *A Glossary of Botanic Terms* by B. D. Jackson *and A Dictionary of Botany* by G. Usher.

abaxial (of a lateral organ): facing away from the axis, normally the lowerside
abortive (spore): incompletely developed
acropetal: produced in succession towards the apex
acroscopic (pinnule; side of pinna): facing towards the apex of the frond; (pinnulet; side of pinnule): facing towards the apex of the pinna
acuminate: having a gradually diminishing drawn out point
acute: sharp-pointed but not drawn out, the angle at less than a right angle
adaxial (of a lateral organ): facing the axis, normally the upperside
adnate (of a lateral organ): if unstalked, attached by its whole width at its base to the axis
adpressed: lying flat for the whole of its length
adventitious: developing in an abnormal position
aggregate: closely packed together
alien: an introduced taxon that has become naturalized in a country or geographical area
alternate (segments): arising in succession on opposite sides of the rachis each one at a higher level than the last
annulus (pl. -li): a line of thick-walled cells (usually in the shape of a ring) or group of cells which brings about the bursting of the sporangium
antheridium (pl.-ia): male sexual organ producing motile sperm (antherozoids)
antherozoid (spermatozoid): a motile sperm produced by the antheridium
apex (adj. apical; pl. apices): tip; end furthest from the point of attachment
apiculate: furnished with a short sharp point
apogamous: growth of sporophyte from gametophyte without any fusion of sexual cells
appressed: pressed closely against another organ
archegonium (pl. -ia): female sexual organ producing the unfertilized egg cell
articulate: jointed
ascending: curving upwards from an oblique base
Asia Minor: the peninsula extending westwards from the Caspian Sea to the Mediterranean including Turkey (in Asia), Syria, Armenia, northern Iraq and northern Iran
attenuate: narrowed, tapered
auriculate: with an ear-like appendage
Australasia: comprising Australia, New Zealand, New Guinea, New Britain and certain smaller islands in their immediate vicinity

awl-shaped: narrow and tapering to a point; subulate
axillary: arising in the axil of a leaf or sporophyll
axis (adj. axial; pl. axes): the central portion of a plant or organ to which the outer portions are attached, as leaves to stem or pinnae to rachis
basal cells: the cells separating the first indurated cells of the annulus from the sporangial stork
base-number (chromosomes): the basic number of chromosomes (x) being the basic set of genetic information carried by the plant
basipetal: produced in succession from the apex downwards
basiscopic (pinnule; side of pinna): facing towards the base of the frond (pinnulet; side of pinnule): facing towards the base of the pinna
bifid: cleft into two
biflagellate: having two whip-like processes (flagella)
bipinnate (blade): pinnate with pinnate pinnae
bipinnatifid (blade): pinnatifid with primary divisions again pinnatifid
bipolar: occurring in the North Temperate and/or Arctic and also in the South Temperate and/or Antarctic zones but not in the Tropics
bistelic: having two strands of specialised tissue (vascular bundles)
blade: the flat expanded portion of the frond thus excluding the stipe
bloom: the bluish-white covering, which can be easily rubbed off, of some leaves such as cabbage or of some fruit such as plum
brindled: streaked with coloured stripes
Britain: also called Great Britain, comprising England, Scotland and Wales (cf. British Isles *and* United Kingdom)
British Isles: England, Scotland, Wales and the whole of Ireland, as well as the offshore islands including the Isle of Man and the Channel Islands (cf. Great Britain *and* United Kingdom)
bulbil: a small vegetative bulb usually axillary
caespitose: growing in tufts
calcicole: a lime-loving plant, i.e. preferring alkaline (base-rich) conditions
calcifuge: a lime-hating plant, i.e. preferring acidic conditions
campanulate: bell-shaped
capsule: the head of a sporangium, inside which the spores develop
carinal canals (in *Equisetum*): inner ring of canals on the inner side of the xylem inside the stem, each opposite a ridge on the surface of the stem
cat's ears (of pinnule or pinna segment): small upward-pointing projections at the junction of the two side-margins with the apex
chlorophyll: green pigment in plants
chromosome: specialized portion of cell nucleus, composed of a compound of basic protein with nucleic acid, which replicates hereditary characters
ciliate (margin): fringed with hairs
circinate (fronds in the bud): crozier-like
circumboreal: around the high latitudes of the northern hemisphere in the boreal zone, which is the one dominated by conifers lying next to the Arctic zone

circumpolar: occurring either throughout the North Temperate and Arctic zones or throughout the South Temperate and Antarctic zones

cleft: cut or split about half-way to the middle or base

coenosorus (pl. -ri): an apparent single sorus formed by several sori having grown fused together

commissure: vein linking together two other veins of a higher degree of importance

concave: having a hollowed out shape, like the upper (inner) surface of a saucer (cf. convex)

cone: the strobilus, being a group of closely packed sporophylls bearing sporangia arranged round a central axis, as found in the clubmosses and horsetails

confluent: blending; running into one another

conspecific: belonging to the same species

contiguous: touching each other at the edges

convex: having a rounded shape, like the underside of a saucer (cf. concave)

cordate: heart-shaped

coriaceous: of a leathery texture, sometimes with shiny surface

corm-like: a bulb-like flesh base to the stem

cortex: the tissue between the stele and the epidermis

cosmopolitan: occurring throughout the world

costa (pl. -as): the midrib of a pinna

crenate (margin): with rounded teeth or notches; scalloped (diminutive: crenulate)

crenate-pinnatifid: shallowly pinnatifid with the segments rounded at their apices

crisped: curled

crozier: the coiled end of the young leaf of a fern

cuneate: wedge-shaped; tapering

cuspidate: abruptly acuminate

deciduous: falling in season

decumbent: with the base reclining but the summit ascending

decurrent (base of leaf segment): running down the axis below the point of attachment

decussate: in pairs, alternately at right angles

deflexed: bent outwards and downwards

dehiscence (of sporangium): spontaneous opening of the sporangial wall to release the spores

deltoid: shaped like an equilateral triangle

dentate (margin): toothed, especially when teeth directed forward (diminutive: denticulate)

dichotomous: forked into two

dichotomy: the state of being repeatedly forked

dimorphic: having two forms

diploid: having twice the basic number of chromosomes (x) present in each cell, i.e. 2n (=2x), as at the sporophytic phase from the zygote through to the mature plant (cf. having the haploid number (n) present in the gametophytic phase)

disarticulate: to separate at a joint, as many leaves do in autumn

dissected: deeply divided

distant: widely spaced

divergent: radiating outwards like an open fan

dorsal: pertaining to the back or outward surface of an organ in relation to the axis, as in the lower surface of a leaf

dorsiventral: used of an organ if possessing dorsal (lower) and ventral (upper) surfaces

elater (in *Equisetum*): four clubbed hygroscopic bands attached to the spores which aid dispersal

elliptic: a flat shape widest in the middle and about 1.2 -3 times as long as wide

elliptic-lanceolate: intermediate between elliptic and lanceolate

elliptic-oblong: intermediate between elliptic and oblong

endogenous: growing from within

entire: without teeth, even, undivided

ephemeral: rapidly being shed

epidermal: of the outer layer of cells

epiphytic: growing upon but not parasitic on other plants

equal: symmetrical

etiolated: lengthened or deprived of colour by lack of light

eutrophic: rich in nutritive salts

exospore: the outer covering of a spore but inside the perispore. It is the opaque region seen under the microscope. See under *Asplenium trichomanes* (Fig. 47)

fastigiate: with branches erect and nearly parallel

fertile: bearing sporangia

filiform: thread-like

fimbriate (indusium): fringed with long slender processes

flaccid: limp, flabby

flagellum (pl. -lla): whip-like organ of propulsion

flange-like: with a projecting edge

fleshy: succulent, juicy or pulpy

free (veins): not joined up, ending blindly

frond: the fern leaf including the stipe

gamete: a sexual cell; the union of an antherozoid and an egg cell producing a fertile egg cell (zygote)

gametophyte: the sexual generation, as in the prothallus, producing gametes not spores

gemma (pl. -ae): bud-like organ capable of reproducing the plant vegetatively

glabrous: devoid of hairs
gland: organ of secretion, often capping a hair
glandular: possessing glands
glaucous: being covered with a thin bluish-white waxy covering (bloom) that often can easily be rubbed off, as on some leaves such as cabbage or on some fruit such as plum
globose: nearly spherical
glochidium (pl. -ia): a barbed hair such as those that cover the massulae of *Azolla* species
gradate (sorus): developing its sporangia in basipetal succession
Great Britain: also called Britain, comprising England, Scotland and Wales (cf. British Isles *and* United Kingdom)
haploid: having the basic number of chromosomes (x) present in each cell, i.e. n (= x), as in the gametophytic phase, with only one member of each chromosome pair present (cf. having the normal diploid number (2n) present, as at the sporophytic phase from the zygote through to the mature plant)
hastate: with the basal pair of lobes enlarged and turned outwards
hectad: a 10-km square, measuring 10 kilometres by 10 kilometres
heterosporous: having spores of two kinds (megaspores and microspores)
hexaploid: having six times the basic number of chromosomes (x), so $2n = 6x$
homosporous: having spores of one kind only
hybrid: a plant resulting from the crossing of sperm from one species with the unfertilized egg cell of another species
hybrid formula: the hybrid written by naming the parents linked by the '×' sign
hybrid name: the hybrid written by naming the genus followed by the assigned name linked by the '×' sign. In intergeneric hybrids the '×' sign precedes the new generic name
hybridization: where cross fertilization occurs between a sperm and an egg cell from prothalli of different species
hygroscopic: susceptible of swelling or shrinking on the application or removal of water
imbricate: with overlapping edges
incised: deeply cut
incurved: turned inwards
indurated: thickened
indusium (pl. -ia): the membrane (usually from brown to translucent white) covering the sorus in many species of fern, often withering as the sporangium matures and dehisces to release the spores
inferior (indusium): situated below the sorus
inflexed: turned or bent inwards and upwards
internode: length of stem between two nodes
keel: a ridge like the keel of a boat

lacerated: torn or irregularly cleft
laciniate: cut into narrow lobes
lamina: the leafy flat portion of the blade of a frond excluding any midribs and stalks
lanceolate: narrow, tapering at both ends and somewhat broadened about one-third from the base
lappet: special fertile lobe of the blade margin often folded under covering the sorus, such as in *Adiantum*
latticed (hair or scale): cross-barred
lax: loose, distant
leaf: the frond, thus comprising the blade and the stipe
leaf-blade: the flat expanded portion of the frond thus excluding the stipe
ligule (adj. ligulate): a small scale or flap borne on the upper surface of a leaf as in *Selaginella* and *Isoetes*
linear: narrow, several times longer than wide
linear-lanceolate: several times longer than wide, tapering to the apex but not necessarily tapered at the base
lingulate: tongue-shaped
lobe: a shallow segment
longitudinal: in the direction of the length
Macaronesia: the biogeographical collective term for the north Atlantic islands of the Azores, the Canary Islands, the Cape Verde Islands, Madeira and associated smaller islands
macroscopic: able to be seen by the naked eye without the aid of a microscope
Malaysia: comprising the Malay states of the Malay Peninsula, and the states in north Borneo
marginal: situated on the margin or edge
Mascarene Islands: the collective term for the islands of Réunion, Mauritius, and Rodrigues
massula (pl. -ae): little clumps of microspores in heterosporous ferns such as *Azolla*.
median: placed midway or nearly midway
megasporangium: sporangium containing megaspores
megaspore: the larger kind of spore in the heterosporous pteridophytes that forms the archegonia
megasorus: sorus giving rise to megasporangia in the heterosporous pteridophytes
meiosis: a reduction division in which the chromosome number is halved by two successive divisions, the nuclei dividing twice but the chromosomes only once. In Pteridophyta this takes place during the production of the spores from the spore mother cell (cf. mitosis)
micron (µm): one thousandth of a millimetre, i.e. one millionth of a metre
Micronesia: a group of small tropical islands mostly north of the Equator in the western region of the Pacific including Guam, the Caroline, Marshall, and Gilbert Islands, and Tuvalu (formerly called the Ellice Islands)

microsporangium: sporangium containing microspores
microspore: the smaller kind of spore, in the heterosporous pteridophytes, that forms an antheridium
microsorus: sorus giving rise to microsporangia in the heterosporous pteridophytes
midrib: distinct central vein of a pinnately veined simple blade or of a segment of a divided blade
mid-vein: when less distinct than a midrib, the primary or main vein of a leaf, or more usually of one of its segments
mitosis: the normal process of cell-division in which the chromosomes duplicate themselves into two separate nuclei, and ultimately into two new daughter cells. In the course of cell division, each chromosome divides along its length, one longitudinal half going to one daughter cell and the other half going to the other daughter cell. It takes place in the sporophyte phase, i.e. from the zygote to the development of the mature plant, and also in the gametophytic phase with the growth of the prothallus (cf. meiosis)
mixed (sorus): developing its sporangia promiscuously, not simultaneously or in sequence
monopodial: having the stem of a single and continuous axis
monotypic: of a genus with one species
morphotype: *see under* taxonomic rank
motile: capable of motion
mucilage: vegetable gelatin, sticky or slimy when wet
mucronate: ending in a short and straight point
multiciliate: many haired
mycorrhiza (pl. -ae): the symbiotic union of a fungus with cells of a plant
naturalized: of foreign origin, but established and reproducing itself as though native
New World: North, Central and South America and the West Indies
node: a joint on a stem where one or more leaves arise
nothomorph: *see under* taxonomic rank
noticeable: easily seen without the aid of a lens
oblanceolate: lanceolate with the widest part above the middle
oblique: slanting; of unequal sides
oblique (annulus): in a plane inclined at an angle to the long axis
oblong: longer than broad with more or less parallel sides
oblong-elliptic: intermediate between oblong and elliptic
obovate: inversely ovate with the widest part above the middle
obovoid: inversely ovoid with the widest part above the middle
obtuse: blunt, with the angle at more than a right angle
Old World: Europe, Asia and Africa
oligotrophic: poor in nutritive (basic) salts
olive-green: yellowish-green darkened with black
opposite (segments): arising in succession in pairs on opposite sides of the rachis

orbicular: with a circular outline
oval: broadly elliptic
ovate: shaped like a longitudinal section of a hen's egg, being broader towards the base
ovate-acuminate: ovate in outline with the apex drawn out into a gradually diminishing point
ovate-deltoid: broadly ovate with the greatest width very near the base
ovate-lanceolate: ovate in shape but two or three times as long as broad
ovate-oblong: ovate with the sides nearly parallel
ovoid: egg-shaped being broadest below the middle
palmate (blade): compound with the divisions all running towards and attached directly to the stipe (stalk)
pan-temperate: throughout most of the temperate regions of the world
pan-tropical: throughout most of the tropical regions of the world
papillose: bearing small knob-like tubercles (papillae)
parallelogram-shaped: four-sided with opposite sides equal and parallel to each other
paraphysis (pl. paraphyses): sterile hair, branched or not, growing among sporangia
peduncle: stalk of inflorescence; used here for the stalk of the cone in clubmosses and horsetails
pedunculate: stalked
peltate: shield-like, roundish with a more or less central attachment
pentagonal: five-sided in outline
perispore: the outer membrane surrounding a spore, sometimes with a winged edge that appears translucent under a microscope
Permian Period: a geological interval of time from about 250 to 290 million years ago
persistent (indusium): at least until the spores are shed
phloem: the food conducting tissue of vascular plants
pinna (pl. -ae): a primary division of a pinnate blade, thus attached to the rachis
pinna-pair: two pinnae attached opposite each other on the rachis
pinna segment: any division into which a pinna is cleft whether completely to the midrib or not
pinnate: branched or divided like a feather with distinct leaflets borne to right and left of a central axis (the rachis)
pinnate (1-pinnate): divided into two rows along a central axis (the rachis); bipinnate (2-pinnate): when the primary divisions of a pinnate blade are themselves pinnate, and so on
pinnatifid: deeply cut but the segments not fully separated from adjoining ones, being connected by some lamina
pinnatisect: pinnately divided almost to the midrib
pinnule (pl. -es): a division of a pinna when quite distinct from adjacent divisions, connected to adjacent divisions only by the midrib of the pinna

pinnule segment: any division into which a pinnule is cleft whether completely to the midrib or not

pinnulet: a division of a pinnule when quite distinct; tertiary pinnae

polymorphous: variable in form

Polynesia: the name used to cover a huge triangular area of the east-central Pacific with New Zealand, Hawaii (Hawaiian Islands) and Easter Island at its corners

primary (segments): of the first order

process: any projecting appendage

procumbent: spread loosely over the surface of the ground

prothallus (pl. -lia): the gametophytic or sexual generation of a pteridophyte arising from the germination of a spore

pubescent: clothed with soft hairs or down

rachis (rhachis; pl. rachises or rachides): the axis (midrib) of a compound frond from its apex down to the base of the blade, thus excluding the stipe

radially (indusium): in a line from the perimeter to the centre of the indusium

receptacle: the part of the frond (usually raised above the general surface) to which the sorus is affixed

recurved: curved backwards or downwards

reflexed: abruptly bent or turned downwards or backwards

refracted: bent sharply from the base backwards

reniform: kidney-shaped

reticulate (veins): forming a network

rhizoid: simple root-like organ

rhizome: stem persisting for more than one season on or under the surface and usually more or less horizontal

rhizophore: a leafless 'rooting' branch in *Selaginella,* which itself emits true roots

rhomboidal: four-sided with the opposite lateral angles obtuse; 'diamond-shaped'

root: a specialized multicellular organ for anchoring the plant and for absorbing water and mineral salts, normally arising from the basal or underground portion of the stem

rugose: wrinkled

runner: an elongated lateral shoot rooting at intervals

saprophytic: deriving its food from dead organic matter

scale: a small thin-textured usually semi-transparent appendage. These are concentrated towards the base of the stipe and usually get fewer higher up

scarious: thin, dry, chaffy appearance

secondary (segments): of the second order

secondary (veins): branching from the midrib or mid-vein

segment: any division into which a blade is cleft whether completely to the midrib or not

sensu lato (*s.l.*): in a broad sense, with a wide or general interpretation
sensu stricto (*s.s.*): in a narrow sense
septate: divided by a partition
serrate: regularly toothed like a saw (diminutive: serrulate)
sessile: stalkless
sheath: a tubular or enrolled part of an organ
shuttlecock (frond): when fronds are arranged in more or less a full circle round the base of the plant, then radiating upwards and outwards
shuttlecock-shaped (indusium): curved upwards and outwards from a central point, as in some indusia just before they are shed
side-margin (of pinnule or pinna segment): the two sides, thus excluding the apical area which links them round the edge
simple (leaf-blade): undivided; (sorus): developing all its sporangia simultaneously
sinuate: wavy margin
sinuate-crenate (margin): waved, with the projections resembling large convex teeth
solenostele: a tubular stele with internal and external phloem
sorus (pl. sori): a cluster of sporangia; in some species initially covered by an indusium
South-east Asia: comprising those continental margins and offshore archipelagos of Asia lying south of China, north of Australia and east of India
spathulate: spatula- or paddle-shaped
sperm: a male reproductive cell
spike: a narrow form of fructification in which the sporangia are sessile
spinulose: with small spines or spinules
spinulose-mucronate: tipped with a spine-like point
sporangiophore: sporangia-bearing organ which is not leaf-like
sporangium (pl. -ia): capsule containing spores and making up the sorus
spore: minute pollen-like body which on germination gives rise to the prothallus; spores look powdery to the eye but are unicellular when seen under the microscope
sporocarp (Marsileaceae - Pillwort family only): a globose body containing sporangia
sporophyll: fertile leaf or homologous structure bearing sporangia
sporophyte: spore-bearing, asexual generation, from the zygote to the mature pteridophyte
stalk: any lengthened (usually narrowed) support of an organ
stalk (frond): the stipe, being the axis of a frond from its base to the base of the blade, thus excluding the rachis
stele: vascular system
stem: the main supporting organ of most plants producing leaves apically and roots basally

stipe: the stalk of a frond, being the axis of a frond from its base to the base of the blade, thus excluding the rachis

stipules: paired leafy growths at the base of the frond

stock: relatively short stout stem, mostly underground, to which the frond bases are attached in some species

stolon: runner or basal branch which roots independently

stoma (pl. stomata): pore in the epidermis of the blade or stipe for gaseous exchange

stomatal guard-cells: two cells which open or close the stoma by their greater or lesser distension caused by their water content

stomium: point on the sporangium where it opens to liberate spores

strobilus (pl. strobili): the cone, being a group of closely packed sporophylls bearing sporangia arranged round a central axis, as found in the clubmosses and horsetails

sub- (prefix): approaching; nearly; rather less than

substrate: the surface or material on which any particular plant grows

subulate: narrow and tapering to a point; awl-shaped (three-dimensional)

succulent: fleshy and juicy or pulpy

superficial (sori): on the surface (in contradistinction to marginal)

superior (indusium): situated above the sorus

symbiont: an individual existing in symbiosis

symbiosis: the living together of dissimilar organisms, with benefit to one only, or both

taxon (pl. taxa): any taxonomic grouping without specifying whether it is, for example, a genus, species, subspecies, variety or hybrid. 'Taxon' is often used when the taxonomic rank of a plant is uncertain. 'Taxa' is commonly used when one wishes to refer collectively to plants which are of various taxonomic ranks

taxonomic rank: a level in the hierarchical scale of classifying organisms, e.g. Family, Genus, Species, Subspecies, Variety, Form. In plants, some intermediate levels, such as 'Morphotype' and 'Nothomorph' are not accepted under the International Code of Botanical Nomenclature

tooth (pl. teeth): small marginal lobe

terete: circular in cross-section

ternate (blade): divided into three more or less equal divisions

terrestrial: growing on the ground

tertiary (segments): of the third order

tertiary (veins): branching from the secondary veins

tetrad: group of four

tetraploid: having four times the basic number of chromosomes (x), so $2n = 4x$

tomentose: covered with dense often matted hairs

tomentum: covering of dense often matted hairs

tooth (pl. teeth): small marginal lobe

trabecula (pl.-ae): sterile tissue haphazardly crossing the sporangia of *Isoetes*

translucent: semi-transparent, diffusing light but not revealing definite contours of an object underneath
transverse: at right angles to the axis
triangular: broadest at the base and narrowing to the apex with more or less straight sides
triangular-lanceolate: tapering to the apex and widening at the base
triangular-ovate (outline): ovate tending to triangular
trident-like: three-pronged
trifid: cleft into three
trigonous: triangular in cross-section
tripartite: divided into three parts
tripinnate (blade): pinnate with pinnate pinnules
truncate: terminating abruptly as if cut off square, at right angles to the midrib or axis
tubercle (rhizome): a small thickened portion of a rhizome, as in *Equisetum*; (surface): a small more or less spherical or ellipsoid swelling on a surface
tuberculate: bearing tubercles; with a surface texture covered in tubercles
ultimate segments: the final divisions of a compound blade
ultrabasic: of igneous rock containing less than 45% silica and virtually no quartz or feldspar
underside: the lower surface of a flat organ, as opposed to the upperside
unequal: asymmetrical
unilocular: having one cavity
United Kingdom: comprising Northern Ireland as well as England, Scotland and Wales (cf. British Isles *and* Great Britain)
upperside: the upper surface of a flat organ, as opposed to the underside
vallecular canals: outer ring of canals inside an *Equisetum* stem each opposite to a groove or furrow on the surface of the stem
veinlets: secondary strands of strengthening and conducting (vascular) tissue running through the divisions of a blade
velum: veil-like outgrowth of tissue
venation: arrangement of the veins of a blade or segment
ventral: pertaining to the front or inward surface of an organ in relation to the axis, as in the upper surface of a leaf
venule: vein of secondary importance
vernation: the order of unfolding from leaf buds
verrucose: covered with rough, wart-like elevations
wilting: drooping, having
winter-green: retaining green colour in winter
xerophilous: growing in dry places
zygote: the fertile egg, the first cell of the new individual after sexual fusion

INDEX

Latin names are in *italics* except the genus entries which are in CAPITAL ROMAN TYPE
BOLD CAPITALS – recorded from Wales.
CAPITALS – recorded from the rest of the British Isles only.
Lower–case (followed in brackets only by Latin name) – synonym.
Lower–case (followed in brackets only by English name) – *under* Other Alien Species *and with an* English name.
Lower–case not as above – other taxa mentioned in the text.

English and Welsh names are in lower–case Roman type.
English name (followed in brackets only by *Latin name in lower–case italics*) – *under* Other Alien Species *and with an* English name.
English name (followed in brackets only by main English name) – alternative English name.

156	Adder's–tongue (***OPHIOGLOSSUM VULGATUM*** L.; Tafod y neidr)
155	Least, (*OPHIOGLOSSUM LUSITANICUM* L.; Tafod y Neidr Lleiaf)
155	Small, (***OPHIOGLOSSUM AZORICUM*** C.Presl; Tafod y Neidr Bach)
154	Adder's–tongues (***OPHIOGLOSSUM*** L.; Tafodau y Neidr)
78	**ADIANTUM** L. (Maidenhair Ferns; Brigerau Gwener)
78	***CAPILLUS–VENERIS*** L. (Maidenhair Fern; Briger Gwener)
79	*PEDATUM* L. (American Maidenhair Fern; Briger Gwener America)
201	reniforme L.
200	venustum D.Don (Evergreen Maidenhair)
116	Alsophila tricolor (Colenso) Tryon (*CYATHEA DEALBATA* (G.Forst.) Sw.)
80	ANOGRAMMA Link (Jersey Ferns; Rhedyn Jersey)
80	*LEPTOPHYLLA* (L.) Link (Jersey Fern; Rhedynen Fach Jersey)
81	**ASPLENIUM** L. (Spleenworts; Duegredyn)
82	***ADIANTUM-NIGRUM*** L. (Black Spleenwort; Duegredynen Ddu)
99	***ADIANTUM–NIGRUM*** × *A. OBOVATUM* = *A.* × *SARNIENSE* Sleep (Guernsey Spleenwort; Duegredynen Guernsey)
99	***ADIANTUM–NIGRUM*** × *A. ONOPTERIS* = *A.* × *TICINENSE* D.E.Mey. (Hybrid Black Spleenwort; Duegredynen Ddu Groesryw)
99	***ADIANTUM–NIGRUM*** × *A. SCOLOPENDRIUM* = *A.* × *JACKSONII* (Alston) Lawalrée (Jackson's Fern; Rhedynen Jackson)
100	***ADIANTUM–NIGRUM*** × *A. SEPTENTRIONALE* = *A.* × *CONTREI* Callé, Lovis & Reichst. (Caernarvonshire Fern; Rhedynen Sir Gaernarfon)
88	adiantum–nigrum subsp. onopteris (L.) Luerss. (*ASPLENIUM ONOPTERIS* L.)
102	× ***ALTERNIFOLIUM*** Wulfen = *A. SEPTENTRIONALE* × *A. TRICHOMANES* subsp. *TRICHOMANES* (Alternate–leaved Spleenwort; Duegredynen Dail Bob yn Ail)

ASPLENIUM (continued)

- 100 × *badense* (D.E.Mey.) Jermy = *A. ceterach* × *A. ruta–muraria*
- 87 *billotii* F.W.Schultz (***ASPLENIUM OBOVATUM*** Viv. subsp. ***LANCEOLATUM*** (Fiori) P.Silva)
- 102 × *breynii* auct. (***A. SEPTENTRIONALE*** × ***A. TRICHOMANES*** subsp. ***TRICHOMANES*** = ***A.*** × ***ALTERNIFOLIUM*** Wulfen)
- **84** **CETERACH** L. (Rustyback; Rhedynen Gefngoch)
- 100 *ceterach* × *A. ruta–muraria* = *A.* × *badense* (D.E.Mey.) Jermy
- 101 × *CLERMONTIAE* Syme = *A. RUTA–MURARIA* × *A. TRICHOMANES* subsp. *QUADRIVALENS* (Lady Clermont's Spleenwort; Duegredynen y Fonesig)
- 101 × *CONFLUENS* (T. Moore ex Lowe) Lawalrée = *A. SCOLOPENDRIUM* × *A. TRICHOMANES* subsp. *QUADRIVALENS* (Confluent Maidenhair Spleenwort; Duegredynen Gwallt y Forwyn Gydlifol)
- **100** × ***CONTREI*** Callé, Lovis & Reichst. = ***A. ADIANTUM–NIGRUM*** × ***A. SEPTENTRIONALE*** (Caernarvonshire Fern; Rhedynen Sir Gaernarfon)
- 85 *cuneifolium* Viv. (Serpentine Black Spleenwort; Duegredynen Ddu Sarff–faen)
- 85 *fontanum* (L.) Bernh. (Smooth Rock–spleenwort; Duegredynen Lefn y Creigiau)
- 102 *germanicum* auct. (***A. SEPTENTRIONALE*** × ***A. TRICHOMANES*** subsp. ***TRICHOMANES*** = ***A.*** × ***ALTERNIFOLIUM*** Wulfen)
- 99 × *JACKSONII* (Alston) Lawalrée = *A. ADIANTUM–NIGRUM* × *A. SCOLOPENDRIUM* (Jackson's Fern; Rhedynen Jackson)
- 87 *lanceolatum* Hudson, non Forssk. (***ASPLENIUM OBOVATUM*** Viv. subsp. ***LANCEOLATUM*** (Fiori) P.Silva)
- 104 × *lusaticum* D.E.Mey. (***A. TRICHOMANES*** subsp. ***QUADRIVALENS*** × subsp. ***TRICHOMANES*** = ***A. TRICHOMANES*** L. nothossp. ***LUSATICUM*** (D.E.Mey.) Lawalrée)
- **86** ***MARINUM*** L. (Sea Spleenwort; Duegredynen Arfor)
- 100 × *MICRODON* (T.Moore) Lovis & Vida = *A. OBOVATUM* × *A. SCOLOPENDRIUM* (Guernsey Fern; Rhedynen Guernsey)
- 101 × *MURBECKII* Dörfl. = *A. RUTA–MURARIA* × *A. SEPTENTRIONALE* (Murbeck's Spleenwort; Duegredynen Murbeck)
- 100 *OBOVATUM* × *A. SCOLOPENDRIUM* = *A.* × *MICRODON* (T.Moore) Lovis & Vida (Guernsey Fern; Rhedynen Guernsey)
- **87** ***OBOVATUM*** Viv. subsp. ***LANCEOLATUM*** (Fiori) P.Silva (Lanceolate Spleenwort; Duegredynen Reiniolaidd)
- 100 *OBOVATUM* subsp. *LANCEOLATUM* × *A. TRICHOMANES* = *A.* × *REFRACTUM* (T.Moore) Lowe
- 88 *obovatum* Viv. subsp. *obovatum*
- 88 ***ONOPTERIS*** L. (Irish Spleenwort; Duegredynen Wyddelig)
- 100 × *REFRACTUM* (T.Moore) Lowe = *A. OBOVATUM* subsp. *LANCEOLATUM* × *A. TRICHOMANES*
- **89** ***RUTA–MURARIA*** L. (Wall–rue; Duegredynen y Muriau)

ASPLENIUM (*continued*)
- *101* *RUTA–MURARIA* × *A. SEPTENTRIONALE* = *A.* × *MURBECKII* Dörfl. (Murbeck's Spleenwort; Duegredynen Murbeck)
- *101* *RUTA–MURARIA* × *A. TRICHOMANES* subsp. *QUADRIVALENS* = *A.* × *CLERMONTIAE* Syme (Lady Clermont's Spleenwort; Duegredynen y Fonesig)
- *99* × *SARNIENSE* Sleep = *A. ADIANTUM–NIGRUM* × *A. OBOVATUM* (Guernsey Spleenwort; Duegredynen Guernsey)
- *90* *SCOLOPENDRIUM* L. (Hart's–tongue; Tafod yr Hydd)
- *101* *SCOLOPENDRIUM* × *A. TRICHOMANES* subsp. *QUADRIVALENS* = *A.* × *CONFLUENS* (T. Moore ex Lowe) Lawalrée (Confluent Maidenhair Spleenwort; Duegredynen Gwallt y Forwyn Gydlifol)
- *92* *SEPTENTRIONALE* (L.) Hoffm. (Forked Spleenwort; Duegredynen Fforchog)
- *102* *SEPTENTRIONALE* × *A. TRICHOMANES* subsp. *TRICHOMANES* = *A.* × *ALTERNIFOLIUM* Wulfen (Alternate–leaved Spleenwort; Duegredynen Dail Bob yn Ail)
- *85* serpentini Tausch (*Asplenium cuneifolium* Viv.)
- *99* × *TICINENSE* D.E.Mey. = *A. ADIANTUM–NIGRUM* × *A. ONOPTERIS* (Hybrid Black Spleenwort; Duegredynen Ddu Groesryw)
- *93* *TRICHOMANES* L. (Maidenhair Spleenwort; Duegredynen Gwallt y Forwyn)
- *94* trichomanes L. subsp. inexpectans Lovis
- *104* *TRICHOMANES* L. nothossp. *LUSATICUM* (D.E.Mey.) Lawalrée = *A. TRICHOMANES* subsp. *QUADRIVALENS* subsp. *TRICHOMANES* (Hybrid Maidenhair Spleenwort; Duegredynen Gwallt y Forwyn Groesryw)
- *95* *TRICHOMANES* L. subsp. *PACHYRACHIS* (H.Christ) Lovis & Reichst. (Lobed Maidenhair Spleenwort; Duegredynen Gwallt y Forwyn Labedog)
- *103* *TRICHOMANES* subsp. *PACHYRACHIS* × subsp. *QUADRIVALENS* = *A. TRICHOMANES* L. nothossp. *STAUFFERI* Lovis & Reichst.
- *96* *TRICHOMANES* L. subsp. *QUADRIVALENS* D.E.Mey. emend. Lovis (Common Maidenhair Spleenwort; Duegredynen Gwallt y Forwyn Gyffredin)
- *104* *TRICHOMANES* subsp. *QUADRIVALENS* × subsp. *TRICHOMANES* = *A. TRICHOMANES* L. nothossp. *LUSATICUM* (D.E.Mey.) Lawalrée (Hybrid Maidenhair Spleenwort; Duegredynen Gwallt y Forwyn Groesryw)
- *103* *TRICHOMANES* L. nothossp. *STAUFFERI* Lovis & Reichst. = *A. TRICHOMANES* subsp. *PACHYRACHIS* × subsp. *QUADRIVALENS*
- *97* *TRICHOMANES* L. subsp. *TRICHOMANES* (Delicate Maidenhair Spleenwort; Duegredynen Gwallt y Forwyn Gain)
- *97* *TRICHOMANES–RAMOSUM* L. (Green Spleenwort; Duegredynen Werdd)
- *97* viride L. (*ASPLENIUM TRICHOMANES–RAMOSUM* L.)

INDEX

- *100* × *Asplenoceterach badense* D.E.Mey. (*A. ceterach* × *A. ruta–muraria* = *A.* × *badense* (D.E.Mey.) Jermy)
- *101* × *Asplenophyllitis confluens* (T.Moore ex Lowe) Alston (*A. SCOLOPENDRIUM* × *A. TRICHOMANES* subsp. *QUADRIVALENS* = *A.* × *CONFLUENS* (T. Moore ex Lowe) Lawalrée)
- 99 × *Asplenophyllitis jacksonii* Alston (*A. ADIANTUM–NIGRUM* × *A. SCOLOPENDRIUM* = *A.* × *JACKSONII* (Alston) Lawalrée)
- *100* × *Asplenophyllitis microdon* (T.Moore) Alston (*A. OBOVATUM* × *A. SCOLOPENDRIUM* = *A.* × *MICRODON* (T.Moore) Lovis & Vida)
- **104 ATHYRIUM** Roth (Lady–ferns; Rhedyn Mair)
- *105 alpestre* auct., non Clairv. (*ATHYRIUM DISTENTIFOLIUM* Tausch ex Opiz var. *DISTENTIFOLIUM*)
- *105 DISTENTIFOLIUM* Tausch ex Opiz var. *DISTENTIFOLIUM* (Alpine Lady–fern; Rhedynen Fair Alpaidd)
- *105 DISTENTIFOLIUM* Tausch ex Opiz var. *FLEXILE* (Newman) Jermy (Newman's Lady–fern; Rhedynen Fair Newman)
- *105 FILIX–FEMINA* (L.) Roth (Lady–fern; Rhedynen Fair)
- *105 flexile* (Newman) Druce (*ATHYRIUM DISTENTIFOLIUM* Tausch ex Opiz var. *FLEXILE* (Newman) Jermy)
- 107 Azolla (Water Fern)
- **107 AZOLLA** Lam. (Water Ferns; Rhedyn y Dŵr)
- *109 caroliniana* auct., non Willd.
- *107 caroliniana* Willd., non auct. (***AZOLLA FILICULOIDES*** Lam.)
- **107 *FILICULOIDES*** Lam. (Water Fern; Rhedynen y Dŵr)
- *109 mexicana* C.Presl
- 118 Bladder–fern, Brittle, (**CYSTOPTERIS FRAGILIS** (L.) Bernh.; Rhedynen Frau)
- 117 Dickie's, (*CYSTOPTERIS DICKIEANA* R.Sim; Ffiolredynen Arfor)
- 119 Mountain, (*CYSTOPTERIS MONTANA* (Lam.) Desv.; Ffiolredynen y Mynydd)
- 117 Bladder–ferns (**CYSTOPTERIS** Bernh.; Ffiolredyn)
- 109 **BLECHNUM** L. (Hard–ferns; Gwibredyn)
- *109 chilense* (Kaulf.) Mett. (***BLECHNUM CORDATUM*** (Desv.) Hieron.)
- **109 *CORDATUM*** (Desv.) Hieron. (Chilean Hard–fern; Gwibredynen Chile)
- *110 PENNA–MARINA* (Poir.) Kuhn (Antarctic Hard–fern; Gwibredynen Antarctig)
- **110 *SPICANT*** (L.) Roth (Hard–fern; Gwibredynen)
- **112 BOTRYCHIUM** Sw. (Moonworts; Lloerlysiau)
- *201 lanceolatum* (S.G.Gmel.) Ångstr.
- **112 *LUNARIA*** (L.) Sw. (Moonwort; Lloerlys)
- *112 lunaria* (L.) Sw. var. *subincisa* Roep.
- *201 matricariifolium* A.Braun ex W.D.J.Koch
- *201 multifidum* (S.G.Gmel.) Rupr.

INDEX 243

184	Bracken (***PTERIDIUM AQUILINUM*** (L.) Kuhn; Rhedynen Gyffredin)
187	Atlantic, (***PTERIDIUM AQUILINUM*** (L.) Kuhn subsp. ***ATLANTICUM*** C.N.Page; Rhedynen Gyffredin y Gorllewin)
187	Common, (***PTERIDIUM AQUILINUM*** (L.) Kuhn subsp. ***AQUILINUM***; Rhedynen Gyffredin)
188	Perthshire, (***PTERIDIUM AQUILINUM*** (L.) Kuhn subsp. ***FULVUM*** C.N.Page; Rhedynen Gyffredin Perth)
190	Pinewood, (***PTERIDIUM PINETORUM*** C.N.Page & R.R.Mill; Rhedynen Gyffredin Pîn)
184	Brackens (**PTERIDIUM** Gled. ex Scop.; Rhedyn Cyffredin)
190	Brake, Chinese, (Ladder Brake)
189	Cretan, (Brake Fern)
190	Ladder, (*PTERIS VITTATA* L.; Rhedynen Ruban Ysgolddail)
189	Spider, (*PTERIS INCOMPLETA* Cav.; Rhedynen Ruban y Corryn)
190	Tender, (*PTERIS TREMULA* R.Br.; Rhedynen Ruban Croendenau)
78	Briger Gwener (***ADIANTUM CAPILLUS–VENERIS*** L.; Maidenhair Fern)
79	America (*ADIANTUM PEDATUM* L.; American Maidenhair Fern)
78	Brigerau Gwener (**ADIANTUM** L.; Maidenhair Ferns)
193	Bristle–fern (Killarney Fern)
195	Veiled, (*TRICHOMANES VENOSUM* R.Br.; Rhedynen Wrychog Orchuddiedig)
132	Buckler–fern, Broad, (***DRYOPTERIS DILATATA*** (Hoffm.) A.Gray; Marchredynen Lydan)
131	Crested, (*DRYOPTERIS CRISTATA* (L.) A.Gray; Marchredynen Gribog)
142	Gibby's, (***D. DILATATA*** × ***D. EXPANSA*** = ***D.*** × ***AMBROSEAE*** Fraser–Jenk. & Jermy; Marchredynen Gibby)
123	Hay–scented, (***DRYOPTERIS AEMULA*** (Aiton) Kuntze; Marchredynen Aroglus)
141	Hybrid Fen, (*D. CARTHUSIANA* × *D. CRISTATA* = *D.* × *ULIGINOSA* (A.Braun ex Döll) Kuntze ex Druce; Marchredynen Gribog Groesryw)
141	Hybrid Narrow, (***D. CARTHUSIANA*** × ***D. DILATATA*** = ***D.*** × ***DEWEVERI*** (J.T.Jansen) J.T.Jansen & Wacht.; Marchredynen Gul Groesryw)
142	Kintyre, (*D. CARTHUSIANA* × *D. EXPANSA* = *D.* × *SARVELAE* Fraser–Jenk. & Jermy; Marchredynen Kintyre)
130	Narrow, (***DRYOPTERIS CARTHUSIANA*** (Vill.) H.P.Fuchs; Marchredynen Gul)
134	Northern, (***DRYOPTERIS EXPANSA*** (C.Presl) Fraser–Jenk. & Jermy; Marchredynen y Gogledd)
138	Rigid, (***DRYOPTERIS SUBMONTANA*** (Fraser–Jenk. & Jermy) Fraser–Jenk.; Marchredynen Anhyblyg)
138	Scaly, (*DRYOPTERIS REMOTA* (A.Braun ex Döll) Druce; Marchredynen Gennog)
121	Buckler–ferns (**DRYOPTERIS** Adans.; Marchredyn)

INDEX

144   *Carpogymnia dryopteris* (L.) Á.Löve & D.Löve (***GYMNOCARPIUM DRYOPTERIS*** (L.) Newman)
146   *Carpogymnia robertiana* (Hoffm.) Á.Löve & D.Löve (***GYMNOCARPIUM ROBERTIANUM*** (Hoffm.) Newman)
179   Celynredynen (***POLYSTICHUM LONCHITIS*** (L.) Roth; Holly–fern)
84   CETERACH Willd. (*see under* **ASPLENIUM** L.)
84   *officinarum* Willd. (***ASPLENIUM CETERACH*** L.)
200   Chain–fern, European, (*WOODWARDIA RADICANS* (L.) Sm.; Rhedynen Gadwyn Ewropeaidd)
200   Chain–ferns (**WOODWARDIA** Sm.; Rhedyn Cadwyn)
33   Clubmoss, Alpine, (***DIPHASIASTRUM ALPINUM*** (L.) Holub; Cnwpfwsogl Alpaidd)
40   Common, (Stag's–horn Clubmoss)
36   Fir, (***HUPERZIA SELAGO*** (L.) Bernh. ex Schrank & C.Mart.; Cnwpfwsogl Mawr)
40   Interrupted, (***LYCOPODIUM ANNOTINUM*** L.; Cnwpfwsogl Meinfannau)
35   Issler's, (***DIPHASIASTRUM COMPLANATUM*** (L.) Holub subsp. *ISSLERI* (Rouy) Jermy; Cnwpfwsogl Issler)
43   Kraus's, (***SELAGINELLA KRAUSSIANA*** (Kunze) A.Braun; Cnwpfwsogl Krauss)
44   Lesser, (***SELAGINELLA SELAGINOIDES*** (L.) Link; Cnwpfwsogl Bach)
38   Marsh, (***LYCOPODIELLA INUNDATA*** (L.) Holub; Cnwpfwsogl y Gors)
40   Stag's–horn, (***LYCOPODIUM CLAVATUM*** L.; Cnwpfwsogl Corn Carw)
39   Clubmosses (**LYCOPODIUM** L.; Cnwpfwsoglau)
33   Alpine, (***DIPHASIASTRUM*** Holub; Cnwpfwsoglau Alpaidd)
36   Fir, (**HUPERZIA** Bernh.; Cnwpfwsoglau Ffeinid)
42   Lesser, (**SELAGINELLA** P.Beauv.; Cnwpfwsoglau Lleiaf)
38   Marsh, (**LYCOPODIELLA** Holub; Cnwpfwsoglau y Gors)
     Cnwpfwsogl
33   Alpaidd (***DIPHASIASTRUM ALPINUM*** (L.) Holub; Alpine Clubmoss)
44   Bach (***SELAGINELLA SELAGINOIDES*** (L.) Link; Lesser Clubmoss)
40   Corn Carw (***LYCOPODIUM CLAVATUM*** L.; Stag's–horn Clubmoss)
35   Issler (*DIPHASIASTRUM COMPLANATUM* (L.) Holub subsp. *ISSLERI* (Rouy) Jermy; Issler's Clubmoss)
43   Krauss (***SELAGINELLA KRAUSSIANA*** (Kunze) A.Braun; Kraus's Clubmoss)
36   Mawr (***HUPERZIA SELAGO*** (L.) Bernh. ex Schrank & C.Mart.; Fir Clubmoss)
40   Meinfannau (***LYCOPODIUM ANNOTINUM*** L.; Interrupted Clubmoss)
38   y Gors (***LYCOPODIELLA INUNDATA*** (L.) Holub; Marsh Clubmoss)
39   Cnwpfwsoglau (**LYCOPODIUM** L.; Clubmosses)
33   Alpaidd (**DIPHASIASTRUM** Holub; Alpine Clubmosses)
36   Ffeinid (**HUPERZIA** Bernh.; Fir Clubmosses)
42   Lleiaf (**SELAGINELLA** P.Beauv.; Lesser Clubmosses)
38   y Gors (**LYCOPODIELLA** Holub; Marsh Clubmosses)

115	Coedredyn Arian (CYATHEA Sm.; Silver Tree–ferns)
120	Coedredyn Awstralia (DICKSONIA L'Hér.; Australian Tree–Ferns)
116	Coedredynen Arian (*CYATHEA DEALBATA* (G.Forst.) Sw.; Silver(y) Tree–fern)
121	Coedredynen Awstralia (*DICKSONIA ANTARCTICA* Labill.; Australian Tree–fern)
195	Coredyn (WOODSIA R.Br.; Woodsias)
196	Coredynen Alpaidd (***WOODSIA ALPINA*** (Bolton) Gray; Alpine Woodsia)
198	Coredynen Hirgul (***WOODSIA ILVENSIS*** (L.) R.Br.; Oblong Woodsia)
114	**CRYPTOGRAMMA** R.Br. (Parsley Ferns; Rhedyn Persli)
114	***CRISPA*** (L.) R.Br. (Parsley Fern; Rhedynen Bersli)
200	*Culcita macrocarpa* C.Presl
115	CYATHEA Sm. (Silver Tree–ferns; Coedredyn Arian)
116	*DEALBATA* (G.Forst.) Sw. (Silver(y) Tree–fern; Coedredynen Arian)
116	**CYRTOMIUM** C.Presl (House Holly–ferns; Rhedyn Celynnog)
116	***FALCATUM*** (L.f.) C.Presl (House Holly–fern; Rhedynen Celynnog)
117	**CYSTOPTERIS** Bernh. (Bladder–ferns; Ffiolredyn)
117	*DICKIEANA* R.Sim (Dickie's Bladder–fern; Ffiolredynen Arfor)
118	***FRAGILIS*** (L.) Bernh. (Brittle Bladder–fern; Rhedynen Frau)
119	*MONTANA* (Lam.) Desv. (Mountain Bladder–fern; Ffiolredynen y Mynydd)
120	DAVALLIA Sm. (Hare's–foot Ferns; Rhedyn Troed–yr–Ysgafarnog)
120	*BULLATA* Wall. group (Squirrel's–foot Ferns; Rhedyn Troed–y–Wiwer)
120	*CANARIENSIS* (L.) Sm. (Hare's–foot Fern; Rhedynen Troed–yr–Ysgafarnog)
201	*zetlandia* Colenso (*Leptolepia novae–zelandiae* (Colenso) Mett. ex Diels)
200	*Dennstaedtia punctiloba* (Michx.) T.Moore (Hay–scented Fern)
120	DICKSONIA L'Hér. (Australian Tree–Ferns; Coedredyn Awstralia)
121	*ANTARCTICA* Labill. (Australian Tree–fern; Coedredynen Awstralia)
121	*fibrosa* Colenso
33	**DIPHASIASTRUM** Holub (Alpine Clubmosses; Cnwpfwsoglau Alpaidd)
33	***ALPINUM*** (L.) Holub (Alpine Clubmoss; Cnwpfwsogl Alpaidd)
33	*complanatum* subsp. *alpinum* (L.) Jermy (***DIPHASIASTRUM ALPINUM*** (L.) Holub)
36	*complanatum* (L.) Holub subsp. *complanatum*
35	*COMPLANATUM* (L.) Holub subsp. *ISSLERI* (Rouy) Jermy (Issler's Clubmoss; Cnwpfwsogl Issler)
35	× *issleri* (Rouy) Holub (*DIPHASIASTRUM COMPLANATUM* (L.) Holub subsp. *ISSLERI* (Rouy) Jermy)
33	*Diphasium alpinum* (L.) Rothm. (***DIPHASIASTRUM ALPINUM*** (L.) Holub)
35	*Diphasium issleri* (Rouy) Holub (*DIPHASIASTRUM COMPLANATUM* (L.) Holub subsp. *ISSLERI* (Rouy) Jermy)
121	**DRYOPTERIS** Adans. (Buckler–ferns; Marchredyn)
137	*abbreviata* auct., non DC. (***DRYOPTERIS OREADES*** Fomin)
123	***AEMULA*** (Aiton) Kuntze (Hay–scented Buckler–fern; Marchredynen Aroglus)

DRYOPTERIS (*continued*)
140   *AEMULA* × *D. DILATATA*
140   *AEMULA* × *D. OREADES* = *D.* × *PSEUDOABBREVIATA* Jermy (Mull Fern; Marchredynen Mull)
124   *AFFINIS* (Lowe) Fraser–Jenk. (Scaly Male–fern; Marchredynen Euraid)
140   *AFFINIS* × *D. FILIX–MAS* = *D.* × *COMPLEXA* Fraser–Jenk. (Hybrid Male–fern; Marchredynen Groesryw)
141   *affinis* × *D. oreades*
126   *AFFINIS* (Lowe) Fraser–Jenk. subsp. *AFFINIS* (Yellow Scaly Male–fern; Marchredynen Euraid)
140   *AFFINIS* subsp. *AFFINIS* × *D. FILIX–MAS* = *D.* × *COMPLEXA* Fraser–Jenk. nothossp. *COMPLEXA*
128   *AFFINIS* (Lowe) Fraser–Jenk. subsp. *BORRERI* (Newman) Fraser–Jenk. (Common Scaly Male–fern; Marchredynen Feddal)
140   *AFFINIS* subsp. *BORRERI* × *D. FILIX–MAS* = *D.* × *COMPLEXA* Fraser–Jenk. nothossp. *CRITICA* Fraser–Jenk.
129   *AFFINIS* (Lowe) Fraser–Jenk. subsp. *CAMBRENSIS* Fraser–Jenk. (Narrow Scaly Male–fern; Marchredynen Gulddail)
140   *affinis* subsp. *cambrensis* × *D. filix–mas*
128   *affinis* subsp. *robusta* Oberh. & Tavel ex Fraser–Jenk. (***DRYOPTERIS AFFINIS*** (Lowe) Fraser–Jenk. subsp. ***BORRERI*** (Newman) Fraser–Jenk.)
128   *affinis* subsp. *stilluppensis* (Sabr.) Fraser–Jenk. (***DRYOPTERIS AFFINIS*** (Lowe) Fraser–Jenk. subsp. ***BORRERI*** (Newman) Fraser–Jenk.)
129   *affinis* subsp. *stilluppensis* sensu Fraser–Jenk., non Sabr. (***DRYOPTERIS AFFINIS*** (Lowe) Fraser–Jenk. subsp. ***CAMBRENSIS*** Fraser–Jenk.)
142   × *AMBROSEAE* Fraser–Jenk. & Jermy = *D. DILATATA* × *D. EXPANSA* (Gibby's Buckler–fern; Marchredynen Gibby)
132   *aristata* (Vill.) Druce (***DRYOPTERIS DILATATA*** (Hoffm.) A.Gray)
134   *assimilis* S.Walker (***DRYOPTERIS EXPANSA*** (C.Presl) Fraser–Jenk. & Jermy)
132   *austriaca* (Jacq.) Woyn. (***DRYOPTERIS DILATATA*** (Hoffm.) A.Gray)
128   *borreri* (Newman) ex Oberh. & Tavel (***DRYOPTERIS AFFINIS*** (Lowe) Fraser–Jenk. subsp. ***BORRERI*** (Newman) Fraser–Jenk.)
142   × *BRATHAICA* Fraser–Jenk. & Reichst. = *D. CARTHUSIANA* × *D. FILIX–MAS* (Brathay Fern; Marchredynen Brathay)
130   *CARTHUSIANA* (Vill.) H.P.Fuchs (Narrow Buckler–fern; Marchredynen Gul)
141   *CARTHUSIANA* × *D. CRISTATA* = *D.* × *ULIGINOSA* (A.Braun ex Döll) Kuntze ex Druce (Hybrid Fen Buckler–fern; Marchredynen Gribog Groesryw)
141   *CARTHUSIANA* × *D. DILATATA* = *D.* × *DEWEVERI* (J.T.Jansen) J.T.Jansen & Wacht. (Hybrid Narrow Buckler–fern; Marchredynen Gul Groesryw)

DRYOPTERIS (*continued*)
- *142*   *CARTHUSIANA* × *D. EXPANSA* = *D.* × *SARVELAE* Fraser–Jenk. & Jermy (Kintyre Buckler–fern; Marchredynen Kintyre)
- *142*   *CARTHUSIANA* × *D. FILIX–MAS* = *D.* × *BRATHAICA* Fraser–Jenk. & Reichst. (Brathay Fern; Marchredynen Brathay)
- *126*   *caucasica* (A. Braun) Fraser–Jenk. & Corley
- **140**   × **COMPLEXA** Fraser–Jenk. = ***D. AFFINIS*** × ***D. FILIX–MAS*** (Hybrid Male–fern; Marchredynen Groesryw)
- **140**   × **COMPLEXA** Fraser–Jenk. nothossp. **COMPLEXA** = ***D. AFFINIS*** subsp. ***AFFINIS*** × ***D. FILIX–MAS***
- *141*   × *complexa* Fraser–Jenk. nothossp. *contorta* Fraser–Jenk.
- **140**   × **COMPLEXA** Fraser–Jenk. nothossp. **CRITICA** Fraser–Jenk. = ***D. AFFINIS*** subsp. ***BORRERI*** × ***D. FILIX–MAS***
- *131*   *CRISTATA* (L.) A.Gray (Crested Buckler–fern; Marchredynen Gribog)
- **132**   **DILATATA** (Hoffm.) A.Gray (Broad Buckler–fern; Marchredynen Lydan)
- *142*   *DILATATA* × *D. EXPANSA* = *D.* × *AMBROSEAE* Fraser–Jenk. & Jermy (Gibby's Buckler–fern; Marchredynen Gibby)
- *143*   *dilatata* × *D. filix–mas* = *D.* × *subaustriaca* Rothm.
- *134*   *dilatata* var. *alpina* T.Moore (***DRYOPTERIS EXPANSA*** (C.Presl) Fraser–Jenk. & Jermy)
- **134**   **EXPANSA** (C.Presl) Fraser–Jenk. & Jermy (Northern Buckler–fern; Marchredynen y Gogledd)
- *124*   *filix–mas* auct., non (L.) Schott (***DRYOPTERIS AFFINIS*** (Lowe) Fraser–Jenk.)
- **5, 135**   **FILIX–MAS** (L.) Schott (Male–fern; Marchredynen)
- **143**   **FILIX–MAS** × ***D. OREADES*** = ***D.*** × ***MANTONIAE*** Fraser–Jenk. & Corley (Manton's Male–fern; Marchredynen Manton)
- *137*   *filix–mas* var. *abbreviata* (DC.) Newman (***DRYOPTERIS OREADES*** Fomin)
- *130*   *lanceolato–cristata* (Hoffm.) Alston (***DRYOPTERIS CARTHUSIANA*** (Vill.) H.P.Fuchs)
- *144*   *linnaeana* C.Chr. (***GYMNOCARPIUM DRYOPTERIS*** (L.) Newman)
- *143*   × ***MANTONIAE*** Fraser–Jenk. & Corley = ***D. FILIX–MAS*** × ***D. OREADES*** (Manton's Male–fern; Marchredynen Manton)
- *128*   *mediterranea* Fomin (***DRYOPTERIS AFFINIS*** (Lowe) Fraser–Jenk. subsp. ***BORRERI*** (Newman) Fraser–Jenk.)
- **137**   **OREADES** Fomin (Dwarf Male–fern; Marchredynen Fach y Mynydd)
- *158*   *oreopteris* (Ehrh.) Maxon (***OREOPTERIS LIMBOSPERMA*** (Bellardi ex All.) Holub)
- *124*   *paleacea* auct. (***DRYOPTERIS AFFINIS*** (Lowe) Fraser–Jenk.)
- *128*   *paleacea* (D.Don) Hand.–Mazz. pro parte, non (Sw.) C.Chr. (***DRYOPTERIS AFFINIS*** (Lowe) Fraser–Jenk. subsp. ***BORRERI*** (Newman) Fraser–Jenk.)
- **140**   × ***PSEUDOABBREVIATA*** Jermy = *D. AEMULA* × *D. OREADES* (Mull Fern; Marchredynen Mull)

DRYOPTERIS (*continued*)
- *126*   *pseudomas* sensu (Woll.) Holub & Pouzar pro parte (***DRYOPTERIS AFFINIS*** (Lowe) Fraser–Jenk. subsp. ***AFFINIS***)
- *138*   *REMOTA* (A.Braun ex Döll) Druce (Scaly Buckler–fern; Marchredynen Gennog)
- *142*   × *SARVELAE* Fraser–Jenk. & Jermy = *D. CARTHUSIANA* × *D. EXPANSA* (Kintyre Buckler–fern; Marchredynen Kintyre)
- *130*   *spinulosa* (O.F.Müll.) Kuntze (***DRYOPTERIS CARTHUSIANA*** (Vill.) H.P.Fuchs)130
- *143*   × *subaustriaca* Rothm. = *D. dilatata* × *D. filix–mas*
- **138**   ***SUBMONTANA*** (Fraser–Jenk. & Jermy) Fraser–Jenk. (Rigid Buckler–fern; Marchredynen Anhyblyg)
- *128*   *tavelii* Rothm. (***DRYOPTERIS AFFINIS*** (Lowe) Fraser–Jenk. subsp. ***BORRERI*** (Newman) Fraser–Jenk.)
- *140*   × *tavelii* auct., non Rothm. (***D. AFFINIS*** × ***D. FILIX–MAS*** = *D.* × ***COMPLEXA*** Fraser–Jenk.)
- *191*   *thelypteris* (L.) A.Gray (***THELYPTERIS PALUSTRIS*** Schott)
- *141*   × *ULIGINOSA* (A.Braun ex Döll) Kuntze ex Druce = *D. CARTHUSIANA* × *D. CRISTATA* (Hybrid Fen Buckler–fern; Marchredynen Gribog Groesryw)
- *139*   *villarii* (Bellardi) Woyn. ex Schinz & Thell.
- *138*   *villarii* (Bellardi) Woyn. ex Schinz & Thell. subsp. *submontana* Fraser–Jenk. & Jermy (***DRYOPTERIS SUBMONTANA*** (Fraser–Jenk. & Jermy) Fraser–Jenk.)
- *128*   *woynarii* auct., non Rothm. (***DRYOPTERIS AFFINIS*** (Lowe) Fraser–Jenk. subsp. ***BORRERI*** (Newman) Fraser–Jenk.)
- *138*   *woynarii* Rothm. (***DRYOPTERIS REMOTA*** (A.Braun ex Döll) Druce)
- 81   Duegredyn (**ASPLENIUM** L.; Spleenworts)

Duegredynen
- 86   Arfor (***ASPLENIUM MARINUM*** L.; Sea Spleenwort)
- 102   Dail Bob yn Ail (*A. **SEPTENTRIONALE*** × *A. **TRICHOMANES*** subsp. ***TRICHOMANES*** = *A.* × ***ALTERNIFOLIUM*** Wulfen; Alternate–leaved Spleenwort)
- 82   Ddu (***ASPLENIUM ADIANTUM–NIGRUM*** L.; Black Spleenwort)
- 99   Ddu Groesryw (*A. ADIANTUM–NIGRUM* × *A. ONOPTERIS* = *A.* × *TICINENSE* D.E.Mey.; Hybrid Black Spleenwort)
- 85   Ddu Sarff–faen (*Asplenium cuneifolium* Viv.; Serpentine Black Spleenwort)
- 92   Fforchog (***ASPLENIUM SEPTENTRIONALE*** (L.) Hoffm.; Forked Spleenwort)
- 99   Guernsey (*A. ADIANTUM–NIGRUM* × *A. OBOVATUM* = *A.* × *SARNIENSE* Sleep; Guernsey Spleenwort)
- 93   Gwallt y Forwyn (***ASPLENIUM TRICHOMANES*** L.; Maidenhair Spleenwort)
- 97   Gwallt y Forwyn Gain (***ASPLENIUM TRICHOMANES*** L. subsp. ***TRICHOMANES***; Delicate Maidenhair Spleenwort)

Duegredynen (*continued*)
- 104　Gwallt y Forwyn Groesryw (**A. TRICHOMANES** subsp. **QUADRIVALENS** × subsp. **TRICHOMANES** = **A. TRICHOMANES** L. nothossp. **LUSATICUM** (D.E.Mey.) Lawalrée; Hybrid Maidenhair Spleenwort)
- 101　Gwallt y Forwyn Gydlifol (**A. SCOLOPENDRIUM** × **A. TRICHOMANES** subsp. **QUADRIVALENS** = **A.** × **CONFLUENS** (T. Moore ex Lowe) Lawalrée; Confluent Maidenhair Spleenwort)
- 96　Gwallt y Forwyn Gyffredin (**ASPLENIUM TRICHOMANES** L. subsp. **QUADRIVALENS** D.E.Mey. emend. Lovis; Common Maidenhair Spleenwort)
- 95　Gwallt y Forwyn Labedog (**ASPLENIUM TRICHOMANES** L. subsp. **PACHYRACHIS** (H.Christ) Lovis & Reichst.; Lobed Maidenhair Spleenwort)
- 85　Lefn y Creigiau (*Asplenium fontanum* (L.) Bernh.; Smooth Rock–spleenwort)
- 101　Murbeck (**A. RUTA–MURARIA** × **A. SEPTENTRIONALE** = **A.** × **MURBECKII** Dörfl; Murbeck's Spleenwort)
- 87　Reiniolaidd (**ASPLENIUM OBOVATUM** Viv. subsp. **LANCEOLATUM** (Fiori) P.Silva; Lanceolate Spleenwort)
- 97　Werdd (**ASPLENIUM TRICHOMANES–RAMOSUM** L.; Green Spleenwort)
- 88　Wyddelig (**ASPLENIUM ONOPTERIS** L.; Irish Spleenwort)
- 101　y Fonesig (**A. RUTA–MURARIA** × **A. TRICHOMANES** subsp. **QUADRIVALENS** = **A.** × **CLERMONTIAE** Syme; Lady Clermont's Spleenwort)
- 89　y Muriau (**ASPLENIUM RUTA–MURARIA** L.; Wall–rue)
- 55　Dutch–rush (Rough Horsetail)
- **49　EQUISETUM** L. (Horsetails; Marchrawn)
- 52　**ARVENSE** L. (Field Horsetail; Marchrhawn yr Ardir)
- 65　**ARVENSE** × **E. FLUVIATILE** = **E.** × **LITORALE** Kühlew ex Rupr. (Shore Horsetail; Marchrawn Croesryw)
- 66　**ARVENSE** × **E. PALUSTRE** = **E.** × **ROTHMALERI** C.N.Page (Ditch Horsetail; Marchrawn y Ffos)
- 68　× **BOWMANII** C.N.Page = **E. SYLVATICUM** × **E. TELMATEIA** (Bowman's Horsetail; Marchrawn Bowman)
- 67　× **DYCEI** C.N.Page = **E. FLUVIATILE** × **E. PALUSTRE** (Dyce's Horsetail; Marchrawn Dyce)
- 54　**FLUVIATILE** L. (Water Horsetail; Marchrawn yr Afon)
- 67　**FLUVIATILE** × **E. PALUSTRE** = **E.** × **DYCEI** C.N.Page (Dyce's Horsetail; Marchrawn Dyce)
- 67　**FLUVIATILE** × **E. TELMATEIA** = **E.** × **WILLMOTII** C.N.Page (Willmot's Horsetail; Marchrawn Willmot)
- 68　× **FONT–QUERI** Rothm. = **E. PALUSTRE** × **E. TELMATEIA** (Font–Quer's Horsetail; Marchrawn Font–Quer)
- 55　**HYEMALE** L. (Rough Horsetail; Marchrawn y Gaeaf)
- 67　**HYEMALE** × **E. RAMOSISSIMUM** = **E.** × **MOOREI** Newman (Moore's Horsetail; Marchrawn Moore)

EQUISETUM (*continued*)
- 68   *HYEMALE* × *E. VARIEGATUM* = *E.* × *TRACHYODON* A.Braun (Mackay's Horsetail; Marchrawn Mackay)
- 54   *LIMOSUM* L. (***EQUISETUM FLUVIATILE*** L.)
- 65   × ***LITORALE*** Kühlew ex Rupr. = ***E. ARVENSE*** × ***E. FLUVIATILE*** (Shore Horsetail; Marchrawn Croesryw)
- 61   *maximum* auct. (***EQUISETUM TELMATEIA*** Ehrh.)
- 68   × *MILDEANUM* Rothm. = *E. PRATENSE* × *E. SYLVATICUM* (Milde's Horsetail; Marchrawn Milde)
- 67   × *MOOREI* Newman = *E. HYEMALE* × *E. RAMOSISSIMUM* (Moore's Horsetail; Marchrawn Moore)
- 57   *PALUSTRE* L. (Marsh Horsetail; Marchrawn y Gors)
- 68   *PALUSTRE* × *E. TELMATEIA* = *E.* × *FONT–QUERI* Rothm. (Font–Quer's Horsetail; Marchrawn Font–Quer)
- 58   *palustre* L. var. *polystachyum* Weigel
- 59   *PRATENSE* Ehrh. (Shady Horsetail; Marchrawn y Cysgod)
- 68   *PRATENSE* × *E. SYLVATICUM* = *E.* × *MILDEANUM* Rothm. (Milde's Horsetail; Marchrawn Milde)
- 59   *RAMOSISSIMUM* Desf. (Branched Horsetail; Marchrawn Brigol)
- 66   × *ROTHMALERI* C.N.Page = *E. ARVENSE* × *E. PALUSTRE* (Ditch Horsetail; Marchrawn y Ffos)
- 60   ***SYLVATICUM*** L. (Wood Horsetail; Marchrawn y Coed)
- 68   *SYLVATICUM* × *E. TELMATEIA* = *E.* × *BOWMANII* C.N.Page (Bowman's Horsetail; Marchrawn Bowman)
- 61   ***TELMATEIA*** Ehrh. (Great Horsetail; Marchrawn Mawr)
- 63   *telmateia* Ehrh. var. *serotinum* A.Br.
- 68   × *TRACHYODON* A.Braun = *E. HYEMALE* × *E. VARIEGATUM* (Mackay's Horsetail; Marchrawn Mackay)
- 63   ***VARIEGATUM*** Schleich. (Variegated Horsetail; Marchrawn Amlywiol)
- 67   × *WILLMOTII* C.N.Page = *E. FLUVIATILE* × *E. TELMATEIA* (Willmot's Horsetail; Marchrawn Willmot)
- 110   Fern, Alpine Water, (Antarctic Hard–fern)
- 79   American Maidenhair, (*ADIANTUM PEDATUM* L.; Briger Gwener America)
- 153   Bead, (Sensitive Fern)
- 162   Beech, (***PHEGOPTERIS CONNECTILIS*** (Michx.) Watt; Rhedynen y Graig)
-    Bladder–, (*see* Bladder–fern)
- 189   Brake, (*PTERIS CRETICA* L.; Rhedynen Creta)
- 142   Brathay, (*D. CARTHUSIANA* × *D. FILIX–MAS* = *D.* × *BRATHAICA* Fraser–Jenk. & Reichst.; Marchredynen Brathay)
-    Buckler–, (*see* Buckler–fern)
- 100   Caernarvonshire, (***A. ADIANTUM–NIGRUM*** × ***A. SEPTENTRIONALE*** = *A.* × *CONTREI* Callé, Lovis & Reichst.; Rhedynen Sir Gaernarfon)

INDEX

Fern (*continued*)
- 200   Chain–, (*see* Chain–fern)
- 201   Christmas, (*Polystichum acrostichoides* (Michx.) T.Moore)
-       Filmy–, (*see* Filmy–fern)
- 100   Guernsey, (*A. OBOVATUM* × *A. SCOLOPENDRIUM* = *A.* × *MICRODON* (T.Moore) Lovis & Vida; Rhedynen Guernsey)
- 120   Hare's–foot, (*DAVALLIA CANARIENSIS* (L.) Sm.; Rhedynen Troed–yr–Ysgafarnog)
-       Hard–, (*see* Hard–fern)
- 200   Hay–scented, (*Dennstaedtia punctiloba* (Michx.) T.Moore)
-       Holly–, (*see* Holly–fern)
- 201   Interrupted, (*Osmunda claytoniana* L.)
- 99   Jackson's, (*A. ADIANTUM–NIGRUM* × *A. SCOLOPENDRIUM* = *A.* × *JACKSONII* (Alston) Lawalrée; Rhedynen Jackson)
- 80   Jersey, (*ANOGRAMMA LEPTOPHYLLA* (L.) Link; Rhedynen Fach Jersey)
- 164   Kangaroo, (*PHYMATODES DIVERSIFOLIA* (Willd.) Pic.Serm.; Rhedynen Cangarŵ)
- 193   Killarney, (**TRICHOMANES SPECIOSUM** Willd.; Rhedynen Wrychog)
- 201   Lace, (*Leptolepia novae–zelandiae* (Colenso) Mett. ex Diels)
-       Lady–, (*see* Lady–fern)
- 201   Leathery Shield, (*Rumohra adiantiformis* (G.Forst.) Ching)
- 158   Lemon–scented, (**OREOPTERIS LIMBOSPERMA** (Bellardi ex All.) Holub; Marchredynen y Mynydd)
- 146   Limestone, (**GYMNOCARPIUM ROBERTIANUM** (Hoffm.) Newman; Llawredynen y Calchfaen)
- 78   Maidenhair, (**ADIANTUM CAPILLUS–VENERIS** L.; Briger Gwener)
-       Male–, (*see* Male–fern)
- 191   Marsh, (**THELYPTERIS PALUSTRIS** Schott; Marchredynen y Gors)
- 158   Mountain, (Lemon–scented Fern)
- 140   Mull, (*D. AEMULA* × *D. OREADES* = *D.* × *PSEUDOABBREVIATA* Jermy; Marchredynen Mull)
- 144   Oak, (**GYMNOCARPIUM DRYOPTERIS** (L.) Newman; Llawredynen y Derw)
- 153   Ostrich, (**MATTEUCCIA STRUTHIOPTERIS** (L.) Tod.; Rhedynen Estrys)
- 114   Parsley, (**CRYPTOGRAMMA CRISPA** (L.) R.Br.; Rhedynen Bersli)
- 189   Ribbon, (Brake Fern)
- 201   Rock Felt, (*Pyrrosia rupestris* (R.Br.) Ching)
- 161   Royal, (**OSMUNDA REGALIS** L.; Rhedynen Gyfrdwy)
- 153   Sensitive, (**ONOCLEA** L.; Rhedyn Croendenau)
- 153   Sensitive, (**ONOCLEA SENSIBILIS** L.; Rhedynen Groendenau)
-       Shield–, (*see* Shield–fern)
- 201   Soft Ground, (*Hypolepis tenuifolia* (G.Forst.) Bernh.)
- 164   Tongue, (Kangaroo Fern)
-       Tree–, (*see* Tree–fern)
- 107   Water, (**AZOLLA FILICULOIDES** Lam.; Rhedynen y Dŵr)

INDEX

162	Ferns, Beech, (**PHEGOPTERIS** (C.Presl) Fée; Rhedyn y Graig)
117	Bladder–, (**CYSTOPTERIS** Bernh.; Ffiolredyn)
189	Brake, (PTERIS L.; Rhedyn Ruban)
121	Buckler–, (**DRYOPTERIS** Adans.; Marchredyn)
200	Chain–, (WOODWARDIA Sm.; Rhedyn Cadwyn)
148	Filmy–, (**HYMENOPHYLLUM** Sm.; Rhedynach Teneuwe)
109	Hard–, (**BLECHNUM** L.; Gwibredyn)
120	Hare's–foot, (DAVALLIA Sm.; Rhedyn Troed–yr–Ysgafarnog)
116	House Holly–, (**CYRTOMIUM** C.Presl; Rhedyn Celynnog)
80	Jersey, (ANOGRAMMA Link; Rhedyn Jersey)
164	Kangaroo, (PHYMATODES C.Presl; Rhedyn Cangarŵ)
193	Killarney, (**TRICHOMANES** L.; Rhedyn Gwrychog)
104	Lady–, (**ATHYRIUM** Roth; Rhedyn Mair)
158	Lemon–scented, (**OREOPTERIS** Holub; Marchredyn y Mynydd)
78	Maidenhair, (**ADIANTUM** L.; Brigerau Gwener)
190	Marsh, (**THELYPTERIS** Schmidel; Marchredyn y Gors)
143	Oak, (**GYMNOCARPIUM** Newman; Llawredyn y Derw)
152	Ostrich, (**MATTEUCCIA** Tod.; Rhedyn Estrys)
114	Parsley, (**CRYPTOGRAMMA** R.Br.; Rhedyn Persli)
160	Royal, (**OSMUNDA** L.; Rhedyn Cyfrdwy)
177	Shield–, (**POLYSTICHUM** Roth; Gwrychredyn)
120	Squirrel's–foot, (*DAVALLIA BULLATA* Wall. group; Rhedyn Troed–y–Wiwer)
	Tree–, (*see* Tree–ferns)
107	Water (**AZOLLA** Lam.; Rhedyn y Dŵr)
117	Ffiolredyn (**CYSTOPTERIS** Bernh.; Bladder–ferns)
117	Ffiolredynen Arfor (*CYSTOPTERIS DICKIEANA* R.Sim; Dickie's Bladder–fern)
119	Ffiolredynen y Mynydd (*CYSTOPTERIS MONTANA* (Lam.) Desv.; Mountain Bladder–fern)
201	Filmy–fern, Shiny, (*Hymenophyllum flabellatum* Labill.)
149	Tunbridge, (**HYMENOPHYLLUM TUNBRIGENSE** (L.) Sm.; Rhedynach Teneuwe)
151	Wilson's, (**HYMENOPHYLLUM WILSONII** Hook.; Rhedynach Teneuwe Wilson)
148	Filmy–ferns (**HYMENOPHYLLUM** Sm.; Rhedynach Teneuwe)
201	Fork–ferns (*Tmesipteris* Bernh. species)
46	Gwair Merllyn (**ISOETES** L.; Quillworts)
48	Gwair Merllyn (**ISOETES LACUSTRIS** L.; Quillwort)
46	Bychan (**ISOETES ECHINOSPORA** Durieu; Spring Quillwort)
47	y Tir (*ISOETES HISTRIX* Bory; Land Quillwort)
109	Gwibredyn (**BLECHNUM** L.; Hard–ferns)
110	Gwibredynen (**BLECHNUM SPICANT** (L.) Roth; Hard–fern)
110	Antarctig (*BLECHNUM PENNA–MARINA* (Poir.) Kuhn; Antarctic Hard–fern)
109	Chile (**BLECHNUM CORDATUM** (Desv.) Hieron.; Chilean Hard–fern)

INDEX

177	Gwrychredyn (**POLYSTICHUM** Roth; Shield–ferns) Gwrychredynen
181	Feddal (***POLYSTICHUM SETIFERUM*** (Forssk.) Woyn.; Soft Shield–fern)
177	Galed (***POLYSTICHUM ACULEATUM*** (L.) Roth; Hard Shield–fern)
181	Gledd Orllewinol (*POLYSTICHUM MUNITUM* (Kaulf.) C.Presl; Western Sword–fern)
183	Groesryw Alpaidd (*P. ACULEATUM* × *P. LONCHITIS* = *P.* × *ILLYRICUM* (Borbás) Hahne; Alpine Hybrid Shield–fern)
184	Groesryw y Gorllewin (*P. LONCHITIS* × *P. SETIFERUM* = *P.* × *LONCHITIFORME* (Halácsy) Bech.; Atlantic Hybrid Shield–fern)
183	Groesryw yr Iseldir (***P. ACULEATUM*** × ***P. SETIFERUM*** = ***P.*** × ***BICKNELLII*** (H.Christ) Hahne; Lowland Hybrid Shield–fern)
143	**GYMNOCARPIUM** Newman (Oak Ferns; Llawredyn y Derw)
144	*DRYOPTERIS* (L.) Newman (Oak Fern; Llawredyn y Derw)
146	*ROBERTIANUM* (Hoffm.) Newman (Limestone Fern; Llawredynen y Calchfaen)
110	Hard–fern (***BLECHNUM SPICANT*** (L.) Roth; Gwibredynen)
110	Antarctic, (*BLECHNUM PENNA–MARINA* (Poir.) Kuhn; Gwibredynen Antarctig)
109	Chilean, (***BLECHNUM CORDATUM*** (Desv.) Hieron.; Gwibredynen Chile)
109	Hard–ferns (**BLECHNUM** L.; Gwibredyn)
90	Hart's–tongue (***ASPLENIUM SCOLOPENDRIUM*** L.; Tafod yr Hydd)
55	*Hippochaete hyemalis* (L.) Bruhin (***EQUISETUM HYEMALE*** L.)
59	*Hippochaete ramosissima* (Desf.) Börner (*EQUISETUM RAMOSISSIMUM* Desf.)
63	*Hippochaete variegata* (Schleicher ex Weber & Mohr) Bruhin (***EQUISETUM VARIEGATUM*** Schleich.)
179	Holly–fern (***POLYSTICHUM LONCHITIS*** (L.) Roth; Celynredynen)
116	Holly–fern, House, (***CYRTOMIUM FALCATUM*** (L.f.) C.Presl; Rhedynen Celynnog)
116	Holly–ferns, House, (**CYRTOMIUM** C.Presl; Rhedyn Celynnog)
68	Horsetail, Bowman's, (*E. SYLVATICUM* × *E. TELMATEIA* = *E.* × *BOWMANII* C.N.Page; Marchrawn Bowman)
59	Branched, (*EQUISETUM RAMOSISSIMUM* Desf.; Marchrawn Brigol)
52	Common, (Field Horsetail)
66	Ditch, (*E. ARVENSE* × *E. PALUSTRE* = *E.* × *ROTHMALERI* C.N. Page; Marchrawn y Ffos)
67	Dyce's, (***E. FLUVIATILE*** × ***E. PALUSTRE*** = ***E.*** × ***DYCEI*** C.N. Page; Marchrawn Dyce)
52	Field, (***EQUISETUM ARVENSE*** L.; Marchrhawn yr Ardir)
68	Font–Quer's, (***E. PALUSTRE*** × ***E. TELMATEIA*** = ***E.*** × ***FONT–QUERI*** Rothm.; Marchrawn Font–Quer)
61	Great, (***EQUISETUM TELMATEIA*** Ehrh.; Marchrawn Mawr)

INDEX

68 Mackay's, (*E. HYEMALE* × *E. VARIEGATUM* = *E.* × *TRACHYODON* A.Braun; Marchrawn Mackay)
57 Marsh, (***EQUISETUM PALUSTRE*** L.; Marchrawn y Gors)
68 Milde's, (*E. PRATENSE* × *E. SYLVATICUM* = *E.* × *MILDEANUM* Rothm.; Marchrawn Milde)
67 Moore's, (*E. HYEMALE* × *E. RAMOSISSIMUM* = *E.* × *MOOREI* Newman; Marchrawn Moore)
55 Rough, (***EQUISETUM HYEMALE*** L.; Marchrawn y Gaeaf)
59 Shady, (*EQUISETUM PRATENSE* Ehrh.; Marchrawn y Cysgod)
65 Shore, (*E. ARVENSE* × *E. FLUVIATILE* = *E.* × *LITORALE* Kühlew ex Rupr.; Marchrawn Croesryw)
63 Variegated, (***EQUISETUM VARIEGATUM*** Schleich.; Marchrawn Amlywiol)
54 Water, (***EQUISETUM FLUVIATILE*** L.; Marchrawn yr Afon)
67 Willmot's, (*E. FLUVIATILE E. TELMATEIA* = *E. WILLMOTII* C.N.Page; Marchrawn Willmot)
60 Wood, (***EQUISETUM SYLVATICUM*** L.; Marchrawn y Coed)
49 Horsetails (**EQUISETUM** L.; Marchrawn)
36 **HUPERZIA** Bernh. (Fir Clubmosses; Cnwpfwsoglau Ffeinid)
36 *SELAGO* (L.) Bernh. ex Schrank & C.Mart. (Fir Clubmoss; Cnwpfwsogl Mawr)
148 **HYMENOPHYLLUM** Sm. (Filmy–ferns; Rhedynach Teneuwe)
201 *flabellatum* Labill. (Shiny Filmy–fern)
151 *peltatum* auct. (***HYMENOPHYLLUM WILSONII*** Hook.)
149 *TUNBRIGENSE* (L.) Sm. (Tunbridge Filmy–fern; Rhedynach Teneuwe)
151 *WILSONII* Hook. (Wilson's Filmy–fern; Rhedynach Teneuwe Wilson)
201 *Hypolepis tenuifolia* (G.Forst.) Bernh. (Soft Ground Fern)
46 **ISOETES** L. (Quillworts; Gwair Merllyn)
46 *ECHINOSPORA* Durieu (Spring Quillwort; Gwair Merllyn Bychan)
47 *HISTRIX* Bory (Land Quillwort; Gwair Merllyn y Tir)
48 *LACUSTRIS* L. (Quillwort; Gwair Merllyn)
46 *setacea* Lam. (***ISOETES ECHINOSPORA*** Durieu)
105 Lady–fern (***ATHYRIUM FILIX–FEMINA*** (L.) Roth; Rhedynen Fair)
105 Alpine, (*ATHYRIUM DISTENTIFOLIUM* Tausch ex Opiz var. *DISTENTIFOLIUM*; Rhedynen Fair Alpaidd)
105 Newman's, (*ATHYRIUM DISTENTIFOLIUM* Tausch ex Opiz var. *FLEXILE* (Newman) Jermy; Rhedynen Fair Newman)
104 Lady–ferns (**ATHYRIUM** Roth; Rhedyn Mair)
123 *Lastrea aemula* (Aiton) Brack. (***DRYOPTERIS AEMULA*** (Aiton) Kuntze)
131 *Lastrea cristata* (L.) C.Presl (***DRYOPTERIS CRISTATA*** (L.) A.Gray)
132 *Lastrea dilatata* (Hoffm.) C.Presl (***DRYOPTERIS DILATATA*** (Hoffm.) A.Gray)
5,135 *Lastrea filix–mas* (L.) C.Presl (***DRYOPTERIS FILIX–MAS*** (L.) Schott)
123 *Lastrea foenisecii* (Lowe) H.C.Watson (***DRYOPTERIS AEMULA*** (Aiton) Kuntze)

INDEX

Lastrea (continued)
- 158   *Lastrea oreopteris* (Ehrh.) Bory (***OREOPTERIS LIMBOSPERMA*** (Bellardi ex All.) Holub)
- 137   *Lastrea propinqua* Woll., non J.Sm. (***DRYOPTERIS OREADES*** Fomin)
- 126   *Lastrea pseudomas* Woll. (***DRYOPTERIS AFFINIS*** (Lowe) Fraser–Jenk. subsp. ***AFFINIS***)
- 142   *Lastrea remota* sensu T.Moore excl. syn. (*D. CARTHUSIANA* × *D. FILIX–MAS* = *D.* × *BRATHAICA* Fraser–Jenk. & Reichst.)
- 138   *Lastrea rigida* (Sw.) C.Presl (***DRYOPTERIS SUBMONTANA*** (Fraser–Jenk. & Jermy) Fraser–Jenk.)
- 191   *Lastrea thelypteris* (L.) Bory (***THELYPTERIS PALUSTRIS*** Schott)
- 38   *Lepidotis inundata* (L.) Opiz (***LYCOPODIELLA INUNDATA*** (L.) Holub)
- 201   *Leptolepia novae–zelandiae* (Colenso) Mett. ex Diels (Lace Fern)
- 143   Llawredyn y Derw (**GYMNOCARPIUM** Newman; Oak Ferns)
- 167   Llawredyn y Fagwyr (**POLYPODIUM** L.; Polypodies)

Llawredynen
- 176   Font–Quer (***P. CAMBRICUM*** × ***P. VULGARE*** = ***P.*** × ***FONT–QUERI*** Rothm.; Font–Quer's Polypody)
- 170   Gymreig (***POLYPODIUM CAMBRICUM*** L.; Southern Polypody)
- 176   Manton (***P. INTERJECTUM*** × ***P. VULGARE*** = ***P.*** × ***MANTONIAE*** Rothm. & U.Schneid.; Manton's Polypody)
- 172   Rymus (***POLYPODIUM INTERJECTUM*** Shivas; Intermediate Polypody)
- 175   Shivas (***P. CAMBRICUM*** × ***P. INTERJECTUM*** = ***P.*** × ***SHIVASIAE*** Rothm.; Shivas' Polypody)
- 146   y Calchfaen (***GYMNOCARPIUM ROBERTIANUM*** (Hoffm.) Newman; Limestone Fern)
- 144   y Derw (***GYMNOCARPIUM DRYOPTERIS*** (L.) Newman; Oak Fern)
- 173   y Fagwyr (***POLYPODIUM VULGARE*** L.; Polypody)
- 112   Lloerlys (**BOTRYCHIUM LUNARIA** (L.) Sw.; Moonwort)
- 112   Lloerlysiau (**BOTRYCHIUM** Sw.; Moonworts)
- **38**   **LYCOPODIELLA** Holub (Marsh Clubmosses; Cnwpfwsoglau y Gors)
- **38**   ***INUNDATA*** (L.) Holub (Marsh Clubmoss; Cnwpfwsogl y Gors)
- **39**   **LYCOPODIUM** L. (Clubmosses; Cnwpfwsoglau)
- 33   *alpinum* L. (***DIPHASIASTRUM ALPINUM*** (L.) Holub)
- 35   *alpinum* var. *decipiens* Syme (***DIPHASIASTRUM COMPLANATUM*** (L.) Holub subsp. ***ISSLERI*** (Rouy) Jermy)
- *40*   ***ANNOTINUM*** L. (Interrupted Clubmoss; Cnwpfwsogl Meinfannau)
- *40*   ***CLAVATUM*** L. (Stag's–horn Clubmoss; Cnwpfwsogl Corn Carw)
- 35   *complanatum* auct. brit. (***DIPHASIASTRUM COMPLANATUM*** (L.) Holub subsp. ***ISSLERI*** (Rouy) Jermy)
- 38   *inundatum* L. (***LYCOPODIELLA INUNDATA*** (L.) Holub)
- 35   *issleri* (Rouy) Lawalrée (***DIPHASIASTRUM COMPLANATUM*** (L.) Holub subsp. ***ISSLERI*** (Rouy) Jermy)
- 36   *selago* L. (***HUPERZIA SELAGO*** (L.) Bernh. ex Schrank & C.Mart.)
- 200   Maidenhair, Evergreen, (*Adiantum venustum* D.Don)

INDEX

5,135 Male–fern (***DRYOPTERIS FILIX–MAS*** (L.) Schott; Marchredynen)
128   Common Scaly, (***DRYOPTERIS AFFINIS*** (Lowe) Fraser–Jenk. subsp. ***BORRERI*** (Newman) Fraser–Jenk.; Marchredynen Feddal)
137   Dwarf, (***DRYOPTERIS OREADES*** Fomin; Marchredynen Fach y Mynydd)
140   Hybrid, (***D. AFFINIS*** × ***D. FILIX–MAS*** = ***D.*** × ***COMPLEXA*** Fraser–Jenk.; Marchredynen Groesryw)
143   Manton's, (***D. FILIX–MAS*** × ***D. OREADES*** = ***D.*** × ***MANTONIAE*** Fraser–Jenk. & Corley; Marchredynen Manton)
129   Narrow Scaly, (***DRYOPTERIS AFFINIS*** (Lowe) Fraser–Jenk. subsp. ***CAMBRENSIS*** Fraser–Jenk.; Marchredynen Gulddail)
124   Scaly, (***DRYOPTERIS AFFINIS*** (Lowe) Fraser–Jenk.; Marchredynen Euraid)
126   Yellow Scaly, (***DRYOPTERIS AFFINIS*** (Lowe) Fraser–Jenk. subsp. ***AFFINIS***; Marchredynen Euraid)
49   Marchrawn (**EQUISETUM** L.; Horsetails)
63   Amlywiol (***EQUISETUM VARIEGATUM*** Schleich.; Variegated Horsetail)
68   Bowman (***E. SYLVATICUM*** × ***E. TELMATEIA*** = ***E.*** × ***BOWMANII*** C.N.Page; Bowman's Horsetail)
59   Brigol (***EQUISETUM RAMOSISSIMUM*** Desf.; Branched Horsetail)
65   Croesryw (***E. ARVENSE*** × ***E. FLUVIATILE*** = ***E.*** × ***LITORALE*** Kühlew ex Rupr.; Shore Horsetail)
67   Dyce (***E. FLUVIATILE*** × ***E. PALUSTRE*** = ***E.*** × ***DYCEI*** C.N.Page; Dyce's Horsetail)
68   Font–Quer (***E. PALUSTRE*** × ***E. TELMATEIA*** = ***E.*** × ***FONT–QUERI*** Rothm.; Font–Quer's Horsetail
68   Mackay (***E. HYEMALE*** × ***E. VARIEGATUM*** = ***E.*** × ***TRACHYODON*** A.Braun; Mackay's Horsetail)
61   Mawr (***EQUISETUM TELMATEIA*** Ehrh.; Great Horsetail)
68   Milde (***E. PRATENSE*** × ***E. SYLVATICUM*** = ***E.*** × ***MILDEANUM*** Rothm.; Milde's Horsetail)
67   Moore (***E. HYEMALE*** × ***E. RAMOSISSIMUM*** = ***E.*** × ***MOOREI*** Newman; Moore's Horsetail)
67   Willmot (***E. FLUVIATILE*** × ***E. TELMATEIA*** = ***E.*** × ***WILLMOTII*** C.N.Page; Willmot's Horsetail)
60   y Coed (***EQUISETUM SYLVATICUM*** L.; Wood Horsetail)
59   y Cysgod (***EQUISETUM PRATENSE*** Ehrh.; Shady Horsetail)
66   y Ffos (***E. ARVENSE*** × ***E. PALUSTRE*** = ***E.*** × ***ROTHMALERI*** C.N.Page; Ditch Horsetail)
55   y Gaeaf (***EQUISETUM HYEMALE*** L.; Rough Horsetail)
57   y Gors (***EQUISETUM PALUSTRE*** L.; Marsh Horsetail)
54   yr Afon (***EQUISETUM FLUVIATILE*** L.; Water Horsetail)
121   Marchredyn (**DRYOPTERIS** Adans.; Buckler–ferns)
190   y Gors (**THELYPTERIS** Schmidel; Marsh Ferns)
158   y Mynydd (**OREOPTERIS** Holub; Lemon–scented Ferns)

INDEX

5,135 Marchredynen (***DRYOPTERIS FILIX–MAS*** (L.) Schott; Male–fern)
138 Anhyblyg (***DRYOPTERIS SUBMONTANA*** (Fraser–Jenk. & Jermy) Fraser–Jenk.; Rigid Buckler–fern)
123 Aroglus (***DRYOPTERIS AEMULA*** (Aiton) Kuntze; Hay–scented Buckler–fern)
142 Brathay (*D. CARTHUSIANA* × *D. FILIX–MAS* = *D.* × *BRATHAICA* Fraser–Jenk. & Reichst.; Brathay Fern)
124 Euraid (***DRYOPTERIS AFFINIS*** (Lowe) Fraser–Jenk.; Scaly Male–fern)
126 Euraid (***DRYOPTERIS AFFINIS*** (Lowe) Fraser–Jenk. subsp. ***AFFINIS***; Yellow Scaly Male–fern)
137 Fach y Mynydd (***DRYOPTERIS OREADES*** Fomin; Dwarf Male–fern)
128 Feddal (***DRYOPTERIS AFFINIS*** (Lowe) Fraser–Jenk. subsp. ***BORRERI*** (Newman) Fraser–Jenk.; Common Scaly Male–fern)
138 Gennog (*DRYOPTERIS REMOTA* (A.Braun ex Döll) Druce; Scaly Buckler–fern)
142 Gibby (***D. DILATATA*** × ***D. EXPANSA*** = ***D.*** × ***AMBROSEAE*** Fraser–Jenk. & Jermy; Gibby's Buckler–fern)
131 Gribog (*DRYOPTERIS CRISTATA* (L.) A.Gray; Crested Buckler–fern)
141 Gribog Groesryw (*D. CARTHUSIANA* × *D. CRISTATA* = *D.* × *ULIGINOSA* (A.Braun ex Döll) Kuntze ex Druce; Hybrid Fen Buckler–fern)
140 Groesryw (***D. AFFINIS*** × ***D. FILIX–MAS*** = ***D.*** × ***COMPLEXA*** Fraser–Jenk.; Hybrid Male–fern)
130 Gul (***DRYOPTERIS CARTHUSIANA*** (Vill.) H.P.Fuchs; Narrow Buckler–fern)
141 Gul Groesryw (***D. CARTHUSIANA*** × ***D. DILATATA*** = ***D.*** × ***DEWEVERI*** (J.T.Jansen) J.T.Jansen & Wacht.; Hybrid Narrow Buckler–fern)
129 Gulddail (***DRYOPTERIS AFFINIS*** (Lowe) Fraser–Jenk. subsp. ***CAMBRENSIS*** Fraser–Jenk.; Narrow Scaly Male–fern)
142 Kintyre (*D. CARTHUSIANA* × *D. EXPANSA* = *D.* × *SARVELAE* Fraser–Jenk. & Jermy; Kintyre Buckler–fern)
132 Lydan (***DRYOPTERIS DILATATA*** (Hoffm.) A.Gray; Broad Buckler–fern)
143 Manton ((***D. FILIX–MAS*** × ***D. OREADES*** = ***D.*** × ***MANTONIAE*** Fraser–Jenk. & Corley; Manton's Male–fern)
140 Mull (*D. AEMULA* × *D. OREADES* = *D.* × *PSEUDOABBREVIATA* Jermy; Mull Fern)
134 y Gogledd (***DRYOPTERIS EXPANSA*** (C.Presl) Fraser–Jenk. & Jermy; Northern Buckler–fern)
191 y Gors (***THELYPTERIS PALUSTRIS*** Schott; Marsh Fern)
158 y Mynydd (***OREOPTERIS LIMBOSPERMA*** (Bellardi ex All.) Holub; Lemon–scented Fern)

INDEX

52	Marchrhawn yr Ardir (*EQUISETUM ARVENSE* L.; Field Horsetail)
152	**MATTEUCCIA** Tod. (Ostrich Ferns; Rhedyn Estrys)
153	*STRUTHIOPTERIS* (L.) Tod. (Ostrich Fern; Rhedynen Estrys)
164	Microsorium diversifolium (Willd.) Copel. (*PHYMATODES DIVERSIFOLIA* (Willd.) Pic.Serm.)
112	Moonwort (***BOTRYCHIUM LUNARIA*** (L.) Sw.; Lloerlys)
112	Moonworts (**BOTRYCHIUM** Sw.; Lloerlysiau)
153	**ONOCLEA** L. (Sensitive Fern; Rhedyn Croendenau)
153	*SENSIBILIS* L. (Sensitive Fern; Rhedynen Groendenau)
154	**OPHIOGLOSSUM** L. (Adder's–tongues; Tafodau y Neidr)
155	*AZORICUM* C.Presl (Small Adder's–tongue; Tafod y Neidr Bach)
155	*LUSITANICUM* L. (Least Adder's–tongue; Tafod y Neidr Lleiaf)
156	*VULGATUM* L. (Adder's–tongue; Tafod y neidr)
155	*vulgatum* subsp. *ambiguum* (Coss. & Germ.) E.F.Warb. (***OPHIOGLOSSUM AZORICUM*** C.Presl)
155	*vulgatum* subsp. *polyphyllum* auct., non A.Braun (***OPHIOGLOSSUM AZORICUM*** C.Presl)
158	**OREOPTERIS** Holub (Lemon–scented Ferns; Marchredyn y Mynydd)
158	*LIMBOSPERMA* (Bellardi ex All.) Holub (Lemon–scented Fern; Marchredynen y Mynydd)
160	**OSMUNDA** L. (Royal Ferns; Rhedyn Cyfrdwy)
201	*claytoniana* L. (Interrupted Fern)
161	*REGALIS* L. (Royal Fern; Rhedynen Gyfrdwy)
164	Pelenllys (**PILULARIA** L.; Pillworts)
165	Pelenllys (***PILULARIA GLOBULIFERA*** L.; Pillwort)
116	PHANEROPHLEBIA C.Presl (**CYRTOMIUM** C.Presl)
116	*falcata* (L.f.) Copel. (***CYRTOMIUM FALCATUM*** (L.f.) C.Presl)
162	**PHEGOPTERIS** (C.Presl) Fée (Beech Ferns; Rhedyn y Graig)
162	*CONNECTILIS* (Michx.) Watt (Beech Fern; Rhedynen y Graig)
144	*dryopteris* (L.) Fée (***GYMNOCARPIUM DRYOPTERIS*** (L.) Newman)
162	*polypodioides* Fée (***PHEGOPTERIS CONNECTILIS*** (Michx.) Watt)
146	*robertiana* (Hoffm.) A. Braun (***GYMNOCARPIUM ROBERTIANUM*** (Hoffm.) Newman)
90	PHYLLITIS Hill (*see under* **ASPLENIUM** L.)
90	*scolopendrium* (L.) Newman (***ASPLENIUM SCOLOPENDRIUM*** L.)
164	PHYMATODES C.Presl (Kangaroo Ferns; Rhedyn Cangarŵ)
164	*DIVERSIFOLIA* (Willd.) Pic.Serm. (Kangaroo Fern; Rhedynen Cangarŵ)
165	Pillwort (***PILULARIA GLOBULIFERA*** L.; Pelenllys)
164	Pillworts (**PILULARIA** L.; Pelenllys)
164	**PILULARIA** L. (Pillworts; Pelenllys)
165	*GLOBULIFERA* L. (Pillwort; Pelenllys)
195	*Polyphlebium venosum* (R.Br.) Copel. (***TRICHOMANES VENOSUM*** R.Br.)
167	Polypodies (**POLYPODIUM** L.; Llawredyn y Fagwyr)

INDEX

167 **POLYPODIUM** L. (Polypodies; Llawredyn y Fagwyr)
170   *australe* Fée (***POLYPODIUM CAMBRICUM*** L.)
171   *australe* Fée 'Cambricum' (***POLYPODIUM CAMBRICUM*** L. **'CAMBRICUM'**)
171   *cambricum* L. (***POLYPODIUM CAMBRICUM*** L. **'CAMBRICUM'**)
170   ***CAMBRICUM*** L. (Southern Polypody; Llawredynen Gymreig)
175   ***CAMBRICUM*** × *P. INTERJECTUM* = *P.* × *SHIVASIAE* Rothm. (Shivas' Polypody; Llawredynen Shivas)
176   ***CAMBRICUM*** × *P. VULGARE* = *P.* × *FONT–QUERI* Rothm. (Font–Quer's Polypody; Llawredynen Font–Quer)
171   ***CAMBRICUM*** L. **'CAMBRICUM'** (Welsh Polypody)
176   × ***FONT–QUERI*** Rothm. = *P. CAMBRICUM* × *P. VULGARE* (Font–Quer's Polypody; Llawredynen Font–Quer)
172   ***INTERJECTUM*** Shivas (Intermediate Polypody; Llawredynen Rymus)
176   ***INTERJECTUM*** × *P. VULGARE* = *P.* × *MANTONIAE* Rothm. & U.Schneid. (Manton's Polypody; Llawredynen Manton)
176   × ***MANTONIAE*** Rothm. & U.Schneid. = *P. INTERJECTUM* × *P. VULGARE* (Manton's Polypody; Llawredynen Manton)
162   *phegopteris* L. (***PHEGOPTERIS CONNECTILIS*** (Michx.) Watt)
175   × *rothmaleri* Shivas (*P. CAMBRICUM* × *P. INTERJECTUM* = *P.* × *SHIVASIAE* Rothm.)
175   × ***SHIVASIAE*** Rothm. = *P. CAMBRICUM* × *P. INTERJECTUM* (Shivas' Polypody; Llawredynen Shivas)
173   ***VULGARE*** L. (Polypody; Llawredynen y Fagwyr)
171   *vulgare* var. *cambricum* (L.) Lightf. (***POLYPODIUM CAMBRICUM*** L. **'CAMBRICUM'**)
172   *vulgare* subsp. *prionodes* (Asch.) Rothm. (***POLYPODIUM INTERJECTUM*** Shivas)
170   *vulgare* var. *serratum* Willd. (***POLYPODIUM CAMBRICUM*** L.)
170   *vulgare* subsp. *serrulatum* Arcang. (***POLYPODIUM CAMBRICUM*** L.)
173   Polypody (***POLYPODIUM VULGARE*** L.; Llawredynen y Fagwyr)
173   Common, (Polypody)
176   Font–Quer's, (*P. CAMBRICUM* × *P. VULGARE* = *P.* × *FONT–QUERI* Rothm.; Llawredynen Font–Quer)
172   Intermediate, (***POLYPODIUM INTERJECTUM*** Shivas; Llawredynen Rymus)
146   Limestone, (Limestone Fern)
176   Manton's, (*P. INTERJECTUM* × *P. VULGARE* = *P.* × *MANTONIAE* Rothm. & U.Schneid.; Llawredynen Manton)
175   Shivas' (*P. CAMBRICUM* × *P. INTERJECTUM* = *P.* × *SHIVASIAE* Rothm.; Llawredynen Shivas)
170   Southern, (***POLYPODIUM CAMBRICUM*** L.; Llawredynen Gymreig)
171   Welsh, (***POLYPODIUM CAMBRICUM*** L. **'CAMBRICUM'**)

177	**POLYSTICHUM** Roth (Shield–ferns; Gwrychredyn)
201	*acrostichoides* (Michx.) T.Moore (Christmas Fern)
177	*ACULEATUM* (L.) Roth (Hard Shield–fern; Gwrychredynen Galed)
183	*ACULEATUM* × *P. LONCHITIS* = *P.* × *ILLYRICUM* (Borbás) Hahne (Alpine Hybrid Shield–fern; Gwrychredynen Groesryw Alpaidd)
183	*ACULEATUM* × *P. SETIFERUM* = *P.* × *BICKNELLII* (H.Christ) Hahne (Lowland Hybrid Shield–fern; Gwrychredynen Groesryw yr Iseldir)
179	*ACULEATUM* (L.) Roth forma *CAMBRICUM*
179	*aculeatum* (L.) Roth var. *cambricum* (Gray) Hyde & A.E.Wade (*POLYSTICHUM ACULEATUM* (L.) Roth forma *CAMBRICUM*)
179	*aculeatum* (L.) Roth var. *lonchitidoides* (Hook.) Deakin (*POLYSTICHUM ACULEATUM* (L.) Roth forma *CAMBRICUM*)
181	*angulare* (Willd.) C.Presl (*POLYSTICHUM SETIFERUM* (Forssk.) Woyn.)
183	× *BICKNELLII* (H.Christ) Hahne = *P. ACULEATUM* × *P. SETIFERUM* (Lowland Hybrid Shield–fern; Gwrychredynen Groesryw yr Iseldir)
201	*braunii* (Spenn.) Fée
116	*falcatum* (L.f.) Diels (*CYRTOMIUM FALCATUM* (L.f.) C.Presl)
183	× *ILLYRICUM* (Borbás) Hahne = *P. ACULEATUM* × *P. LONCHITIS* (Alpine Hybrid Shield–fern; Gwrychredynen Groesryw Alpaidd)
177	*lobatum* (Huds.) Chevall. (*POLYSTICHUM ACULEATUM* (L.) Roth)
184	× *LONCHITIFORME* (Halácsy) Bech. = *P. LONCHITIS* × *P. SETIFERUM* (Atlantic Hybrid Shield–fern; Gwrychredynen Groesryw y Gorllewin)
179	*LONCHITIS* (L.) Roth (Holly–fern; Celynredynen)
184	*LONCHITIS* × *P. SETIFERUM* = *P.* × *LONCHITIFORME* (Halácsy) Bech. (Atlantic Hybrid Shield–fern; Gwrychredynen Groesryw y Gorllewin)
181	*MUNITUM* (Kaulf.) C.Presl (Western Sword–fern; Gwrychredynen Gledd Orllewinol)
181	*SETIFERUM* (Forssk.) Woyn. (Soft Shield–fern; Gwrychredynen Feddal)
116	Ponga (Silver(y) Tree–fern)
184	**PTERIDIUM** Gled. ex Scop. (Brackens; Rhedyn Cyffredin)
184	*AQUILINUM* (L.) Kuhn (Bracken; Rhedynen Gyffredin)
187	*AQUILINUM* (L.) Kuhn subsp. *AQUILINUM* (Common Bracken; Rhedynen Gyffredin)
187	*AQUILINUM* (L.) Kuhn subsp. *ATLANTICUM* C.N.Page (Atlantic Bracken; Rhedynen Gyffredin y Gorllewin)
188	*AQUILINUM* (L.) Kuhn subsp. *FULVUM* C.N.Page (Perthshire Bracken; Rhedynen Gyffredin Perth)
188	*aquilinum* (L.) Kuhn subsp. *latiusculum* (Desv.) C.N.Page (*PTERIDIUM PINETORUM* C.N.Page & R.R.Mill)
186	*latiusculum* (Desv.) Hieron.

PTERIDIUM (*continued*)
- *188*   *PINETORUM* C.N.Page & R.R.Mill (Pinewood Bracken; Rhedynen Gyffredin Pîn)
- *188*   *PINETORUM* C.N.Page & R.R.Mill subsp. *OSMUNDACEUM* (H.Christ) C.N.Page
- *188*   *PINETORUM* C.N.Page & R.R.Mill subsp. *PINETORUM*
- 189   PTERIS L. (Brake Ferns; Rhedyn Ruban)
- *184*   *aquilina* L. (**PTERIDIUM AQUILINUM** (L.) Kuhn)
- *189*   *CRETICA* L. (Brake Fern; Rhedynen Creta)
- *189*   *INCOMPLETA* Cav. (Spider Brake; Rhedynen Ruban y Corryn)
- *190*   *longifolia* auct., non L. (*PTERIS VITTATA* L.)
- *201*   *longifolia* L.
- *189*   *palustris* Poir. (*PTERIS INCOMPLETA* Cav.)
- *189*   *serrulata* Forssk., non L.f. (*PTERIS INCOMPLETA* Cav.)
- *190*   *TREMULA* R.Br. (Tender Brake; Rhedynen Ruban Croendenau)
- *190*   *VITTATA* L. (Ladder Brake; Rhedynen Ruban Ysgolddail)
- *201*   *Pyrrosia rupestris* (R.Br.) Ching (Rock Felt Fern)
- 48   Quillwort (**ISOETES LACUSTRIS** L.; Gwair Merllyn)
- 47   Land, (**ISOETES HISTRIX** Bory; Gwair Merllyn y Tir)
- 46   Spring, (**ISOETES ECHINOSPORA** Durieu; Gwair Merllyn Bychan)
- 46   Quillworts (**ISOETES** L.; Gwair Merllyn)

Rhedyn
- 200   Cadwyn (WOODWARDIA Sm.; Chain–ferns)
- 164   Cangarŵ (PHYMATODES C.Presl; Kangaroo Ferns)
- 116   Celynnog (**CYRTOMIUM** C.Presl; House Holly–ferns)
- 153   Croendenau (**ONOCLEA** L.; Sensitive Fern)
- 184   Cyffredin (**PTERIDIUM** Gled. ex Scop.; Brackens)
- 160   Cyfrdwy (**OSMUNDA** L.; Royal Ferns)
- 152   Estrys (**MATTEUCCIA** Tod.; Ostrich Ferns)
- 193   Gwrychog (**TRICHOMANES** L.; Killarney Ferns)
- 80   Jersey (ANOGRAMMA Link; Jersey Ferns)
- 104   Mair (**ATHYRIUM** Roth; Lady–ferns)
- 114   Persli (**CRYPTOGRAMMA** R.Br.; Parsley Ferns)
- 189   Ruban (PTERIS L.; Brake Ferns)
- 120   Troed–y–Wiwer (*DAVALLIA BULLATA* Wall. group; Squirrel's–foot Ferns)
- 120   Troed–yr–Ysgafarnog (DAVALLIA Sm.; Hare's–foot Ferns)
- 107   y Dŵr (**AZOLLA** Lam.; Water Ferns)
- 162   y Graig (**PHEGOPTERIS** (C.Presl) Fée; Beech Ferns)
- 148   Rhedynach Teneuwe (**HYMENOPHYLLUM** Sm.; Filmy–ferns)
- 149   Rhedynach Teneuwe (**HYMENOPHYLLUM TUNBRIGENSE** (L.) Sm.; Tunbridge Filmy–fern)
- 151   Wilson (**HYMENOPHYLLUM WILSONII** Hook.; Wilson's Filmy–fern)

Rhedynen
- 114   Bersli (***CRYPTOGRAMMA CRISPA*** (L.) R.Br.; Parsley Fern)
- 164   Cangarŵ (*PHYMATODES DIVERSIFOLIA* (Willd.) Pic.Serm.; Kangaroo Fern)

Rhedynen (*continued*)
- 116 Celynnog (***CYRTOMIUM FALCATUM*** (L.f.) C.Presl; House Holly–fern)
- 189 Creta (*PTERIS CRETICA* L.; Brake Fern)
- 190 Croendenau (*PTERIS TREMULA* R.Br.; Tender Brake)
- 153 Estrys (***MATTEUCCIA STRUTHIOPTERIS*** (L.) Tod.; Ostrich Fern)
- 80 Fach Jersey (*ANOGRAMMA LEPTOPHYLLA* (L.) Link; Jersey Fern)
- 105 Fair (***ATHYRIUM FILIX–FEMINA*** (L.) Roth; Lady–fern)
- 105 Fair Alpaidd (*ATHYRIUM DISTENTIFOLIUM* Tausch ex Opiz var. *DISTENTIFOLIUM*; Alpine Lady–fern)
- 105 Fair Newman (*ATHYRIUM DISTENTIFOLIUM* Tausch ex Opiz var. *FLEXILE* (Newman) Jermy; Newman's Lady–fern)
- 118 Frau (***CYSTOPTERIS FRAGILIS*** (L.) Bernh.; Brittle Bladder–fern)
- 200 Gadwyn Ewropeaidd (*WOODWARDIA RADICANS* (L.) Sm.; European Chain–fern)
- 84 Gefngoch (***ASPLENIUM CETERACH*** L.; Rustyback)
- 153 Groendenau (***ONOCLEA SENSIBILIS*** L.; Sensitive Fern)
- 100 Guernsey (*A. OBOVATUM* × *A. SCOLOPENDRIUM* = *A.* × *MICRODON* (T.Moore) Lovis & Vida; Guernsey Fern)
- 184 Gyffredin (***PTERIDIUM AQUILINUM*** (L.) Kuhn; Bracken)
- 187 Gyffredin (***PTERIDIUM AQUILINUM*** (L.) Kuhn subsp. *AQUILINUM*; Common Bracken)
- 188 Gyffredin Perth (*PTERIDIUM AQUILINUM* (L.) Kuhn subsp. *FULVUM* C.N.Page; Perthshire Bracken)
- 188 Gyffredin Pîn (*PTERIDIUM PINETORUM* C.N.Page & R.R.Mill; Pinewood Bracken)
- 187 Gyffredin y Gorllewin (***PTERIDIUM AQUILINUM*** (L.) Kuhn subsp. *ATLANTICUM* C.N.Page; Atlantic Bracken)
- 161 Gyfrdwy (***OSMUNDA REGALIS*** L.; Royal Fern)
- 99 Jackson (*A. ADIANTUM–NIGRUM* × *A. SCOLOPENDRIUM* = *A.* × *JACKSONII* (Alston) Lawalrée; Jackson's Fern)
- 190 Ruban Croendenau (*PTERIS TREMULA* R.Br.; Tender Brake)
- 189 Ruban y Corryn (*PTERIS INCOMPLETA* Cav.; Spider Brake)
- 190 Ruban Ysgolddail (*PTERIS VITTATA* L.; Ladder Brake)
- 100 Sir Gaernarfon (***A. ADIANTUM–NIGRUM*** × ***A. SEPTENTRIONALE*** = ***A.*** × ***CONTREI*** Callé, Lovis & Reichst.; Caernarvonshire Fern)
- 120 Troed–yr–Ysgafarnog (*DAVALLIA CANARIENSIS* (L.) Sm.; Hare's–foot Fern)
- 193 Wrychog (***TRICHOMANES SPECIOSUM*** Willd.; Killarney Fern)
- 195 Wrychog Orchuddiedig (*TRICHOMANES VENOSUM* R.Br.; Veiled Bristle–fern)
- 189 y Corryn (*PTERIS INCOMPLETA* Cav.; Spider Brake)
- 107 y Dŵr (***AZOLLA FILICULOIDES*** Lam.; Water Fern)
- 162 y Graig (***PHEGOPTERIS CONNECTILIS*** (Michx.) Watt; Beech Fern)
- 190 Ysgolddail (*PTERIS VITTATA* L.; Ladder Brake)

85	Rock–spleenwort, Smooth, (*Asplenium fontanum* (L.) Bernh.; Duegredynen Lefn y Creigiau)
201	*Rumohra adiantiformis* (G.Forst.) Ching (Leathery Shield Fern)
84	Rustyback (***ASPLENIUM CETERACH*** L.; Rhedynen Gefngoch)
90	*Scolopendrium vulgare* Sm. (***ASPLENIUM SCOLOPENDRIUM*** L.)
42	**SELAGINELLA** P.Beauv. (Lesser Clubmosses; Cnwpfwsoglau Lleiaf)
201	*denticulata* (L.) Link
201	*helvetica* Link
43	***KRAUSSIANA*** (Kunze) A.Braun (Kraus's Clubmoss; Cnwpfwsogl Krauss)
44	***SELAGINOIDES*** (L.) Link (Lesser Clubmoss; Cnwpfwsogl Bach)
183	Shield–fern, Alpine Hybrid, (*P. ACULEATUM* × *P. LONCHITIS* = *P.* × *ILLYRICUM* (Borbás) Hahne; Gwrychredynen Groesryw Alpaidd)
184	Atlantic Hybrid, (*P. LONCHITIS* × *P. SETIFERUM* = *P.* × *LONCHITIFORME* (Halácsy) Bech.; Gwrychredynen Groesryw y Gorllewin)
177	Hard, (***POLYSTICHUM ACULEATUM*** (L.) Roth; Gwrychredynen Galed)
183	Lowland Hybrid, (***P. ACULEATUM*** × ***P. SETIFERUM*** = *P.* × ***BICKNELLII*** (H.Christ) Hahne; Gwrychredynen Groesryw yr Iseldir)
181	Soft, (***POLYSTICHUM SETIFERUM*** (Forssk.) Woyn.; Gwrychredynen Feddal)
177	Shield–ferns (**POLYSTICHUM** Roth; Gwrychredyn)
102	Spleenwort, Alternate–leaved, (***A. SEPTENTRIONALE*** × *A. TRICHOMANES* subsp. ***TRICHOMANES*** = *A.* × ***ALTERNIFOLIUM*** Wulfen; Duegredynen Dail Bob yn Ail)
82	Black, (***ASPLENIUM ADIANTUM–NIGRUM*** L.; Duegredynen Ddu)
96	Common Maidenhair, (***ASPLENIUM TRICHOMANES*** L. subsp. ***QUADRIVALENS*** D.E.Mey. emend. Lovis; Duegredynen Gwallt y Forwyn Gyffredin)
101	Confluent Maidenhair, (*A. SCOLOPENDRIUM* × *A. TRICHOMANES* subsp. *QUADRIVALENS* = *A.* × *CONFLUENS* (T. Moore ex Lowe) Lawalréc; Duegredynen Gwallt y Forwyn Gydlifol)
97	Delicate Maidenhair, (***ASPLENIUM TRICHOMANES*** L. subsp. ***TRICHOMANES***; Duegredynen Gwallt y Forwyn Gain)
92	Forked, (***ASPLENIUM SEPTENTRIONALE*** (L.) Hoffm.; Duegredynen Fforchog)
97	Green, (***ASPLENIUM TRICHOMANES–RAMOSUM*** L.; Duegredynen Werdd)
99	Guernsey, (*A. ADIANTUM–NIGRUM* × *A. OBOVATUM* = *A.* × *SARNIENSE* Sleep; Duegredynen Guernsey)
99	Hybrid Black, (*A. ADIANTUM–NIGRUM* × *A. ONOPTERIS* = *A.* × *TICINENSE* D.E.Mey.; Duegredynen Ddu Groesryw)

Spleenwort (*continued*)
104   Hybrid Maidenhair, (*A. TRICHOMANES* subsp. *QUADRIVALENS* × subsp. *TRICHOMANES* = *A. TRICHOMANES* L. nothossp. *LUSATICUM* (D.E.Mey.) Lawalrée; Duegredynen Gwallt y Forwyn Groesryw)
88   Irish, (*ASPLENIUM ONOPTERIS* L.; Duegredynen Wyddelig)
101   Lady Clermont's, (*A. RUTA–MURARIA* × *A. TRICHOMANES* subsp. *QUADRIVALENS* = *A.* × *CLERMONTIAE* Syme; Duegredynen y Fonesig)
87   Lanceolate, (*ASPLENIUM OBOVATUM* Viv. subsp. *LANCEOLATUM* (Fiori) P.Silva; Duegredynen Reiniolaidd)
95   Lobed Maidenhair, (*ASPLENIUM TRICHOMANES* L. subsp. *PACHYRACHIS* (H.Christ) Lovis & Reichst.; Duegredynen Gwallt y Forwyn Labedog)
93   Maidenhair, (*ASPLENIUM TRICHOMANES* L.; Duegredynen Gwallt y Forwyn)
101   Murbeck's, (*A. RUTA–MURARIA* × *A. SEPTENTRIONALE* = *A.* × *MURBECKII* Dörfl; Duegredynen Murbeck)
86   Sea, (*ASPLENIUM MARINUM* L.; Duegredynen Arfor)
85   Serpentine Black, (*Asplenium cuneifolium* Viv.; Duegredynen Ddu Sarff–faen)
81   Spleenworts (**ASPLENIUM** L.; Duegredyn)
181   Sword–fern, Western, (*POLYSTICHUM MUNITUM* (Kaulf.) C.Presl; Gwrychredynen Gledd Orllewinol)
46   *Stylites* Amstutz
156   Tafod y neidr (**OPHIOGLOSSUM VULGATUM** L.; Adder's–tongue)
155   Bach (**OPHIOGLOSSUM AZORICUM** C.Presl; Small Adder's–tongue)
155   Lleiaf (**OPHIOGLOSSUM LUSITANICUM** L.; Least Adder's–tongue)
90   Tafod yr Hydd (**ASPLENIUM SCOLOPENDRIUM** L.; Hart's–tongue)
154   Tafodau y Neidr (**OPHIOGLOSSUM** L.; Adder's–tongues)
190   **THELYPTERIS** Schmidel (Marsh Ferns; Marchredyn y Gors)
144   *dryopteris* (L.) Sloss. (***GYMNOCARPIUM DRYOPTERIS*** (L.) Newman)
158   *limbosperma* (All.) H.P.Fuchs (***OREOPTERIS LIMBOSPERMA*** (Bellardi ex All.) Holub)
158   *oreopteris* (Ehrh.) Sloss. (***OREOPTERIS LIMBOSPERMA*** (Bellardi ex All.) Holub)
191   ***PALUSTRIS*** Schott (Marsh Fern; Marchredynen y Gors)
162   *phegopteris* (L.) Sloss. (***PHEGOPTERIS CONNECTILIS*** (Michx.) Watt)
146   *robertiana* (Hoffm.) Sloss. (***GYMNOCARPIUM ROBERTIANUM*** (Hoffm.) Newman)
191   *thelypteroides* Michx. subsp. *glabra* Holub (***THELYPTERIS PALUSTRIS*** Schott)
201   *Tmesipteris* Bernh. species (Fork–ferns)

121	Tree–fern, Australian, (*DICKSONIA ANTARCTICA* Labill.; Coedredynen Awstralia)
116	Silver(y), (*CYATHEA DEALBATA* (G.Forst.) Sw.; Coedredynen Arian)
121	Soft, (Australian Tree–fern)
120	Tree–Ferns, Australian, (DICKSONIA L'Hér.; Coedredyn Awstralia)
115	Silver, (CYATHEA Sm.; Coedredyn Arian)
193	**TRICHOMANES** L. (Killarney Ferns; Rhedyn Gwrychog)
193	*radicans* auct. (***TRICHOMANES SPECIOSUM*** Willd.)
193	***SPECIOSUM*** Willd. (Killarney Fern; Rhedynen Wrychog)
195	*VENOSUM* R.Br. (Veiled Bristle–fern; Rhedynen Wrychog Orchuddiedig)
193	Vandenboschia radicans (Sw.) Copel. (***TRICHOMANES SPECIOSUM*** Willd.)
89	Wall–rue (***ASPLENIUM RUTA–MURARIA*** L.; Duegredynen y Muriau)
195	**WOODSIA** R.Br. (Woodsias; Coredyn)
196	***ALPINA*** (Bolton) Gray (Alpine Woodsia; Coredynen Alpaidd)
198	***ILVENSIS*** (L.) R.Br. (Oblong Woodsia; Coredynen Hirgul)
196	Woodsia, Alpine, (***WOODSIA ALPINA*** (Bolton) Gray; Coredynen Alpaidd)
198	Oblong, (***WOODSIA ILVENSIS*** (L.) R.Br.; Coredynen Hirgul)
195	Woodsias (**WOODSIA** R.Br.; Coredyn)
200	WOODWARDIA Sm. (Chain–ferns; Rhedyn Cadwyn)
200	*RADICANS* (L.) Sm. (European Chain–fern; Rhedynen Gadwyn Ewropeaidd)